RUTLAND BOUGHTON AND THE GLASTONBURY FESTIVALS

Rutland Boughton and the Glastonbury Festivals

MICHAEL HURD

CLARENDON PRESS · OXFORD
1993

Oxford University Press, Walton Street, Oxford OX2 6DP
Oxford New York Toronto
Delhi Bombay Calcutta Madras Karachi
Kuala Lumpur Singapore Hong Kong Tokyo
Nairobi Dar es Salaam Cape Town
Melbourne Auckland Madrid
and associated companies in
Berlin Ibadan

Oxford is a trade mark of Oxford University Press

Published in the United States
by Oxford University Press Inc., New York

British Library Cataloguing in Publication Data
Data available
ISBN 0–19–816316–9

Library of Congress Cataloging in Publication Data
Hurd, Michael.
Rutland Boughton and the Glastonbury festivals / Michael Hurd.
Rev. and enl. ed. of: The immortal hour. 1962.
Discography:
Includes bibliographical references and index.
1. Boughton, Rutland, 1878–1960. 2. Composers—England—
Biography. I. Hurd, Michael. Immortal hour. II. Title.
ML410.B772H9 1993 780'92–dc20
ISBN 0–9–816316–9

Set by Hope Services (Abingdon) Ltd.
Printed in Great Britain
on acid-free paper by
Biddles Ltd
Guildford and King's Lynn

The most vital drama is that which takes place in the secret hearts of us all–the strife which is not completely known even to our nearest and dearest–the strife between us and our destiny.

Rutland Boughton: Parsifal, 1920.

ACKNOWLEDGEMENTS

Rutland Boughton and the Glastonbury Festivals is a radically revised and significantly enlarged version of my first Boughton biography, which Routledge & Kegan Paul published in 1962 under the title *Immortal Hour*. Thus there lie embedded in its text many passages which, though often much changed in their detail, are also to be found in the 1962 volume. Virtually unchanged are the many letters and newspaper reports for whose use in the original version permission was obtained from the appropriate copyright holders. In 1962, however, it was possible thus to acknowledge the help of people who were living, and in many cases known to me. These included members of the Boughton family, in particular Mrs Kathleen Boughton, Mr and Mrs Arthur Boughton, and Joy Boughton and her husband Christopher Ede, Adolph Borsdorf, Mr and Mrs Roger Clark, Gordon Craig, Sir Eugene Goossens, Kingsley Martin, Lord Morrison of Lambeth, Charles Kennedy Scott, Sir Steuart Wilson, and Frederick Woodhouse. A similar debt of gratitude was acknowledged to the Trustees, Executors, and Descendants of Frederic Austin, Sir Granville Bantock, Sir Arnold Bax, Sir Thomas Beecham, Reginald Ramsden Buckley, Kenneth Curwen, Professor Edward J. Dent, Sir Edward Elgar, Hamilton Fyfe, Gustav Holst, Laurence Housman, the Marchioness of Londonderry, Professor Gilbert Murray, Ernest Newman, Dame Ethel Smyth, and Ralph Vaughan Williams for permission to quote letters and other material. I was, and am, particularly indebted to the Society of Authors, acting on behalf of the Bernard Shaw Estate, the Trustees of the British Museum, the Governors and Guardians of the National Gallery of Ireland, and the Royal Academy of Dramatic Art, for permission to print Shaw's delightful correspondence with Rutland Boughton, including unpublished material, © 1993. Without their co-operation the present volume would be no more possible than its 1962 progenitor. A similar debt was, and still is, due to the following newspapers and periodicals for permission to use extracts from articles, letters, and critical reviews, the source of which is recorded in the text: *The Times, Daily Telegraph, Birmingham Post, Daily Mirror, Daily Worker, Western Daily Press, Daily News* (Associated Press), *Sunday Times, Observer,* and *Nation* (by permission of the *New Statesman*); Novello & Co. for quotations from the *Musical Times* (now published by the Orpheus Press Ltd.), and J. Curwen & Sons for material originally published in the *Musical*

News and Herald and the *Sackbut*. Special thanks must go to the proprietors of the *Central Somerset Gazette* (now part of the Mid Somerset Series of Newspapers, and the Westminster Press)—that newspaper having proved a mine of invaluable information. Extracts from *Music Drama of the Future* were included by kind permission of William Reeves Ltd., and permission to quote music examples was readily granted by Novello & Co., J. Curwen & Sons, and Stainer & Bell. These dispensations have been renewed for the present volume.

In revising and expanding the original biography new debts have been contracted, and it is a pleasure to acknowledge the help freely given by the late Desirée Ames (Mrs Arthur Newell), Miss Kathleen Beer, Gwydion Brooke, Alan Bush, A. T. K. Crozier, Kathleen Dillon (Mrs Angus Morrison), Sir John Gielgud, the Baroness von Kühne, Dan H. Laurence, George Lloyd, Percy Lovell, Alan G. Melville, Dr Tania Rose, Dr K. M. Tomlinson, Mrs Ursula Vaughan Williams, and Dame Ninette de Valois. The surviving members of Rutland Boughton's family, in particular his children Peter, Brian, and Jennifer, and his grandchildren Ian, Shelagh, Caroline Abraham, Gillian Weeks, and Robin Moore Ede, have given me their unreserved co-operation and constant encouragement. For this I am deeply grateful; without it I could not have completed the task I set myself. I had hoped to be able to place a copy of the new biography in Dame Gwen Ffrangcon-Davies's hands, but after 101 years of vibrant living this was perhaps too much to much to ask. The memory of the interview she granted me in 1987 remains an unforgettable experience. I owe a very special debt of gratitude to Mrs Jackie Flowers for her painstaking research into Boughton's journalism, which both complemented and added to my own researches. My thanks go also to Mrs Pamela Burrough, Librarian of the Street branch of the Somerset County Library; to Jim Hastie, Administrator of the Margaret Morris Movement, Glasgow; and to the Company Archivist (then Ms Annabel Ainsworth) of C. & J. Clark Ltd., Street. I am grateful to Lewis Foreman for his helpful comments at the manuscript stage, to the staff of the British Library for their unfailing courtesy, to A. J. Clark for his photographic skills, and to Bruce Phillips of the Oxford University Press for his constant encouragement.

As I have been actively involved with Boughton's extraordinary life and achievements for nearly forty years, there must inevitably be people who provided me with valuable information and insights into his character and career who I have now forgotten. I trust they will accept my apologies for the unintentional oversight. Finally, it was my pleasure to thank Mai Zetterling, David Hughes, and the late Peter Price for the support they gave me during the writing of the original biography: it is an equal pleasure, thirty years later, to recall the enduring influence of their friendship.

M.H.

West Liss, Hampshire
April 1992

CONTENTS

Contents

LIST OF PLATES

1. William and Grace Boughton, with their children (left to right) Rutland, Muriel, and Edward, *c*.1896
2. Rutland Boughton during his Birmingham days, *c*.1906
3. Florence Hobley, the first Mrs Boughton, on the eve of her departure for Canada, 1911
4. Christina Walshe, the second 'Mrs Boughton', 1911
5. Boughton about to enter the Glastonbury Assembly Rooms, *c*.1923
6. Kathleen Davis, the third 'Mrs Boughton', as Mary in *Bethlehem*, 1922
7. A dancing class in the Victoria Rooms, Glastonbury, *c*.1916
8. Margaret Morris as the Wicked Queen in *Snow White and the Seven Little Dwarfs*, 1916
9. *The Birth of Arthur*, Glastonbury Assembly Rooms, 1921
10. *The Immortal Hour*, Act 2, Glastonbury. Gwen Ffrangcon-Davies as Etain, and Sheerman Hand as Eochaidh
11. The Regent Theatre, King's Cross, November 1923
12. Gwen Ffrangcon-Davies as Etain, 1923
13. *Alkestis*, Glastonbury 1922: Kathleen Davies, Handmaiden (left), Astra Desmond, Alkestis, (centre), Steuart Wilson, Admetus, (right)
14. *The Queen of Cornwall*, Glastonbury 1924: King Mark (Frederick Woodhouse) stabs Tristram (Frank Phillips) before the horrified gaze of Queen Iseult (Gladys Fisher)
15. *Bethlehem*, Bristol Folk Festival School 1922: the Angel Gabriel (Kathleen Beer) greets Mary (Kathleen Davis)
16. The modern-dress *Bethlehem*, Church House, Westminster, 1926: Herod, Edward Nichol (centre): Herodias, Dorothy D'Orsay (right); Dancer, Penelope Spencer (left)
17. Kilcot, *c*.1930. The lady is the Boughtons' friend Miss Agnes Thomas
18. Kathleen Boughton, *c*.1950
19. Rutland Boughton, passport photograph, *c*.1950

LIST OF FIGURES

Rutland Boughton
His Life

1

1878–1899

Aylesbury, the county town of Buckinghamshire, lies at the foot of the Chiltern hills. At the end of the nineteenth century it was a pleasant place, neat and compact—its life centring round a fine market square where people would gather, as their forefathers had done, to buy and sell and hear the latest news. Something of this still remains, dimmed by neon signs and plate-glass windows: a shadow of the days when William Rutland Boughton asked a young schoolmistress from the nearby village of Askett to be his wife. He offered her the comfort of his easygoing nature, and the security of a small grocery business ('W. R. & C. A. Boughton: Tea Dealers, Family Grocers, and Provision Merchants') which he ran from 37 Buckingham Street in partnership with his brother. The young schoolmistress, Grace Martha Bishop, accepted him.

Like most Victorians of their class they were conservative by inclination, religious by conviction, and puritan by nature. They accepted without question both good times and bad, secure in the knowledge that such things were ordained in a system of just rewards and punishments. Life, for the most part, meant work; and pleasure, if enjoyed at all, was to be enjoyed soberly. Indeed, pleasure might even be the handmaiden of religion, as when they sang together in the chorus of some Handel or Mendelssohn oratorio for the Aylesbury Sacred Harmonic Society, or when the young Mrs Boughton felt moved to play hymns on the family harmonium or compose religious verse. The wider aspects of culture meant very little to them, and in this they were at one with the majority of their fellow countrymen.

Of their six children only three survived the hazards of Victorian childbirth. The second son, born in one of the newly built and as yet unnumbered terraced houses in the Bicester Road on 23 January 1878, was the first to live. He was given a distinctive Christian name: Rutland. It was, in fact, a family name inherited from Ambrose Rutland, a farmer in Newton Longville who had died in 1765 leaving two sons and a daughter. The sons remained single, but the daughter, Ann Rutland, married John Boughton (1772–1838) the composer's great-grandfather. Throughout the eighteenth and most of the nineteenth centuries the Boughtons and Rutlands had figured as yeoman farmers in Buckinghamshire, and it is only with William Rutland Boughton's generation that the pattern changed. Ancestry on the Bishop side was also largely concerned with farming; and, typical of the very circumscribed possibilities of the time, it is interesting to find that two of the four Boughton brothers, William (1841-1905), the composer's father, and Cecil, married two of the Bishop sisters, Grace (1845–1928), the composer's mother, and Sarah. The other brothers, John and Edward, followed suit and also married two sisters, Polly and Nellie Hedges.[1]

William and Grace's three children, Rutland, Edward, and Muriel, were brought up in such comfort as a lower middle-class family could afford. Life was modest enough, yet it was evident that there were others even less happily placed. Accordingly, a portion of each Sunday lunch, the week's best meal, was taken to some poorer home before the Boughton family sat down to enjoy its own slender good fortune. As a lesson in practical socialism it was not easily forgotten. But if Sunday was given over to piety and good works, there was always Saturday when, free from school, the children could join their cousins, who lived over the shop, in 'helping' their parents—blending the tea, cutting the huge sugar cones into manageable lumps, weighing the spices, and keeping a watchful eye on such oranges as might, after a helpful squeeze, be thought too far gone for selling. Sometimes, as a mark of special favour, one or other of them would be allowed to go with their father in the pony trap as he visited the outlying villages with the week's grocery orders, stopping off at some wayside inn for the midday meal, washed down with cold ginger beer in summer, hot peppermint in winter.

And then there was school itself. First, a small Dame School run by two old ladies, and then the Aylesbury 'Grammar' or 'Latin' School. Here the pupils were divided into an 'Upper' order of those whose parents could afford a modest fee, and a 'Lower' one for the less affluent. The two sections were segregated within doors, arrived and left at different times, but mixed happily and without prejudice as soon as lessons were over. Despite its rather grand names the school was a simple one: two rooms, two large classes, two teachers; the three 'R's', a smattering of French, and list upon list of historical dates. Little was offered the pupils and little accepted. The elderly clergyman who functioned as headmaster carried out his duties more often than not in a doze from which his assistant was too tactful to wake him—his very owlishness delightfully endorsed by his name: the Reverend Mr Howells.

But if the Grammar School in St Mary's Square did little to arouse the young Rutland Boughton's interest, the music that played so prominent a part in his mother's religious observance was altogether a different matter. He listened entranced as she played the harmonium, and joined in bravely with the many relatives who came round to sing hymns after the Sunday evening service. When he was old enough he became a chorister at the parish church, and in due course joined his parents in the activities of the Sacred Harmonic Society. His first actual music lessons came from two girl cousins who taught him to play the piano which his mother had arranged to buy on the instalment plan, despite the strain it placed on her housekeeping money. Throughout his childhood it was she who encouraged his obvious gift for music, and she who started his career as a composer by the simple act of giving him a small, leather-bound manuscript book in which her father had written a few anthems and hymns of his own composition. The blank pages made imitation irresistible and the boy soon set about the business of filling them up.

The grandfather whose music book was thus defiled was Francis Bishop, remembered by Boughton as 'a cheerful red-robin of a man', the owner of a small farm deep in the Chilterns. The farmhouse did double duty, for it was also an inn; its scrubbed tables, gleaming copper pans, and freshly sanded floors, the slow-turning spit, the clatter and bustle, and general good humour remaining a special memory of childhood and a sharp contrast with the earnest ways of the Aylesbury household. Like his brother Charles, Francis Bishop had a strong love of music and enough talent to be able to compose simple anthems, hymns, and chants for the local church choir. Charles, who became a teacher at Lydbrook in the Forest of Dean—a few miles south of what was to be Rutland Boughton's own Gloucestershire home—showed signs of a more dramatic talent. A 'May-Day Cantata', which he composed for the villagers so that each man might have something to represent his own special trade, remained popular until long after his death. It is thus from his mother's family that Rutland Boughton inherited his musical gifts and a concern for his fellow men.

His own earliest pieces were inspired by the religious music he heard as part of everyday life. Ironically, as it now seems, it was the marriage service that attracted his first efforts: 'it had never been done and seemed to need attention badly'. He quickly fell into difficulties and turned for guidance to a book on Figured Bass. Neither he nor his mother could make anything of it. The mysteries of chord progression were incomprehensible when reduced to such bewildering mathematics, so he carried on without them, his ear a surer guide.

The pile of manuscripts grew and it was not long before he felt obliged to share the results of his labours. Gathering together a 'children's orchestra', he announced that rehearsals would take place in his mother's kitchen until such time as they were fit for a public appearance. The forces were unusual enough: a battery of school friends on penny whistles and home-made zithers, a small choir of very mixed voices, a violin, and Boughton himself as conductor, composer, and impresario. At last it was felt that perfection was indeed at hand. The vicar agreed to loan the parish hall, and a date was announced. In order that things might be done in style, a local printer was talked into producing suitable programmes—though style received a grievous blow when the number of misprints were eventually totted up, for on being shown the proofs Boughton thought he was simply being invited to admire the printer's handiwork and politely did just that. But this was a trifling matter compared with the thrill of appearing, in a bower of potted plants, as composer and conductor before your first real audience. Perhaps the most genuinely admired item was his own performance as a minstrel, complete with blackened face and home-made banjo, in imitation of a concert party he had seen at Ramsgate one summer holiday—though whether it was realized that his efforts were intended as comic 'relief' has not been recorded.

Despite these excursions into public life Boughton was a solitary boy, preferring always the company of books and music and the solace of his own imagination. At home there was little for such a mind to feed on. Bunyan and the Bible were soon exhausted and *The Confessions of St Augustine* held little attraction. He turned instead to the penny-dreadful adventures of Deadwood Dick and a bowdlerized version of *The Arabian Nights*. His mother was horrified and quickly referred the matter to the curate, who suggested that the historical novels of Harrison Ainsworth might satisfy her son's taste for adventure and at the same time calm her own fears—each earnest volume carrying, as it were, a guarantee of genuine, irreproachable literature. In the end, however, it was his school that provided the first really decisive literary influence. At the age of 12 he received a prize: a volume of Shakespeare's plays.

The idea of drama aroused an immediate response and he was soon scribbling away at music for the Witches in *Macbeth*. But fresh problems arose. The only day which offered a really generous measure of free time was Sunday. To use this for composing music to such obviously profane words would be, inevitably, to provoke the wrath of God. With commendable ingenuity he evolved a solution that would satisfy both the Lord and himself, and incidentally quell any doubts his parents might harbour. He had for some time felt that the mere singing of oratorio was inadequate, and it now struck him that if a choir were to be grouped around three sides of a raised platform the soloists might act out the drama as it unfolded in the music. With all the abandon of a Stockhausen he devised a cycle of fourteen music dramas on the life of Christ and began to work through each Sunday with a clear conscience. Although it was never completed, and was eventually destroyed, Boughton's 'Jesus' cycle anticipated to a surprising degree that Arthurian dramas that were to be his life's work. That such an idea could have occurred to him at all is remarkable enough when one considers he had never been to a theatre and did not know that opera existed. But he had seen a seaside concert party perform on a raised platform, heard a few oratorios, and read some of Shakespeare's plays. These disparate elements combined to give him that peculiarly English form: the Choral Drama.

As has already been observed, Boughton's musical talents came to him through his mother's side. So also must have come the will to succeed. The merest glance at family photographs is sufficient to endorse his memory of his father as a mild, somewhat ineffectual man, and his mother as a proud, determined spirit. She was not a woman to be trifled with. Her will power seems also have affected the second son, Edward (1881-1971), who when he died left a fortune of nearly three million pounds, earned through his skills as an engineer and designer (he helped design the first tank, and later designed the Ruston Hornsby car, and in 1920 was co-founder of the Automotive Products Group, the company that brought Lockheed's hydraulic brakes into Europe). Only Muriel, Rutland's sister, enjoyed no obvious talent other than a voice that enabled her to take part in the first performance of *The Immortal Hour*. Even in the matter of looks she cannot be said to have offered any competition to her brothers, taking after her father as fatally as they took after their mother.

By 1892 Boughton's boyhood dreams had to give way to stern realities. He was now 14, and as head boy of the school had received the presentation Bible and could expect no more. Further education was out of the question: he had to earn a living. At first his parents had no idea what to do. The only thing that interested him was outside their experience, but they consulted the local organist who suggested they might do worse than advertise in the *Musical Times* offering the usual premium of £25 to apprentice a likely boy to some musical trade.[2] They scraped together the money, placed the advertisement, and in October 1892 Rutland Boughton set out for London and the office of Cecil Barth's newly founded Concert Agency.

After the gentle rhythms of Aylesbury, London must have seemed a frightening and lonely place to a young boy away from home for the first time. 'The sky seemed so meagre,' he remembered, 'and the people so aggressive.' Some things he never forgot: 'When, a spare, shy boy, I once asked for a glass of milk and some cake the waitress looked so horribly, frighteningly superior that for a long time I dared not show my face in any sort of eating house.' But there were some friends, and his employer proved to be one of them.

Barth was not his real name. He was in fact a member of the Walenn family, well known in London musical circles as singers and instrumentalists. Advertisements for his concert agency at 2 Oxford Mansions, Oxford Street, make their first appearance in the *Musical Times* for September 1892, so it would seem that Rutland Boughton joined the firm at a moment when a £25 premium might have been very welcome. By November that year Barth had acquired a partner and was able to advertise as 'Messrs Barth & Black—A Musical Agency under Professional Direction.' In later years he turned to the business management of plays and musical comedies, abandoning the 'high-class Vocalists & Instrumentalists [for] Concerts, Oratorio Performances, Garden Parties, At Homes [and] Public Dinners' that had been the modest backbone of his fortunes. According to Boughton, who once appeared in an introductory promotion campaign at Olympia, Barth was largely responsible for introducing his fellow countrymen to the possibilities of confetti as a means of expressing nuptial approval. Apart from this lapse he seems to have been tactful, generous, and good-natured. He took a liking to his inefficient office-boy and dealt gently with him when he proved capable of 'posting the latch-key and bringing back the letters'. He soon discovered the reason for the boy's dreaminess.

One day, returning from lunch a little earlier than usual, he caught Boughton in the act of concealing something beneath his blotting-pad. Investigation made it plain that the office-boy had been using the firm's time for private composition. The music in question was commandeered, and next morning a worried Rutland Boughton was summoned into his employer's office. Instead of getting the sack he expected and deserved, he was told that arrangements had been made for lessons in harmony and counterpoint with an elderly musician, a Mr Smith-Webster, in Camden Town. Barth went further, arranging for the use of a small piano and passing on all the spare concert and opera tickets that came his way. Boughton was thus able to attend Henry Wood's Promenade Concerts at the recently built Queen's Hall (including the first, on 10 August 1895), and sit in the Gallery at Covent Garden to hear Mozart and Meyerbeer—not to mention Gilbert and Sullivan at the Savoy Theatre.

Had it not been for Barth's generosity there would have been no concerts and certainly no lessons, for Boughton was hardly wealthy. He had started work at 6 shillings a week, rising, by easy stages, to a guinea. To begin with he lived with his mother's sister in Kilburn, sometimes walking, sometimes taking the horse-bus to Oxford Street, saving all he could for music. 'I do not think, dear,' wrote his mother 'that one or two days you have had sufficient lunch. A penny bun is not enough to go eight hours on. Do not let the music tempt you to deny your body.'

At Kilburn the atmosphere was even more restrictive than it had been at home. His uncle and aunt were strong Methodists and quick to assume the worst when they realized that to their nephew music meant concerts and opera and not just hymns on a Sunday. Only partly convinced by his assurance that these delights were nothing like as dangerous as the theatre, they allowed him to continue, but insisted on having him publicly 'saved' at the local Mission Hall. 'How wicked I was in every atom of my being I had never known until the pressure of that religious meeting brought it to my knowledge in a flood of tears. My aunt insisted that she could easily recognize the difference in my character.' But the emotional confidence trick had no lasting effect. When at last the chance came he was glad to escape their kindly but stifling attentions. An unexpected addition to his uncle's family, and the conviction that he could support himself on the 15 shillings he was now earning each week, led him into lodgings in West Kensington.

The move to St Helen's Gardens proved to be a happy one. Thanks to Barth he had his own piano (a Cramer upright) and his landladies, the Misses Abrams, encouraged him to play. The journey to Oxford Street (by Metropolitan Line from Latimer Road to Great Portland Street) was that much easier, and above all he could come and go as he pleased, with no one to question the direction his thoughts were now taking. One of the first results of this freedom was the formation of a piano trio which seemed, at least to its members, good enough to undertake professional engagements. Life in the concert agency had underlined the need for effective advertisement and the value of aiming high, so Boughton wrote to Johann Strauss for patronage. Rather to his surprise it was granted and the 'Johann Strauss Trio' was born. It did not live long. At a one and only engagement in Hull the audience made it quite clear that there had been some mistake. Thereafter Boughton and his friends (one was the neighbourhood milkman) contented themselves with the private study of chamber music, the Misses Abrams their only audience.

By now Boughton was beginning to feel his way as a composer. Barth had given him a tremendous leg-up in 1893 by publishing a song called 'Passing Joys'. It was his first work to appear in print and he dedicated it, proudly, to his mother. A year or so later a *Japanese Suite* was played by one of the Guards' Bands, and in 1897 Messrs Weekes brought out his 'Hungarian Fantasia' for piano duet. In the following year the firm of Kene & Co. published a piano solo entitled, for no very obvious reason, 'Sentry's March'. At the age of 20 Rutland Boughton was on his way.

None of these works, however, have any artistic merit, each being typical of Edwardian salon music of the most elementary kind. The most astonishing aspect is the Opus numbers Boughton bestowed on them. The 'Sentry's March' figures as 'Op 70, No 1', while the 'Hungarian Fantasia' weighs in as 'Op 94'! Whether they represent the true tally of his compositions, or merely reflect a desire to impress, is impossible now to say, for he was soon to destroy all his early efforts. But it may have been so, for in later years he claimed to have tried his hand at all manner of chamber works, a Requiem Mass, and settings of seventy of Heine's poems made on reading Bowering's translations.[3] One work came back to haunt him: the *Japanese Suite*, which was issued in piano solo form by a publisher hoping to cash in on the popularity of *The Immortal Hour*—the furious composer having forgotten that the manuscript had never been returned.

Of somewhat greater value is a cycle of twelve songs, *The Passing Year*, which Weekes brought out in 1898 after Arthur Walenn had sung them at a Queen's Hall concert. Copies can still occasionally be found. Although the music has no great distinction, and the words, by Boughton's friend Lizzie Miller Pengelly, even less, it has a fluency that does credit to its young, untrained composer. Traces of Schubert and Schumann jostle with Victorian balladry, but there are also moments of originality that explain the opinion that Ernest Newman formed of them in 1902: 'I know of no other of the younger school in whom there are clear indications of a quite personal style to come. There are certain things that cannot be taught in musical composition, that must come to you, if they come at all, by the grace of God.'[4]

Other pieces, such as a setting of Tennyson's poem *Sir Galahad* for chorus and orchestra, composed in 1898 but eventually published by Curwen in 1910 as Boughton's 'official' Opus 1, have a certain energy—as, indeed, has his setting of Longfellow's *The Skeleton in Armour*, composed in November and December 1898 as a 'Ballad for baritone and Orchestra', but recast in 1908 as a full-blown choral work. The most ambitious work of this period, however, was a Piano Concerto which was performed at the St James's Hall on 1 February 1898. Although it was torn to shreds by the critics and subsequently destroyed by the composer, its history is of special interest for the light it throws upon his character.

One day when Boughton was alone in Barth's office a young

German came in, desperate for work. Even though he was by now almost hardened to the disappointments of an agency, which had come to seem to him more like an 'ante-chamber of hell', he was struck by the man's woebegone condition and offered him the only money he had, a half-sovereign. The young man hesitated for a moment and then gave in. 'I must have it', he stammered, 'I must', and then he rushed out of the office. Boughton naturally assumed that the incident was at an end. But a few weeks later the young man returned. Apparently the gift had seemed to change his luck and he was now beginning to establish himself. He told Boughton that he was soon to conduct an orchestral concert and offered to include in the programme anything he might like to suggest. Rather airily, Boughton suggested a Piano Concerto, and then hastened back to his lodging to write one.

Paul Graener (1872-1944), the young German, kept his word and the concerto was duly performed by Freeman Whatmore, an organist friend of Boughton's from St Albans, in company with Mozart's G minor Symphony and Hamish Hamilton's *Highland Memories Suite*. After the harsh reception Boughton launched the first of his outspoken attacks on the critics and in doing so happened to reveal his age. His feelings were somewhat mollified when Fuller Maitland, *The Times* critic, wrote to apologize saying 'We thought it was the work of a much older man.'

Despite the failure of the concerto, Boughton's association with Paul Graener ripened into a friendship which soon embraced a musical cousin, Georg, secretary to the Naval Attaché at the Austrian Embassy. At one point the three friends planned a joint composition—a symphony trilogy entitled *Hill and Forest* to which each was to contribute a movement—but it seems never to have been completed. Paul Graener himself was to become well-known in Germany as a fluent composer of opera and Lieder in the Strauss-Reger tradition, and ended his career as Vice-President of the Reichsmusikkammer. Boughton's relations with Georg Graener, whom he regarded as one of his closest friends, ended in 1914 with the declaration of war.

By the end of 1898 Boughton's eagerness to take on anything and everything brought about a serious illness and put an end to his career as office-boy. The immediate cause was a competition for a cantata on the life of John the Baptist which Curwen had announced. In order to complete his score in time, Boughton decided to spend the nights composing as well as any part of the day that might be available. Inevitably he collapsed and Cecil Barth suggested that he needed a complete rest and possibly a change of job. Boughton took the hint and returned home to 'shock the neighbours by becoming a parasite on my parents'.

There was now sufficient evidence of a talent to justify their rather anxious toleration, and he himself set about doing what he could to contribute to the family budget. He acquired a number of piano pupils, and gradually got together a choir with the intention of giving concerts in the neighbourhood. One was billed as a 'Farewell Concert'—not because he had any intention of saying goodbye, but because he had learnt a trick or two at the agency. His local reputation rose considerably when Barth sent along the Band of the First Life Guards with a 'March by Rutland Boughton' in their programme,[5] and his own self-esteem was greatly increased when the editor of *Musical Opinion* agreed to take a few articles, though he did not feel able to offer any money for them. At this stage in his career everything was worth doing and no experience too mean to be overlooked. In his spare time he began work on his first music-drama: a setting of Bulwer-Lytton's translation of Schiller's 'The Bride of Messina'. Exactly what this was like we shall never know, for he destroyed it in later years as 'Schummanesque and mostly rubbish'. A copy of a Grand March that had figured in it did, however, survive to justify his verdict. But at the time of composition he thought sufficiently well of the work to attempt a subscription list for its performance and even managed by persuade Lord Battersea to head the promised contributions. Nothing came of the scheme, and many people must have thought it yet another example of the wildness of 'that young Boughton'. Indeed, he had already acquired something of a reputation among the town gossips. Not only did he appear to have rejected any idea of a steady job with prospects, but there was also the business of the girl he had been making such a fuss about.

In reality it was all very innocent. He had bumped into Ethel Gilbert in London and recognized her as someone from his Aylesbury childhood. They met again, held hands, and decided they were in love. His earliest sexual encounters had followed the usual paths: 'a passionate love for the solo boy in the choir, reaching the climax of one secret kiss in the vestry; scares of revulsion when coarser advances were made, first by a school-girl and then, even more alarmingly, by a boy cousin; and a series of child sweethearts who were personifications of all that was modest and gentle'. This time his feelings ran deeper and were ripe for idealization. Following his mother's example, he was already in the habit of expressing himself in verse. Now his emotions flowed over into admonitory love lyrics:

> Words fail, my mouth is stopped,
> Thy beauty who can measure?
> All others are o'erlopped—

Thou givest the greatest pleasure;
And thoughts of thee now comfort me
And fill up all my leisure.
Nay, more than that! When duty
Demands my whole attention,
Thoughts of thee and thy beauty
Oft act as a detention.
When duty calls and love enthralls
They quickly cause dissension.
                    (28 March 1895)

Towards the end of his London career Ethel had returned to Aylesbury, and when at last he was able to join her he found that her father had begun to raise objections to their friendship, largely on the grounds that a musician's future was far from certain and that the daughter of a railway guard might expect to do better. Boughton rose to the drama of the situation and begged the girl to elope with him:

O fly with me and be my bride
And none shall part us ever,
Or chains of love shall sever
With thee, dear, by my side.
Heed not thy father's irate word
For truly we are one;
All sacred 'neath the sun
Doth testify accord.
                    (26 February 1898)

Wisely, Miss Gilbert refused the invitation, whereupon her swain rushed into the garden and promptly demolished the bench on which they had been accustomed to sit, vowing never to see her again. His action merely convinced her that her father had been right and she made haste to put him out of her mind—though not entirely, for it was rumoured that he was wandering the streets of Aylesbury with a revolver determined to shoot himself, or her, or both.

As it happened, he was doing nothing more drastic than resolving his feelings in a bout of Whitmanesque verse:

Come, O thou beautiful Death.
Come with all thy soothing balms and perfumed odours.
Come with thy hand of relief,
Whether it hold the happy influence of unconsciousness
Or the more bitter one of suffering.
Come, I wait for thee—
My soul yearns to break the boundaries

Of its present limited existence
And leap into the great Hereafter.
(28 February 1898)

Fortunately the 'Great Hereafter' had a while to wait, and in the mean time the boundaries of his limited existence were broken in a totally unexpected manner. One day his attention was caught by an article in the *Musician* in which Robin Legge, the music critic of the *Daily Telegraph* asked where the 'New Musical Messiah' that England so badly needed was to be found. Boughton lost no time in telling him.

Despite his surprise at having a rhetorical question answered, Legge took him at his word and asked to see a sample of his work. Again he got rather more than he bargained for, but dutifully read the vast quantity of music that Boughton unloaded on him and was impressed. It was nothing if not ambitious. In addition to 'The Bride of Messina' there were settings of *The Skeleton in Armour* and *The Invincible Armada* (both in their original solo versions), not to mention a symphonic poem, apocalyptically entitled *Lucifer*, which he had completed in May 1898. Legge passed the music to Fuller Maitland for a second opinion, who in turn handed it to Sir Charles Stanford. Stanford decided that the boy had talent and began pulling strings to such effect that the MP for Aylesbury found himself agreeing to pay for lessons at the Royal College of Music. As he was a member of the Rothschild family the undertaking was not particularly onerous, and though he died shortly afterwards, his sister, Lady Battersea, agreed to carry out his wishes. Thus in the autumn of 1899 Rutland Boughton found himself again bound for London, not this time as a nervous schoolboy, but as a proud student of the Royal College of Music.

2

1899–1905

He cut a poor figure as a student. Confident enough of his musical abilities, he was painfully shy when it came to mixing with the others. He felt himself uncouth and awkward, a country bumpkin let loose in polite society. And Stanford can scarcely have improved matters by introducing him as 'a fellow who's been playing Beethoven in a barn'. He looked at the senior students—Gustav Holst and Ralph Vaughan Williams among them—and at the younger men, John Ireland, Frank Bridge, George Dyson, William Henry Harris, and compared his own rudimentary education and lower class background. It must have seemed as if he would never be able to match their knowledge and sophistication. Moreover he was, as Parry once remarked, a 'bad mixer'. During the whole time he spent at the Royal College he made only one close friend: Edgar Bainton.

Bainton (1880–1956) was the son of a Nonconformist clergyman. He had received a good education and was already by Boughton's standards, a man of culture. Although Boughton had been directed by the college into lodgings with a naval officer's widow, apparently in the hope that something of her gentility might rub off on him, he soon decided to throw in his lot with his new friend. They agreed to share lodgings, and tactfully, but firmly, Bainton began to guide his reading and encourage the process of self-education that was to last throughout his life. Once Boughton's mind had been opened to the world of other men's thoughts he never tired of exploration. But like most self-education it began just that much too late to be digested easily, and this, coupled with a natural earnestness and lack of any deep sense of humour, often led him into disastrous errors of judgement. Though he found it easy to revere knowledge, he was never able to take his learning lightly.

Perhaps the most decisive books that came his way during this period were Carlyle's *The French Revolution* and Tolstoy's *What is Art?*—the one turning his thoughts to the problem of social justice, and the other to the idea that art should be the spiritual servant of mankind rather than the plaything of a selfish élite. Both were calculated to appeal to the feelings of personal inadequacy of which student life had made him conscious. He fixed on these as his main lines of thought precisely because they helped him ward off the sense that in a country in which art was regarded as a gentlemanly, amateur pursuit he had, in social terms at least, started on the wrong foot.

In matters of musical achievement his position was equally difficult. He was already a published composer and had enjoyed a number of substantial performances. He had written a great deal of music—much of it ambitious in scale and intention. Yet everything had been done in isolation. There had been no equally gifted friends to poke fun at him and show him his mistakes, no one to test his ideas against. He made a difficult pupil because he already knew too much and yet knew it only through the light of his own limited experience. He had plunged into serious composition before he could swim, and found it difficult to understand that simply because he hadn't drowned there could be anything left to learn.

Nevertheless he was grateful to his teachers and in later years spoke affectionately of both Stanford who had tried to make him a virtuous imitator of Brahms, and Walford Davies who, through the study of Bach and Palestrina, had taught him 'the spiritual value of counterpoint'. In the long run, however, the most valuable lessons may have been those of St John Dykes who, despairing of ever turning his hearty attacks on the piano into an orthodox technique, reached the sensible conclusion that he might do worse than get a thorough grounding in Bach fugues and Beethoven sonatas.

Such was the pressure of college work that only two pieces of any size date from this period: the symphonic suite *The Chilterns* and the symphonic poem *Imperial Elegy*, completed in 1900 and 1901 respectively. The former is a four-movement work lasting approximately half an hour. It was first performed on 31 December 1901 at a concert given by the Incorporated Society of Musicians in the Hotel Cecil—one of seven works, including Josef Holbrooke's *Ode to Victory*, selected from a batch of seventy. It was widely and favourably reviewed. E. A. Baughan, writing in the *Musical Standard* on 4 January 1902, was typical of the general approval: 'Mr Boughton has invention because he has something to say, and he uses his orchestra as if he thought in orchestral colours.' He also noted that the work 'owed something to Wagner and Tchaikovsky'. Stanford, on the other hand, thought it 'curiously like Berlioz' (a dubious compliment in those days) and, according to Boughton, 'improved' his English scherzo tune by 'twisting it into an Irish Jig'. A subsequent performance at the Crystal Palace met with approval, but Boughton eventually withdrew the piece.

The *Imperial Elegy* is of greater interest, partly because it reflects the patriotic emotions brought out in Boughton by the events of the Boer War (in which Edward, his brother, was fighting), and partly because some of the music was used again in his music-drama *Alkestis*. His account of its origins pinpoint the date to 23 January 1901 and a birthday treat abandoned because Queen Victoria had died:

With two other College students I was waiting in the queue for admission to the gallery of Her Majesty's Theatre, when notice was given that there would be no performance that night as the Queen was dead, so we tramped back in the night gloom along by Hyde Park to South Kensington without a word, and the tramp of our feet grew into the rhythm of the work. A week or two later it was tried over by the College orchestra. Dr Stanford told me that it was the ugliest thing he had ever heard apart from the music of Richard Strauss. I tried to look unconcerned, but something of my heart was in that work.

Vaughan Williams, who was in the audience, overheard Stanford's helpful comments and with typical generosity of spirit comforted the composer by observing that 'Some of Strauss's uglinesses are better than Stanford's beauties.' Holst, on the other hand was less approving and, according to Boughton, merely 'goggled his eyes at me as if I were a strange insect'. The first public performance took place at a Henry Wood Promenade Concert on 22 September 1903 in Queen's Hall. By this time Boughton had decided that 'good art ought not to be topical' and had therefore changed the title to *Into the Everlasting*, adding a quotation from Walt Whitman as a clue. Though this only served to confuse the critics, who could not 'see the drift of his symphonic poem, in spite of—or perhaps because of—his motto', they were still able to praise its 'strength and imagination'.[1]

But if student life had its disappointments and difficulties they were nothing compared with those at home. For some years the family business had been in decline. Both partners had tended to borrow from the firm when sudden personal expenses arose. The finances inevitably got into 'an unholy mess', and William Boughton tried to put things right by 'borrowing' from a local Friendly Society of which he was the treasurer. The unsanctioned loan was discovered and he found himself in an awkward predicament. When his son heard of it he took immediate steps to retrieve the situation with a loan from Edgar Bainton (a matter of £30) and thus saved his father from complete disgrace. But the incident proved too much for the business. A meeting of creditors was called and William Boughton was forced to retire from the partnership, handing everything over to his brother who eventually ran the entire venture into the ground. The Boughton family moved to a larger and cheaper house so that rooms might be let out to boarders. Boughton gave up his London lodgings and returned home, travelling up to the college twice a week for lessons, but giving his mother the benefit of his maintenance money.

The new mode of life was in some respects a happier one. Seeing his father wander round the house, tired and dispirited, searching out jobs to occupy his mind, suddenly an old man, Boughton felt in sympathy with him for the first time. He began to realize that what had so often seemed like harshness and bad temper had been as much the result of worry and overwork as anything fundamental to his character. It seemed, too, that his behaviour had sometimes been inspired directly by his wife, whom Boughton now suspected of wishing to keep her children's affections entirely to herself. As he grew nearer in under-standing to his father, he grew more and more critical of his mother and the things she held dear.

One of these was religion. As early as April 1898 he had begun to record his difficulties in verse that contrived to be challenging and deferential at one and the same time:

> I sought Thee in religion—found Thee not.
> Religion maketh Thee a poor weak fool
> As e'er one could imagine to exist.
> (I say it in all reverence of Thee)—
> Religion is too narrow for a God.

Now, his doubts sharpened by argument with his fellow students, fur-ther reading served to undermine his faith altogether:

One day, when looking through the books in a penny tray outside a second-hand bookshop, and casually opening one by Annie Besant, I read 'Jesus Christ was the illegitimate son of a girl called Mary.' That was a bit of a stunner. My own church-going habits had been entirely broken and I felt guilty about them; but if there was any truth in this blasphemy my neglect would become a virtue.

I first tried out the thought on an uncle—not the Methodist, but a bank manager, the youngest and most successful of my father's family. He rebuked me and wrote to my father to enquire if he knew that his son was on the way to perdition. There then ensured a long argument between my mother and myself in which I retailed Annie Besant's theory and my mother expressed her grief. The argument finally petered out, but left us in an entirely new relation-ship.

The truth was that Rutland Boughton had outgrown his family in every possible way and the fact could no longer be disguised.

There were, as it happened, even greater changes on the horizon. The money for his studies at the Royal College had been set aside as a lump sum to be drawn upon as and when expenses arose. Now it had been used up and all attempts to find an alternative source had failed. Boughton left the Royal College of Music in 1901 at the end of the

Spring Term, not because he had grown impatient with the restraints
of technical study, but because there was no more money. He left as
well prepared as any of his contemporaries—that is to say, he had
a considerable technique, but no style of his own. Only time could
remedy that.

The next few years were among the most difficult of his whole life.
Having nothing to contribute to the family budget, and aware that he
was only filling a room that might otherwise be let, he left home and
went into cheap lodgings in Harrow (19a Wellesley Road). He lived for
some time on the very edge of starvation, existing for one dreadful
week on 'a loaf of bread and a small pot of treacle'. Even his landlord
was moved and instead of collecting the rent left a florin on the
mantlepiece. But Boughton was too proud to accept the gift, preferring
to walk in hungry desperation to friends in Hampstead—only to find
their situation almost as bad as his. He set out again for Kensington
where he hoped Edgar Bainton might be able to lend him some money.
On the way he met a tramp who was about to throw away a bag of
crusts that were not to his taste. Boughton was only too glad to eat
them. At one point he applied for the post of organist at a South
London church, hoping that his recollections of the harmonium would
pull him through. He was soon found out, though the clergyman who
interviewed him was moved to stand him a square meal. There were
also more practical attempts at fund-raising, including, on 24 October
1901, an all-Boughton Subscription Concert in the Assembly Room of
Aylesbury's Victoria Club. The programme featured settings of Kipling
poems (*Six Songs of the English*) and *The Chilterns*—played, however,
not by an orchestra but economically in an arrangement for piano and
harmonium. Flowers for the occasion were 'loaned by Mrs Astley P.
Cooper and Mrs A. T.L Hobley'—the latter soon to figure disastrously
in his life.

   Little by little his circumstances began to improve. The music critic
Alfred Kalisch, who had been responsible for handling his college
allowance, put him in the way of several hack jobs: arranging the vocal
score of one of Baron Frederic d'Erlanger's operas, and, in 1903, trans-
posing part of *The Dream of Gerontius* for the benefit of the German
tenor Ludwig Wüllner—'Elgar's approval of the job was the sort of
thing that stiffened my self-respect.' And one day he introduced him
too the great Welsh baritone David Ffrangcon-Davies who was looking
for an accompanist. Ffrangcon-Davies took to Boughton and offered
him the job.[2]

   It proved to be an invaluable experience. Boughton accompanied him
when he gave lessons, watched and listened and learnt for himself.

Eventually Ffrangcon-Davies decided to take his accompanist in hand and train his voice—initially, it seems, because he couldn't stand his strong Buckinghamshire accent. 'You can't go round saying "Beowton",' he growled, 'it just won't do.' And soon his pupil had learnt so well that he was allowed to act as his deputy when he was too busy to teach. On several occasions Boughton also acted as his accompanist in recitals, earning respectful notices from the press—*The Times* (12 November 1903) remarking that he had played 'with sympathy and spirit'.

In addition to training his voice, Ffrangcon-Davies encouraged Boughton's composition. In October 1903 he introduced three of the *Six Songs of the English* at a concert in Queen's Hall, and in the following year sang the complete cycle. Shortly afterwards he bought the exclusive rights for 10 guineas and sang the songs frequently, both in this country and abroad. Like the *Imperial Elegy*, the cycle reflects the mood of intense patriotism Boughton was going through in common with the rest of Boer-War England. The titles tell it all ('Fair is Our Lot', 'The Coastwise Lights', 'Song of the Dead', 'The Price of Admiralty', 'The Deep-Sea Cables', 'Song of the Sons') and the musical style, pre-echoing that of Stanford's *Songs of the Sea*, is appropriately imperious. Two other works are in the same vein: the orchestral 'March of the British' (dedicated 'To Our Brothers and Cousins Overseas') and a setting for baritone and orchestra of Schiller's poem *The Invincible Armada*, dedicated 'To my Brother (At War, 1901)'.

Of rather more interest, in so far as an understanding of Boughton's character and emotions are concerned, is the opera *Eolf* composed between 1901 and 1903. It is the first dramatic piece he wrote to his own libretto, and the first to show a tendency to dramatize his own situation. The story (his own devising) is pure fustian, and the libretto seems to have been concocted under the baleful influence of the Poet Bunn:

Eolf, the son of a nobleman, has been captured in battle and is now a slave in the service of Oscar, a monk and part-time minstrel. Oscar [as at this date the name might lead one to expect] is a villain and particularly jealous of rival minstrels.

The opera opens with Eolf in a mood of depression because he has been forbidden to sing. He is cheered by the arrival of Eda, also one of Oscar's slaves, with whom he is in love even though she is of humble birth. Stimulated by Eda's beauty, Eolf bursts into song and is overheard by a passing nobleman, Ethelred, who offers to take him into his service and train his voice (Ethelred is also a part-time minstrel). Eda urges him to accept the offer:

> O love, O man, O minstrel mine,
> Reject not this deliverance!

Eolf leaves with his new master, promising to return as soon as his voice is trained.

Oscar now rushes in, having heard the sound of unlawful singing, but is too late to stop Eolf. Furious at being thwarted, he gives Eda to his henchmen, fastidiously observing:

> To me she'd be but humble gain,
> Nor will I honour so his choice.
>   Here, lusty varlets—
>   Take yon shrinking thrall.
> Do with her as ye best will know.

In the second act, King Arthur holds a competitive Festival of Song which Eolf, inevitably, wins over all others, including Oscar and Ethelred. His prize-song urges the King to take a step forward in political evolution and free all the slaves. The King agrees. Oscar is furious at this fresh blow to his prestige:

> O foiled, foiled, foiled!
>   O hate! O vengenace!
> Power inferne, to me, to me
> And aid my endangered power!
>   [A flash of lightening. Curtain.]

In the third and last act we are back outside Oscar's castle. Eda, unlike most operatic heroines, has contrived to live through a fate worse than death but now languishes in appropriate despair. Eolf returns in triumph to claim her hand, but Oscar describes the nature of his revenge. After only a moment's hesitation Eolf magnanimously decides to overlook the whole affair and takes Eda to be his wife.

It is not difficult to see in the character of Eolf an unconscious projection of the young Rutland Boughton, determined to win through the power of his music. Oscar, a monk, may be a symbol of corrupt religion (the values, perhaps, of life in Aylesbury); and Ethelred is clearly the generous Ffrangcon-Davies. Eda is most probably a reflection of the unfortunate Ethel Gilbert—though it is difficult to imagine anyone inspiring so vapid a character and it is more likely that she is simply Boughton's idea of what a woman ought to be: faithful unto death or worse, and self-denying to the point of imbecility. It is also apparent that the state of disgrace from which she is raised by Eolf's excessive nobility foreshadows a situation that was soon to develop into the tragicomedy of his own first marriage.

Mixed in with this farrago are Boughton's views on patriotism and socialism, and the notion that art has the power to purify men's souls and bring about the recognition of some ultimate truth. They are views that were to recur again and again in his later works. The music itself is influenced by Wagner, but lacks themes of any distinction. Boughton realized how inadequate the work was, for he never completed the orchestration, even though it was 'forward-looking enough to include a saxophone'. He later destroyed what little he had done, but could not bring himself to jettison the vocal score.

Although the work with Ffrangcon-Davies was both enjoyable and instructive it did not bring in enough money for even a mildly comfortable existence and Boughton was forced to look for other outlets, especially since he was anxious to contribute to the family budget. He had already drawn attention to himself through articles written for the *Musical Opinion* and *Music Trades Review* from as early as July 1898 and as a result E. A. Baughan, its editor, recommended him as music critic to the *Daily Mail*. A certain difficulty arose when he was called for the interview, for he realized he had no presentable clothes. Once again a characteristic mixture of ingenuity, naïvety, and sheer effrontery came to his rescue: he wore the only passable things he could get hold of—a pair of white flannels and a white shirt—trusting that the editor would think he had just come from the tennis courts and had not had time to change. He got the job.

Within a few weeks of joining the paper he came up against a problem that was to recur in all his periods as a journalist. He discovered that he was not always expected to tell the truth and that sometimes his opinions had to be trimmed to suit the paper's editorial bias. This he found repugnant and quite impossible to accept. After several warnings he was summoned to Lord Northcliffe's office on a charge of having written 'Melba won't do'. Northcliffe took one look at his music critic and said 'Boughton won't do', and that was that. The month's salary in lieu of notice was small comfort. But once again his luck held, this time leading him back to Paul Graener who was now conductor of the pit orchestra at the Haymarket Theatre. Hearing of Boughton's plight he immediately offered him a job. The fact that he could play no orchestral instrument caused only a momentary hitch. Graener produced a harmonium and instructed his new recruit to fill in any missing parts.

This became Boughton's working life for the next two years: eight performances a week at 6s. 8d. a performance, rehearsals in the afternoon, and the probability of being kept hanging around until after midnight. It was dispiriting work, but may have taught him something about theatrical effect. At the very least, as he later recalled, it taught him 'what not to do'. Only one piece of music was written as a direct result of his work at the Haymarket: an overture to Cyril Maude's production of *The School for Scandal*. Though Boughton destroyed the score, a study of the orchestral parts suggests a liveliness appropriate to the play even if his assessment of the work as 'little more than a pot-boiler' is accurate enough.

More important for his own development were the long discussions and arguments that went on between the members of the orchestra. As a result his thoughts began to turn more and more to politics and to an acceptance of socialism as the proper alternative to his father's staunch conservatism and the liberalism with which he had been toying as a result of that party's recent pro-Boer stance. The titles of two songs he wrote at this time reflect the direction his political opinions were taking: 'In Prison', to words by William Morris, and 'The Love of Comrades', a setting of Walt Whitman. The writings of both men, together with those of John Ruskin, had made a deep impression on Boughton and form the basis of the political views that were to have such a crucial effect on his life and career. The communism which he eventually embraced proved to be more the child of English nineteenth-century liberal-socialism than anything to do with Karl Marx.

Nor is it surprising that his thoughts should have turned in this direction. His upbringing and experience made his mind peculiarly receptive to socialist theories. Born into an atmosphere of unquestioning religious conviction, he had come to realize how fundamentally the dogma and practice of the Church differed from the actual teaching of Christ. He had experienced poverty and felt the lack of privilege. The need to work out a Christian ethic in practical terms, to see an end to poverty, to feel a part of a caring society—all found answer in the socialism he was beginning to discover.

A further aspect of this development is revealed in the subject-matter of the Symphony he completed in 1905. It bears the sub-title 'Cromwell' and is intended as a character-study of the man he obviously regarded as a hero (not for nothing had John Hampden's statue gazed resolutely over Aylesbury's market square!) The four movements detail first Cromwell's general character; then his changing moods of grief and resolution, as expressed in the letter he wrote to his wife before the Battle of Dunbar (second movement); the turmoil of the Civil War itself (a brisk march movement); and finally his death, including a setting of his last prayer, as recorded in Carlyle's biography. Boughton dedicated the work to his friend Georg Graener, in whose company he had discovered 'the only right aesthetic path' through their mutual admiration of Ruskin. Though chosen for performance at a Patron's Fund concert, and read through by the Royal College of Music's senior orchestra, the symphony seems never to have been performed in public. Boughton eventually withdrew the score, using two of its themes for *Alkestis* (one to represent the heroic nature of Herakles' character). Musically speaking, the style of Boughton's first symphony leans heavily on Wagner. Its themes are bold and effective, though the working-out is somewhat pedestrian. The orchestration, however, is brilliant, as would be the orchestration of all his more important works, early and late. His decision to withdraw it was, in the light of his subsequent development, understandable and wise, but it is interesting to observe how the idea of a hero intent upon leading his people to salvation had crept in—for it is the idea that underlies the Arthurian cycle on which he would shortly begin to work.

But if Boughton's political ideas developed during this period, the same cannot be said for his ideas on other aspects of life. In matters of personal relationships he was as naïve at 25 as he had ever been. Although unable to accept his mother's religious convictions, he had managed to let some of her other beliefs go unchallenged. Among them was the disastrous notion that love came only once in a lifetime. Thus it seemed obvious that he had already squandered his ration of available love on the faithless Ethel Gilbert, and when the daughter of a hard-drinking next-door neighbour began to make herself pleasant to him he offered no very strong show of resistance. He was sorry for the girl. 'I did not love her', he later wrote, 'but admired the way she tried to protect her younger sisters and brothers from their drunken mother's ill temper.' Within a short time they began to look upon themselves as more or less engaged.

And there matters might have rested, for he was in no position to think of marriage, had it not been that the drunken mother became so abusive that the girl ran away from home and threw herself on his mercy. She arrived at the stage-door of the Haymarket in a state of collapse and Boughton took the sensible course of packing her off to stay with his friends, Edmund and Lizzie Pengelly, in Hampstead.

Next day I wrote to her parents, told them what I thought of them and that we intended to get married. Meanwhile the Pengellys, who had a clearer view of what this was going to mean, persuaded Florence to write to her parents asking them to take her back. When they told me what they had done I got it into my silly head that the girl would be going back to a hell and would have to suffer extra for what she had done. So I persuaded her to return with me to my lodgings, where I explained the situation to my very unhappy landlady. After leaving her I went to the theatre as usual. But when the night's show was over my landlord was waiting for me with the news that Flo's mother had turned up at my lodgings in Harrow after going to the Pengellys and finding her gone.

When I alighted at Harrow Station the mother was waiting for me and wanted to know what I had done with her daughter. She reviled me and said that Flo would not have come to me unless there had been physical intimacy between us. Now I had been addicted to regular masturbation for years and was revolted by the idea of sexual relations with a woman, so I denied the accusation and said that her daughter was safe and sound at my lodgings. The old fury said 'Then she shall come home and you will never see her again.' That course would have been the best for all of us, but it seemed to me, in my fool's courage, equivalent to dooming the girl to the protection of a cowardly father and the carving-knife of this old demon. I recanted my previous denials.

The next thing he knew was an umbrella breaking over his head and a torrent of abuse falling about his ears. He was marched back to his lodgings at the point of the umbrella and locked in until the following day when he was allowed out to make the arrangements for a hasty wedding. And so, on 16 December 1903, Florence Elizabeth Hobley became the first Mrs Rutland Boughton. Back home in Aylesbury it was not until the September of the following year had passed without incident that his mother felt able to look the neighbours in the face again.

One more tragedy was to occur before he finally left London. His father had been unwell for some time and by the spring of 1904 it was obvious that he was dying. During the last few weeks he was nursed almost entirely by his son who was profoundly affected not only by the painful way his father had to die, but also by the gross inhumanity of the doctors who seemed only interested in delaying death. In the end he refused to let them administer the brandy that alone seemed to be keeping the cancer-ridden body alive and, sustained by his father's gratitude, remained with him until he died. With the death of William Rutland Boughton the old pattern of life came to an end. Within a few months the Boughton family had left Aylesbury for ever, and Rutland Boughton's own career had begun to move towards a new and exciting future.

3

1905–1907

The opportunity for Boughton to leave London and the grinding routine of the pit orchestra occurred early in 1905. Again it was through a chance encounter, this time with Granville Bantock.[1] The two composers already knew of each other. Boughton had written appreciately of Bantock in the 'Studies in Modern British Music' series he had contributed to the *Musical Standard*, and as Principal of the Birmingham and Midland Institute's School of Music Bantock had been able to form an opinion of the younger composer's worth when his symphonic poem *A Summer Night* was given its first performance by the Halford Concerts Society on 25 November 1902. When he heard how Boughton was having to earn a living, Bantock lost no time in offering him a post. In July 1905 Rutland Boughton left for Birmingham to take up his duties as a teacher of Piano and Rudiments.

They lived at first in Coventry, in the house which his mother and sister were already sharing with his brother Ted who worked in Daimler's drawing office. Initially things went tolerably well. Flo, whose own upbringing had been decidedly erratic, began to learn more economical ways under the eagle eye of her mother-in-law. The first of their children (Ruby) had been born in October 1904 and two more (Estelle and Arthur) were to arrive in March 1907 and January 1910, but Flo's attitude to motherhood seemed to Boughton rather less than satisfactory:

I do not think I am exaggerating when I say I became half mother as well as father to my little daughter because of the limitation in my wife's attitude to her. For example, one evening when a cat got into the baby's bedroom and was itself frightened and wild because it found itself in a strange place, my wife ran shrieking from the room instead of getting rid of it. A trifling occurrence, but the sort of detail much multiplied by circumstance . . .

For a year the household managed to contain three women without undue strain, but in June 1906 it collapsed with the addition of a fourth. Ted married, and his bride soon found herself quite unable to cope with her sister-in-law's vagaries. The two households split up, Boughton departing with his family to live in what was then the quiet country village of Marston Green, a few miles from the centre of Birmingham.

Despite the domestic upheavals it was a period of great excitement for Rutland Boughton. For the first time in his life he felt free and in touch with his innermost being. The sense of elation, at times, approached a profound spiritual experience:

Walking late the other evening through the leafiest Warwickshire lanes, at the hour when all colour deepens into blue and green, I became aware of a very wonderful thing. For the last five years I have been more or less swamped in the vortex of London. The dust of the city has been in my eyes; the noise of the city in my ears; the fog of the city in my throat; the hard stones of city pavements in my heart; the poisonous commercial vapour of the city enwrapping my senses, even as gas enwraps and poisons the senses, physical and spiritual, in the drowsy dens of a city's trade, its offices and shops. In that condition of spiritual coma I have spent five long years. Five long, wasted years! For five long years I have striven to reconcile the spiritual longing and the musical longing of man— they are the same thing—with conditions of life that stifle spiritual and musical longing. For five long years I have allowed my perceptions of life and music to suffocate. For five long years I have lain in a state of spiritual stupor. In that state of stupor I have dreamed of a life and of a noble art that should be ethical because I have willed it so; I have dreamed that my fellows, suffering with me, should be influenced and ennobled by an art made ethical by my own will; I have dreamed that arts should establish definitions of good and evil, wilfully accepted by my fellows because wilfully purposed by me; and I have been mystified when others could not dream with me.

But the other evening as I passed along that lane, losing myself in its gulfs of blue and green, elms reaching from either side with outstretched arms in a large and loving embrace, suddenly, as in a revelation, the city-coma was lifted from me and I was awake, alive, merged in the force and poise of the world itself. In that moment I knew why savages and all simple folk are obliged to invent some idea of a god. In that moment I became aware of what music is, and why it is by all peoples and in all ages bound up with religious observance.

We have all probably known, at some time or other, the curiously impersonal feeling of looking on while other forces strive for possession of us. Some call those forces God and Satan; others, Force and Matter; others, Life and Death; others, Good and Evil; while others regard the struggle as a phase in the evolution of man's nobler self. So far as words go, it is none of these things: so far as meaning goes, they are all true. That struggle affects music because it touches the deeps, the vitals of our whole existence; and during the coma of the last five years that struggle has been strong in my clay. Now it is over. Not because Good or Evil have triumphed. Neither has. But because in that moment of beautiful evening gloom they both slipped outside my conception of life.

Work left little to be desired. The atmosphere at the Midland Institute was stimulating and his pupils were quick to recognize what they had gained. 'He was', wrote one of them,

a torrent of vitality. We were sustained and inspired by his integrity and sincerity, encouraged by his generosity and understanding, and impressed by his fine musicianship. How he simultaneously insisted on and drew out our best efforts! How merciless he was to anything shoddy or mean in our work or in ourselves!—for his influence, like that of any good teacher, was far wider than the mere instruction he gave us.[2]

Boughton himself was a little nervous at first, fearing that his youth and still boyish appearance would tell against him. As a precaution he grew a beard, but only managed to make himself look, if anything, even more vulnerable. He need not have worried. The classes in Rudiments proved such a success that others in Harmony and Sight-Singing were soon added, and when Ernest Newman gave up his post as singing teacher Boughton was invited to take his place. The years with Ffrangcon-Davies began to pay dividends.

As a teacher of singing Boughton based his work on the belief that the human voice can only properly be divided into two categories: male and female. He held that it was pointless to make further divisions in the early stages, and set about training his pupils' voices as a whole. His own range was considerable, running from a low baritone G to the high B natural in the tenor. He proved the point on many occasions, but never more spectacularly than with Arthur Jordan who arrived at the Midland Institute as a baritone and left to sing Siegfried at Covent Garden.

Success as a teacher soon led to other things. People began to talk about this bright-eyed, scurrying little man. They responded to his quick, decisive gestures, the quivering excitement of his high-pitched, thin, clear voice. He bubbled over with vital energy, infectious and irresistible. Towards the end of 1906 he was asked to take over the Birmingham City Choral Society while its conductor, Frederick Beard, was away in Australia. It was a fine choir and he revelled in the opportunity. His first concert, on 21 February 1907, was nothing if not ambitious, consisting as it did of Humperdinck's *The Pilgrimage of Kevelaar*, Coleridge-Taylor's *The Death of Minnehaha*, the 'Grail' scene from *Parsifal*, and Act II, Scene ii, of *Tristan und Isolde*. Ernest Newman, reporting for the *Birmingham Post* (22 February), was very complimentary:

Even in the great finale [of the love duet from Tristan und Isolde], in which Mr Boughton obtained a superb climax from the players, the voices were never lost in the swirling flood of tone that came from the orchestra. The whole concert was a decided triumph for Mr Boughton, on whom the bulk of the evening fell and who did his share of it better than anyone else on the platform. He could not have been set a more difficult task for his first appearance as a concert conductor that last night's programme. He came through the trying ordeal with the utmost credit. He always had his forces well in hand, and showed great breadth and versatility of sympathy in his treatment of music so widely different as that of Tristan and Parsifal.

It would have been too much to expect that every member of the choir would fall under Boughton's sway, and when it was discovered that Mr Beard proposed to stay in Australia the choir divided into factions over their new conductor. Although the pro-Boughtons were in the majority the choir disintegrated under the tension, only to be resurrected by Boughton in 1908 as 'The New Choral Society'. Its first concert, in the Birmingham Town Hall on 10 December 1908, was greeted by an enthusiastic press. 'Nothing better, all things considered, has been done in Birmingham by any choir' declared the *Birmingham Gazette & Daily Express*, 'The choir', wrote Ernest Newman in the *Birmingham Daily Post*, 'has made an excellent beginning . . . There was enjoyment to be had out of all it did . . . It was best, perhaps, in Elgar's impressive *Owls*, here not only were the considerable technical difficulties quite overcome, but there was an intimate understanding of the whole meaning and atmosphere of the piece.' Subsequent concerts saw Boughton conduct an impressive range of work, from Beethoven's Seventh Symphony and Violin Concerto to Edgar Bainton's *The Blessed Damosel* and Algernon Ashton's C minor Symphony—all part of a five-day 'Arts & Crafts' exhibition in September 1910 for which he acted as musical director. Perhaps even more to his taste were the massed voices of 3,500 children as, under his baton and accompanied by Messrs Priestley's Military Band, they sang an alfresco farewell to King Edward and Queen Alexandra after the opening of Birmingham University of 7 July 1909. He was, in short, a force to be reckoned with in Birmingham's musical life.

In its turn, choral conducting led to an increase in Boughton's choral writing, and there followed a series of works that helped to establish his reputation was a composer of importance and through which he began to achieve a style of his own. The most immediately popular were the sets of *Choral Variations on English Folksongs* that appeared between 1907 and 1910. Like many of his contemporaries, Boughton felt that English music needed to speak out boldly with an English voice and saw in folksong the key that might help bring this about. In these ingenious sets of variations the folksong is not so much absorbed into the composer's musical language (as it was in the case of Vaughan Williams) as used as a focal point for a series of elaborate exercises in a style that could only have emerged from late Victorian England. It was to be some time before Boughton can be said to have absorbed the folksong influence. There is, therefore, much in these variations to annoy the purist. But to anyone willing to enjoy them as bluff, good-humoured dramatic sketches, they still have something to offer.

So far as his standing as a composer was concerned, 1909 was a crucial year for Rutland Boughton, and by the end of it he could reasonably claim to be one of the most promising men of his generation—a talent to watch. The process of recognition had begun in November 1908 when a short orchestral work *Tintagel* (in fact the Prelude to the second act of *Uther and Igraine*, the first of his Arthurian music-dramas) was performed by Thomas Beecham at a municipal concert in Leeds. The *Yorkshire Post* (30 November) greeted it with enthusiasm: 'It is rich and sombre in colour, and the moaning and surging of the sea is vividly suggested. The themes have individuality and the music is forceful, possessing genuine poetic feeling.' In January 1909 he completed 'The City', a motet for unaccompanied voices to words by Henry Bryan Binns. It was an unequivocal statement of his hopes for a socialist future:

> I see a city being wrought
> Upon the living rock of human thought.

He was particularly proud of the workmanship and therefore dedicated it 'To Charles Villiers Stanford—My Master, 1899-1901', though what that testy luminary thought is not recorded.

On 16 March Edward Mason's choir gave the first performance of *The Skeleton in Armour*, a work originally designed for solo baritone but now rearranged for full choir and orchestra. The performance, at Queen's Hall, was something of a triumph. Boughton was called to the platform three times and was gratified to be assured in the subsequent issue of the *Musical Standard* that he had written 'music of fibre, music that shows the sure touch of a master—a man who knows what he is about'. The climax came, however, in the autumn. On 5 October his most ambitious choral work, *Midnight*, made a deep impression at the Birmingham Triennial Festival. Ten days later the Southport Triennial Festival heard the first of his sets of Choral Variations ('William and Margaret' and 'Widdicombe Fair'), and on Thursday, 21 October, *The Invincible Armada* received its première at the Newcastle on Tyne Triennial Festival. It must have seemed as if the Midlands and North Country were devoting all their musical efforts to the service of Rutland Boughton.

By far the most important of the new works is the Symphonic Poem for Chorus and Orchestra, *Midnight*. It is a setting of words from Edward Carpenter's *Towards Democracy* that trace the thoughts of one who considers the sleeping city, its miseries and weariness for the moment at rest, and who is reminded by the tolling bells that the thoughts of men are changing little by little and that even now there are those who 'dream the impossible dream' of social justice—'the sound which is not yet on earth'. Boughton had completed his setting in December 1907 and offered it to the Birmingham Festival for the following year. The Committee had agreed to take it if it met with the approval of the principal conductor, Hans Richter. 'I played it to him', Boughton remembered, 'he walking about the while and puffing like a grampus. At the end of it he said "Very well! Very well!" and bundled me out. Such was Richter's way of dealing with young composers!'

Midnight met with a rapturous reception and on 6 October was warmly praised by most of the critics. The *Manchester Courier* declared: 'It is spontaneous music, unforced, direct, yet charged with mystical feeling. But, above all, it is original: it is unlike anything else that has been written in our time, and it is difficult to trace the sources from which it has been derived. Mr Boughton is now a composer to be reckoned with.' While the *Yorkshire Post* found that: 'The orchestra is often finely and subtly used, and helps to convey a sense of the spaciousness and mystery of night. There are one or two really fine climaxes in the work—a superbly brilliant one near the middle . . .'. The *Liverpool Daily Post*, on the other hand, was upset by a passage depicting the clatter and chatter of clocks and bells as they strike the midnight hour—a moment evidently too advanced and dissonant to be passed over without a reprimand: 'The music did not seem in keeping with the sustained dignity of the composition, and the undignified treatment of the voices jarred somewhat . . .'. Not to be outdone, the critic of the *Birmingham Post* took great delight in airing his discovery that six of Boughton's notes (an awkward linking passage—see p. 257) were identical with two bars of Sullivan's *The Golden Legend* (the aria and chorus 'The night is calm and cloudless'), and appeared to think it a hanging matter.

From Boughton's point of view the Newcastle performance of *The Invincible Armada* may have been even more exciting, for Elgar was in attendance to conduct *The Kingdom* and his First Symphony. Since Clara Butt was to sing his *Sea Pictures* in the same programme he may have stayed on to hear Boughton's work (he had missed the Birmingham première by one day), noting, perhaps, the arrival of a new and forceful voice on the British musical scene. Whether he met

Boughton on this occasion is not recorded, though he may have recalled a brief correspondence with him in November 1897 when Boughton wished to write about his music.[3] But whatever Elgar may have thought of Boughton and his music the critics were in no doubt about its value, as the *Morning Post* made plain on 22 October:

The music of The Invincible Armada is as boisterous and vigorous as that of Midnight is impressionistic and intangible. The difference between the two is extraordinary, and since both are equally successful in their several ways the proof of the highly varied gifts they afford is unmistakable. Mr Boughton follows the varying phases of expression in the poem with all possible success. A highly commendable feature is to be found in the legitimacy of the part-writing. It is often daring in tonality, but always laid out with the hand of a craftsman. The piece, therefore, retains all the best features of the British school of vocal composition and, effective in every part, is likely to achieve the immense popularity to which it is entitled.

Novello & Company, who had published all three works, must have felt they had backed a winner, even though in *Midnight* they had been unable to bring themselves to print the words of a passage depicting a young bride, her husband 'worshipping sleepless on her bosom'.

To some papers, however, it seemed that Mr Rutland Boughton was also a composer with definite political views and one or two of them were rather sniffy about it. 'Mr Boughton', wrote the *Daily Telegraph*, 'is very young to have already begun the issuing of political, or economical, or socialist tracts—one of which *Midnight* appears to be.' And Boughton himself had the peculiar pleasure of overhearing, as he left the platform, a scandalized lady whispered to her friend 'He's a Socialist, you know!'

Indeed he was, and it was not very surprising. To a young man of Boughton's temperament Birmingham at the turn of the century was a singularly exciting place to live in. It had already established itself as a go-ahead, socially conscious city under the mayoralty of Joseph Chamberlain and now it was a hotbed of liberal socialism. Under the guidance of Bishop Charles Gore (1853–1932) the Church itself had become vaguely militant on behalf of the working classes. Boughton was invited by Father Adderley (the Honourable and Reverend James Adderley, 1861-1942) to take part in organizing the weekly concerts at his Saltley church—concerts that soon became too popular for the church's capacity. At much the same time he began to appear as a regular speaker at the Sunday Evening Lecture Society which Adderley also ran. His talks proved immensely popular, and he used them to sort out his ideas on music and society and the way life should be. 'Music in the Social Life to Come' (28 November 1909) is a typical subject, cheerful in its denounciation of music as mere entertainment, and hot in its praise of music that is 'an expression of life'.

Nor were Adderley and his friend and co-worker in the socialist vineyards the Reverend Arnold Pinchard the only clergymen to exert an influence on Boughton in those days. He made friends with an equally dedicated socialist, the Reverend George Herbert Davis, and his family. Though he could not know it at the time, one of the Davis children, a delightful girl named Kathleen, was destined to assume a role of the greatest importance in his later life. But in 1909 she was no more than an engaging 9 year old, to be loved in the way he loved all children.

In the mean time there was Cornish's bookshop, stocked high with new experiences, an inexhaustible mine for the curious intelligence. There it was possible to meet with other men and women enthusiastic for the same ideas and visions. John Drinkwater was one of them.[4] Within a short time of their meeting, the two men had begun work on a version of the King Cophetua legend. As it turned out, Boughton found himself unable to agree with Drinkwater's view of the main characters and the work proceeded as a play and not the opera they had intended. But their friendship remained, and when in 1911 Drinkwater was asked to write one of the Cadbury Pageants at Bourneville he insisted that Boughton should be engaged to write the music.

Throughout this period Boughton's own literary efforts took wing. He already wrote regularly for the *Musical Standard* (as its Birmingham correspondent he was able to obtain tickets for any concert he wished), and now he found himself a welcome contributor to other magazines. An extensive series under the title 'Britannia Singing!' completed in November 1905 and amounting to a very personal interpretation of British musical history, may have appeared in magazine form, as did a similar 'Music and Democracy' series (1907), though both seem to have been designed as books. For *Musical Opinion* he wrote about Loewe's ballads (1905) and the music of Algernon Ashton (1906); while for *Music* he wrote articles on Richard Strauss (1903) and a stinging attack on contemporary musical conditions under the provocative title 'The Prostitution of Music' (1903-4). The *Musical Times* fell to his pen in July 1910 with an article on 'English Folksong and English Music', and in between whiles he found time to write a short but extremely perceptive study, *Bach*, for the Bodley Head series 'The Music of the Masters', dedicating it to David Ffrangcon-Davies with the legend:

> My Dear Friend,—to you, whose Faith and Art are one, I dedicate
> this little study of great music. One, also, are Bach's Art and Faith.

Yours in thought and work,

Rutland Boughton

It was to remain in print for twenty years and go through three editions.

But by far the most interesting of his publications at this period is the pamphlet he had privately printed in 1909 under the ingenuous title: *The Self-Advertisement of Rutland Boughton*. Its thirty-two pages amount to a detailed catalogue of his works to date, accompanied by reviews, both good and bad, and his own commentary on the origins of each work. The 'Forward' sets out his artistic credo in terms that are both powerful and moving:

All art should be a free gift. It is an abominable thing that the product of a man's deepest emotions should be carried to market like a pound of butter. But this horrid commercial civilization offers the artist no fair alternative. He must either do this—or weaken and waste his energy in bread-and-cheese work for which he is unfitted—or starve. The more a man is forced from within to paint pictures or write symphonies, the less is he willing for ever to fritter away his strength in making chairs or shoeing horses—work which other people can do far better than he. But, on the other hand, it is well for him and his art if some bodily labour fall to him. It is well for him, because work in field or forest will keep him in fellowship with the aristocracy—with carpenters, field labourers, and all men who are doing true productive work on the material side. It is well for his art because the painter of the studio and the composer of the study have a nasty habit of becoming absurd—of growing pedantically conventional, or pedantically unconventional. If an artist is to do true productive work on the spiritual side, he must live a true life on every side, developing not only the craft of his art, but his physical and intellectual faculties as well. And if these faculties can be developed in immediate intercourse with Father Sky and Mother Earth, so much the better. They are the original source of all energy on this planet. Pictures painted and symphonies composed under such conditions have a solar force behind them. But the difficulty for the modern artist is to get into such a position. The modern artist is hindered on every side, unless, like Thoreau, he is willing to forgo the fellowship of man, and if he forgo that fellowship, farewell to all the best art, for the noblest art is nothing but the spiritual contact of mankind in joy and sorrow. Those false joys and sorrows which arise in our commercialized world have neither binding power or creative force. Whatever is good in modern art comes into being in spite of modern civilization, and in direct opposition to it; but, once in being, it will be still born if the artist refuse to work through the web of commerce in which we are all entangled. This is the dilemma in which William Morris was placed. He loathed the commercial system, and was yet forced to use it. This is the dilemma in which I am placed. If you like to give me a nice little farm near a town on such terms as will allow me to do my musical work when I feel moved to do it, I will give you my music. Of course, I should not feel in the least grateful to you for the gift; only glad that you had sufficient discernment to recognize the value of my work. But you are not likely to do this, so I am forced to sell my soul to you—to advertise my music and ask what price I can get. I am a modest man (even a shy man), and have no desire to force my music upon those poor souls who don't want it; but I am convinced that all sensible people will finally want it if once they can be got to know it. I say finally, because at two of the recent Triennial Festivals the choruses began by disliking my music. However, several of the Birmingham choristers told me that my Midnight grew upon them, while it was obvious to everybody at the Festival that their hearts were in the work. And I heard that the Newcastle chorus, at the first rehearsal of my Armada, thought they were in for some 'awful stuff', but when they had learned it they placed it among their favourite works.

Now, I daresay that a good thing will eventually make its own way without being pushed. But I do not expect to live a century, and in the meantime, as I cannot afford to give you my art, I should like you to recognize my right to a decent livelihood on the strength of it. The fact that you allow this commercialized life to continue proves that you do not object to it, so you cannot logically object to this advertisement. Mr Pears did not wait for posterity to enjoy his very nice soap. Why should I leave to posterity the exclusive enjoyment of my very nice music? I can warrant the feeling which caused it to be true, the workmanship clean and honest, and the result effective.

Few composers, today or at any other time, could find much to argue about in Boughton's declaration of intent. His words hold as good today as they did in 1909, or would have done in any other century. It is a remarkable document.

Another significant preoccupation in Boughton's Birmingham days was his work as conductor and adjudicator for the Clarion Choir competitions. These choirs had been started in 1896 by Montague Blatchford, brother of the more famous Robert Blatchford, as a means of educating and giving heart to the working classes in the North of England. They spread remarkably and, as Boughton soon found, reached out to a high standard of performance. He found in them a people eager for music and responded wholeheartedly. For example, for the November competitions of 1911 he wrote, under the title 'Song of Liberty', a setting of appropriately revolutionary verses by Bantock's wife, Helena. He dedicated the work in confident hope 'To the Awakening Manhood of Britain', and sang his conviction in suitably forthright Elgarian terms. All in all, his experience with the Clarion Choirs, together with his work as Inspector of Singing in the

Birmingham Elementary Schools, convinced him of the immensely civilizing influence of music. He began to feel that music might one day succeed where religion had failed, and it is this belief that underlies the pattern taken by his later career.

1907–1911

As Boughton's fame began to spread, he came into contact with a series of people who were to influence his life in various important ways. Two men in particular left their mark: Edward Carpenter and George Bernard Shaw.

At the turn of the century Edward Carpenter (1844–1929) was a well-known and highly controversial figure. Born into comfortable circumstances and trained for an academic career, he came under the influence of Morris, Ruskin, and Walt Whitman and threw up everything to live in the country, near Sheffield, and devote himself to writing, thinking, and a certain amount of gentle manual labour— small-scale farming, the making of sandals, and the like. His poetry was written under the shadow of Whitman and, like Whitman's, celebrates the brotherhood of mankind and the need for a universal love. Such principles appealed enormously to Boughton, who would probably have been making sandals too had he enjoyed a similar private income. By the time they met, Boughton had already set four of Carpenter's poems to music and was eager to begin work one major choral work, *Midnight*. They got on well together. Boughton became a regular visitor to the house at Millthorpe, where Carpenter lived with his lover George Merrill, and the poet declared that the song-settings had 'an eagle's flight'.

He was not far wrong. Boughton's Carpenter songs have a breadth and passion that is quite striking. Even the critics noticed it. 'Sincerity of purpose and clearness of utterance distinguish them. They treat of subjects worthy of thought, and in a manner that will provoke thought. Entirely devoid of eccentricity, they yet strike the ear as original. There is a similarity of mood, here and there, with Sir Hubert Parry's virile manner of writing': so thought the *Musical Standard* (17 July 1909) as it reviewed their publication. The first group, composed in February 1906 and September 1907, consists of 'To Freedom', a title which speaks for itself; 'The Dead Christ', a grim portrait of humanity crucified down the ages; 'Fly Messenger, Fly', a call to hope; and 'Standing Beyond Time', which advances a mystical apprehension of the power of Universal Love. Boughton's approach was essentially dramatic and very different from the 'chamber' quality of most British songs. There are in these settings a distinct indication of the operatic composer to come. The second set belongs to 1914 and is somewhat more restrained. It consists of 'The Lake of Beauty', a Zen-like plea for wholeness with the Universe; 'Child of the Lonely Heart', a more personal plea for love, whose homosexual underlay Boughton almost certainly failed to recognize; and 'The Triumph of Civilization', a savage indictment of poverty.

Boughton learnt a great deal from Carpenter. Not only of art and democracy, but also of love and sexuality—for Carpenter was a courageous and outspoken champion of its free expression. 'His book *Love's Coming of Age* had its influence on a sexual nature tiresomely insistent and divorced from what the virginal man had conceived as love. His condonation of comparative promiscuity confused a mind at once puritanical and adventurous.' But there were even more confusing revelations to come:

I had grown to look on Carpenter as my own brother and kissed him when we met. But later on I learned that Horatio Bottomley was considering the exposure of Carpenter as an homosexual. The idea, entirely outside my own experience or imagination, seemed absurd and one day I chaffed my friend about the projected attack. Then I learned that it had some basis in fact. That night, for the first time in his house, I bolted my bedroom door.

Later, of course, he realized how foolish and insulting his reaction had been, but by that time it was too late to put their friendship back on the old footing and they gradually drifted apart. But if the friendship cooled, Carpenter's influence remained. 'He was', wrote Boughton, 'the gentlest and most Christ-like being I ever met.'

For George Bernard Shaw, Boughton's friendship and admiration went much deeper and lasted to the end of his life, so that when Shaw died Boughton felt that he had lost a second father. In the beginning, however, such a relationship seemed unlikely. Shaw was already a famous man and not in the least bit impressed when he heard that a young and unknown composer had criticized him in a lecture delivered to the Birmingham Sunday Lecture Society. And later, when the same young man wrote to ask for a comic opera libretto, sending music for the first few pages of *The Admirable Bashville* as a sample of his art, his reply was crushing:

27 May 1908

Dear Sir,

I tried over the Bashville music. It is no good. I could compose Waldweben of that sort by the yard myself. Besides, it spoils my music. What musicians seldom understand is that all artistic literature is itself music—word music.

The Admirable Bashville is written from end to end in the music of blank verse; and the whole comic effect of it depends on this burlesque of the music of the Elizabethans. To smother this up with orchestral tremolandos and six-eight tunes is a piece of philistinism. I do not say it would be impossible to find a sort of music that would incorporate with the word music; but the man who could do this would not waste his time on The Admirable Bashville, and his style would not be the tremolandoish sugary modulatory style that has been compounded out of Gounod and Wagner.

For the present there is nothing to be got out of the Shakespear Memorial. The project is still in fluid condition, and is practically without funds. I greatly mistrust your project of a series of Arthurian music dramas. You will simply produce a second-hand Ring. The Ring itself would never have existed had it not been to Wagner an expression of his strongest religious and social convictions; and unless you have equally strong convictions and an equally deep penetration into the social life that surrounds you, you will only waste a great deal of scoring paper which you might employ far better by trying to deal, as Strauss does (not to mention Elgar) with the modern world in a crisp and powerful style, making a clean sweep of the tremolandos and the sentimental and grandiose modulations of the nineteenth century.

I shall be in London on Friday; but I am sorry to say that I cannot find a spare half hour in which to ask you to call: I am overwhelmed with work and engagements of one sort and another at the moment.

Yours faithfully,
G. Bernard Shaw.

What Shaw can have least expected was a spirited retort which ended with the suggestion that he would be free to rewrite the music as long as Boughton was free to rewrite the words. 'That will be true collaboration and knock out the specialization that makes modern art so absurd.'

Even if he wasn't convinced, Shaw must have enjoyed the Shavian argument and recognized that he was dealing with an unusual man. Sure enough, Boughton persisted in his attempts to enlist his interest in King Arthur. On 17 September 1908 Shaw defended himself with a characteristic postcard:

Arthur be blowed! What on earth do I care about Arthur? and what could I be to you if I did? There are about sixteen communal theatre projects which have reached the stage of assuring me that they are going to materialize. Meanwhile I have lost heavily by endowing the advanced British drama, and am not to be consoled by first acts of King Arthur or anyone else. Purcell is quite good enough for me: he has no tremolandos, anyhow.

G.B.S.

But in his next letter the tone has altered slightly, and though equally crushing it is the kind of letter that could only be written to someone who is understood to be well able to look after himself:

9 November 1908

Dear Mr Boughton

Why do you want to come and play your King Arthur to me? I do not insist on reading my plays to you. Do you suppose that I am an impresario, or that I have influence at the Opera or with music publishers? If so, I assure you you are mistaken. I can do absolutely nothing for you; and your desire to waste my time and your own is not one to be encouraged, especially now that you are married and have a wife to provide for. Get your King Arthur performed if you can; and I will attend the performance if I feel inclined. Until then bend all your energies to achieving the performance; and remember that there are few experiences more trying, even to people whose business it is, than to hear a young composer making a horrible noise at the piano under the impression that he is conveying the beauties of his score to the unfortunate listener. Besides, one knows beforehand by your age and your way of going on that the score will be rubbish, though no doubt the composition of it has helped to educate you, and some scraps of it may come in later on.

Yours faithfully,
Bernard Shaw.

In time he was to write: 'Now that Elgar has gone you have the only original English style on the market . . . I find that I have acquired a strong taste for it.' But that was twenty years later, and in the mean time Shaw was content to tease his new disciple with the kind of banter that clearly afforded him the greatest amusement.

The *King Arthur* that Shaw had begged not to have to listen to was the first of a projected cycle of music-dramas on the Arthurian legends. Boughton's interest in King Arthur began in about 1906: 'I knew nothing then of Malory, and was bored by Tennyson, but the Arthurian ballads in the Percy *Reliques* and some poems of William Morris opened the doors I was looking for.' He saw in them something that would take the place of the 'Jesus' dramas he had dreamt of as a boy and which now seemed irrelevant. He explained the evolution of his thinking in an extended essay he published in 1911 under the forthright title *Music Drama of the Future*: 'I became aware of the truly prophetic nature of all the greatest art, and of the fact that the greatest artists acquire their superhuman power by acting as the expression of the *oversoul* of a people. Then I understood why Wagner had chosen folk subjects which had been produced by that oversoul.' The Arthurian legends seemed to have just this quality and he started to plan a cycle on Wagnerian lines, though differing in the crucial use they were to make of the chorus. In his eyes 'the Wagner drama lacked just that channel of musical expression which is absolutely necessary to the English people'. But the work went badly and he was on the point of giving it up altogether when he received the manuscript of a series of poetic dramas covering the very same ground.

The author was Reginald Ramsden Buckley, a young poet and part-time journalist working as a bank clerk in Halifax, where he lived with his mother. He too had reached much the same conclusions as Boughton, and under the same influences. The account he gave in *Music Drama of the Future* (which he brought out with Boughton) suggests, however, a much more romantic, not to say gushing, element in his approach:

Wagner's Tannhäuser revealed to me a world undreamed of, and I fastened eagerly upon his prose writings, and, through them, was led to a study of Greek art.

The works of Ruskin burned in my veins together with the keenest interest in the world of action. I loved the art of the pen and was ready to do something.

Having read how Milton had contemplated an Arthurian epic; how Shakespeare had conned the Arthurian chronicles; how Dante knew of the French romances; I admired the foresight and reticence of these men in leaving well alone. Tennyson's Memoirs showed that he too had thought of a musical work on those legends, before publishing the Idylls. I understood the reason— that the music of their day could not have entered into union with their poetry, but that, through Wagner, the way had become clear.

A great dream was born within me to make these national scriptures the quarry from which to hew a huge music drama on the lines of Wagner's Ring, with Merlin as Britain's Isaiah, Galahad her Parsifal, Arthur her type of Manhood. Amid all the talk of 'supermen' in an age of philosophic and artistic healers and quacks, I longed for a dramatic and poetic art, wherein the sane, healthy England might bathe, as in the pure rhythmic sea of Cornwall, loved by all flaming souls from the Round Table till now. . . . Alone I worked, putting aside all else, my good mother thinking me a little mad, a Peer Gynt weaving his own crown of straw! But, unlike Åse, she helped me through, praying that out of all his toil some good might come!

In a word, I wrote four poems, showing forth the Coming of the Hero, his Manhood, the Quest of the Grail, and the ultimate fading of those glorious days, deeming their production as popular works of art would move men to the passionate desire of my own heart, that Britain shall become again a joyful garden, a fruitful field for labour, the home of manly toil and the armed and sea-girt hero-land, as long ago.

Buckley had adopted as his motto: 'First to Dream—and then, to Do'. He sent his poem to Elgar. Elgar sent it back. He then sent it to Granville Bantock, and Bantock suggested that Rutland Boughton might be interested. Boughton was, and Buckley, overjoyed, wrote in reply on 17 October 1907: 'Your letter made me feel that the dream of my life was within grasp. Coincidence of ideas is unlikely, but I am prepared if necessary to recast the whole drama.' When it came to the point, however, he didn't at all enjoy seeing his work altered, but Boughton was determined and Buckley had to submit:

The poems were planned for Wagnerian treatment—the passages for purely musical development being represented by what Buckley called a *verbal orchestra*—and the whole thing was so musically suggestive that I jumped at it. But when I came face to face with the composition of them I found that while they were exquisitely musical, they were not dramatic. On my pointing out certain things to my poet, he accepted some of them; but others he fought me upon tooth and nail, so I was forced to the alternative of either accepting his exquisite musical lie, or of forcibly altering, omitting and adding to the libretto so that the characters might behave like the men and women around me and not like the beautiful creatures of romance.[1]

In plain terms Boughton found that the characters were not only life-less, but all resembled Buckley: 'Buckley as Galahad, Buckley as Guenevere, Buckley as Morgan le Fay, Buckley as Lancelot.' But they reached a compromise (or rather, Boughton simply went his own way and did not inform Buckley of the changes until it was too late!) and the music for the first part of the cycle, *Uther and Igraine*, proceeded steadily through 1908 and 1909.

Despite the shortcomings of Buckley's characterizations the poem had one tremendous advantage in Boughton's eyes: it gave him plenty of scope for choral writing:

When I came to the passages which he intended as mere indication of the musical mood—and full of a new poetic technique: the translation of the Wagnerian leitmotive to the art of words—it seemed to me as wanton a shame to banish their loveliness and suggestiveness from the dramatic perfor-mance as it would be to strike out all the most beautiful of the Shakespearean speeches. Then I bethought me of my Christ-Drama with its chorus outside the stage, commenting, elucidating, pointing the climaxes, and expressing that mighty spirit which can only be found in the mass of humanity.[2]

Once the music was under way the two men began laying plans for a production, and as a preliminary issued a challenge in the form of *Music Drama of the Future*. But before publication they had made a certain amount of generalized propaganda:

The first response came from Lady Isabel Margesson, near whom I was living at Alvechurch. She was building a badminton court in the grounds of her home in Barnt Green, and offered to incorporate in it a stage suited to our purpose. It would have been a wise beginning, but I was scandalized by the idea that our theatre should be associated with a trivial game and declined the offer.[3]

The truth was that Boughton and Buckley's aims and expectations were wildly ambitious, and more than a little priggish.

Music Drama of the Future must have caused its readers considerable amusement. It is a brash work, fuelled by youthful enthusiasm and self-assurance, immodest to a degree. Boughton's contribution was an essay, some forty pages long, on the origins and nature of choral drama which, despite its high-flown style and a certain amount of confused thinking, presents a stimulating case. Buckley's share consisted of the libretto of *Uther and Igraine* and an essay entitled 'The Growth of Dreams'. He was also responsible for a decidedly amateur frontispiece—a line drawing of 'The Gateway of the Future' which displays an intriguing set of values, even for the period. William Morris, George Frederick Watts, Michelangelo, and Sophocles are named as the pillars that support an archway composed of Shakespeare, Wagner, and Beethoven. Beyond the gateway is an altar upon which stands what can only be described as a plate of hot mashed potatoes. Doubtless the symbolism was clear in Buckley's mind, but to the general public it can only have seemed rather ludicrous.

The libretto is a curious mixture: passages of some power offset by stretches of utter bathos. It says all too plainly that only the greatest music could hope to redeem its more questionable features. The critics were sceptical and judged the book 'ill-timed and ill-advised'. For Boughton, however, it was a statement of faith from which there could be no going back, and it is impossible not to be moved by the bravery of his words and the nobility of his vision:

We are dreamers. That is undeniable. But without dreams nothing can be done. Wagner's dreams necessitated the building of the Bayreuth theatre. Our dramas necessitate the building of a place which Buckley has fitly forenamed the Temple Theatre. That theatre we are intent upon making the centre of a commune. There have been many communes and they have failed—for lack of a religious centre. Our theatre supplies that. It shall grow out of the municipal life of some civically conscious place if we can get such a place to co-operate with us. Failing that, a new city shall grow around the theatre.

Thus Boughton joins the ranks of those who dream that especially English dream of pantisocracy: the eternal quest for the Celestial City where Art makes all men equal.

Important as Reginald Buckley was in helping to give direction to Boughton's dreams, there appeared in 1910 a person who was to play an even greater part in the development of his art and his life. Her name was Christina Walshe. Early in 1910 Boughton had founded the 'Birmingham Literary and Musical Fellowship', a club which met regularly for concerts, talks, discussions, and anything of artistic interest. Among its members was a young student from the Birmingham School of Art. He soon found himself increasingly anxious for her company. She seemed to understand his hopes and dreams. She believed in everything he held dear, and helped him to have faith in himself and his ideas. She was everything a man of his temperament could want as a companion: equal yet complementary. In her he found something he had never suspected could exist in a woman, and having found it he knew he could never do without it again.

The situation was thrown into relief by the increasingly unsatisfactory nature of his relationship with his wife. Long before he met Christina his marriage had begun to go badly. Flo seemed incapable of running their home in a manner that came anywhere near the standards his mother had led him to expect. She seemed to spend money faster than he could earn it, and in his eyes she was, as he put it, 'an indifferent mother and a bad manager'. Worse, she had no interests of her own and seemed incapable of sharing his. The very fact that he was totally absorbed in something she would never understand made her resentful and ill-tempered. Nor was he inclined to make things easier for her. Instead of accepting her limitations he simply shut her out of all his preoccupations, so that she could only think of herself as an unpaid, unappreciated servant to his domestic needs. Had she married an ordinary man—someone less single-minded and egotistic—she might have settled to a happy, humdrum life. As it was, they were badly matched in every way. What had begun in the farce of his naïve chivalry was rapidly turning into tragedy. Probably he took few steps to conceal his new affections, for he found it difficult to practise deceit. Understandably Flo became jealous and her unhappiness made her act foolishly, aggravating the situation rather than controlling it. In the end he was forced to choose between them and unhesitatingly declared his love for Christina.

At this point Lady Isabel Margesson decided to exercise her talents for managing the crises of other people's lives. She arranged to send Flo to Canada, Christina to Chicago, and for Boughton to take refuge (rather unnecessarily in view of her other arrangements) in her own home. Christina, who had said that if ever Flo left home she would stand by him, declared that London would be a more sensible retreat. Flo decided in favour of Canada (she did not remain there, however) and on 21 April 1911 signed a deed of separation whereby, in return for a sixth part of his gross quarterly income, she gave him custody of the children—reserving for herself the right of access once a month and a more extended visit twice a year. The idea of a complete divorce was, apparently, against her principles (or, more probably, beyond her capacity for forgiveness) and would have cost rather more than he could afford. For the time being the children went to live with his mother and sister at Yardley, while he and Christina began to face up to the realities of their situation. At the end of the spring he asked Christina to make her decision:

she was doubtful, but we took a weekend away together, on her insistence that no intimacy took place. That weekend went as planned, with a nightingale to fill the sleepless hours. Then we went to consult Shaw, I taking with us some of Christina's drawings. 'Holbein!', he said when he saw them, and advised her to have nothing to do with me unless she wanted to sacrifice her art.

But they were not to be parted. From now on Christina would be, to all intents and purposes, Mrs Rutland Boughton.

The scandal was immense. Boughton was immediately presumed to be an unscrupulous adventurer. Acquaintances recalled how he had practised vegetarianism, worn a soft collar, sometimes even gone without a tie. A passing remark about free love was remembered and elaborated upon. Had he not been a champion of the notorious Edward Carpenter? Had he not supported the suffragette movement, spoken in favour of socialism? Parents whose daughters had been his pupils let their imaginations hover round the unspeakable. Nothing was assumed impossible. His fellow teachers began to invent philanderings that would have done credit to the most remorseless Don Juan—until Granville Bantock, sensible as always, forced an inquiry and had the satisfaction of seeing his friend completely vindicated.

And it is true that, whatever anyone may have thought, his relationship with Christina thus far had been entirely innocent. They had turned to each other because each was the mirror of the other's needs. They were perfect comrades. It was not until their motives had been publicly misconstrued that they began to think of themselves in other terms. The step they took was no easy one. Both had been brought up strictly, and in any case the idea of an open relationship was hardly popular in Edwardian England—and certainly not to people in their class. It is one thing to be a free thinker in such matters, but a very different thing actually to brave convention. They did so only when it was plain there could be no other course.

Once the step had been taken it was obviously impossible to remain in Birmingham. Tongues continued to wag, fingers to point. Boughton resigned his post and in July 1911 set off with Christina to try his luck in Berlin.

5

1911–1912

They spent only two months in Germany. Neither had more than a smattering of the language and without it work was more or less impossible. But the Graener family was hospitable, and Boughton's writings on Wagner enabled him to get free passes for Bayreuth. 'Neither art-form, method of production, nor theatre was the sort of thing I wanted for my own work. Only one detail seemed important: the semi-covered orchestra pit from which issued such perfect blending of sound.' He even began work on an opera of his own, but soon decided it was a 'foolish effort' and abandoned it. Christina sketched and painted, including a striking portrait of her lover. In the background she included a 'raven of ill-hap and a castle-in-the-air' as an ironic comment on his ambitions and a prophetic insight into where they might lead.

But such intuitions were for the future. For the present they were happy in each other's company and quite able to shrug off the news from England that her parents wished never to see her again. They visited art galleries, Christina teaching him to see paintings for the first time. They went to concerts. Each was eager to learn the secrets of the other's art. For Boughton it was the beginning of a new way of life: 'Christina's company was very heaven. For the first time a rather starved mind could link up with one of similar but more advanced perceptions.' He celebrated their loving comradeship by turning a group of her suffragette poems into a song-cycle, the *Songs of Womanhood*. The finest of them, 'A Song of Giving', perfectly expresses their feeling for one another:

> I make lament that I so little have to give,
> But all that I possess I offer you
> To be my love.

Whatever society might choose to say, they knew they had made the right decision and that they belonged together.

Christina Anne Stansfield Walshe, the daughter of William Henry Walshe and Agnes Eulalia Collard, had been born at 16 Trumpington Street, Cambridge, on 10 December 1888. On her birth certificate her father is described as 'tutor', which presumably means that he was working as some kind of 'crammer' for anxious undergraduates. He does not appear to have been a member of the University itself. But he was an educated man, having matriculated from Worcester College, Oxford, in 1868. He was also a member of an established Irish family that had borne arms for at least 300 years and, in its various ramifications, was exceptionally well-connected. The earliest traceable ancestor, Theobald (b. 1485) had been Laird of Carrickmain, and the artistic streak can be traced to various eighteenth-century Walshes who were distinguished Dublin goldsmiths. More ominously, from Christina's point of view, was the fact that William Henry and Agnes Eulalia had embraced the faith of the Plymouth Brethren and were therefore wedded to the infallibility of the Scriptures and eager in their anticipation of the Second Coming. Thus, while they may reluctantly have come to terms with her decision to study art and design (1908-11) at Birmingham's College of Art, her relationship with a married man was more than righteous flesh and blood could stand.

Indeed, Christina was typical of very few women at that time. To be 'Ann Veronica' in a novel was bad enough.[1] To put such advanced ideas into practice required an extraordinary amount of courage. For Boughton she was the ideal companion—eager, impulsive, ready to dash off in a whirlwind of ideas, the perfect foil for his own reflective, withdrawn nature. 'She was all for the sun', he wrote, 'and I was all for the moon.' Beyond any doubt it was the impact of her personality that first brought into focus what was original in his.

After two months they returned to London to face the disapproval of certain friends, though even this died down when it was realized how sincere their attachment was. His mother astonished them when she first greeted Christina by sliding her hands over her breasts and saying 'Why! I thought you'd be a woman with more here!' At Christina's suggestion they were now joined by his three children. He was grateful and deeply touched by her action, for she was not particularly maternal by nature, though a certain coolness and whimsicality where emotions were concerned helped to balance his own impatience and frequent anger. For the children she was like a perfect elder sister and they seem to have had no problem accepting her.

But life was difficult at first. Christina was able to earn a little as a freelance art critic for the *Daily Herald*, and this, together with the small amounts that came from the music publishers (the delightful *Three Folk Songs* for string orchestra belong to this period), enabled them to eke out an existence. Even so, they touched the borderlines of starvation—at one time being reduced to a three-day diet of pease-pudding. In the end it was Shaw who came to the rescue by recommending Boughton as music critic to the *Daily Citizen*. Besides acting in this capacity he seems also to have contributed a certain amount of art criticism under the name of J. Jameson, though it is more likely that Christina was the guiding intelligence in this instance.

Such affluence—he was earning up to £10 a week—must have seemed too good to last, and sure enough it came to an end. Boughton had been puzzled by the paper's attitude to the suffragette movement and on asking his editor for an explanation had been told, with a knowing smile, 'We are *ostensibly* in favour.' It was too much. There could be no alternative but to resign. Almost immediately he was offered work on the *Daily Herald* and though it meant a lower salary he accepted rather than see the family go hungry again. Nominally a music critic, he claimed also to have turned his hand to art and drama criticism, and even political commentary, each under a different name. But the effort of being all these personalities proved too great and after a few months he left the paper and, for the next ten years, journalism generally.

Again it was Shaw who came to the rescue, this time with a very unusual suggestion. Among his friends was an elderly Mr Frederick Jackson, a retired solicitor with socialist interests who lived securely on a substantial private income in a large house, Tarn Moor, at Grayshott on the Surrey-Hampshire borders.[2] Music was Jackson's passion and his despair. He felt driven to improvise small pieces at the piano, but was quite unable to write them down. Rather than see his music for ever lost to posterity he decided to engage a professional musician as an amanuensis. Acting on Shaw's advice he gave the job to Rutland Boughton.

The two men got on well, and when Jackson heard about Christina and the children he suggested that the whole family, which he supposed to be entirely orthodox in its constitution, might be happier living together and away from their meagre lodgings in Battersea.[3] He offered them the use of a small cottage in the woods at the far end of his estate. It seemed an admirable idea and they were soon settled in their new home. Each day Boughton would walk through the woods to Jackson's house, notate the few bars the old man felt able to reveal and then return to the cottage, free to work at his own composition. Gradually life began to assume a settled pattern and, relieved of the immediate worry of providing food and shelter for the family, Boughton was able to begin rebuilding his public career.

To begin with he advertised himself as a lecturer, offering his agent a wide choice of subjects from 'Bach', 'Beethoven', and 'Wagner', to 'Modern British Music', 'Folksong', and 'Edward Carpenter'. Then he began to look around for teaching appointments and was delighted when his former Birmingham pupils begged him to return and give private lessons. He found a studio over Crane's music shop and gradually began to win back something of the position he had lost when he had been forced to choose between conventional morality and the imperatives of his own feelings. Few people could hold out against such candour as his. He had never for one moment attempted to deceive his friends, and they in turn soon realized they had nothing to complain of: his private life was, after all, his own.

Yet he was conscious of the demands he had made on other people's tolerance and was grateful when they proved considerate. A letter to Charles Kennedy Scott, written immediately after the Oriana Madrigal Society had sung one of his *Six Spiritual Songs* and he had conducted them in Part VI of Bach's *Christmas Oratorio* at a concert on 17 December 1912, is typical of his attitude and typical of the charm that helped other people forget their reservations:[4]

Now just a word of true thanks to you for your kindness in mixing me up with your people.

First, of them: they sang my piece perfectly (except where I tried to mislead them). They put in all sorts of nuances which I felt freshly and differently at the last moment—it was like playing on an immediately responsive instrument—and we had a real pianissimo! O, it was lovely. Do thank them for me. To make a composer love his own trifles again must surely be due either to exceptional beauty in the rendering, or extreme vanity in the composer. Now I am vain, I know, but not so vain as that. No, it was due to the fact that the dear people just made up their minds to do their best for me and they did it. Bless them!

But the heart and reason of all that? Well, of course, it lies in you. It is a real shouting joy to meet a man like you, especially in such a circle. How can you live in that atmosphere? But you do, and you transfigure it with the sense of musical beauty, which exists in you more strongly than in any other man I have met. I have been inclined to underestimate, even despise, just this power. Indeed, when I heard your people I could have sworn you saw things from my angle, but couldn't make them understand what you were after. Now I know you climb the opposite side of the hill I am climbing, so that when we meet at the top we are on the same ground. The difference is: on my side is the mob with all its folly and power, on your side the elect with all their authority and sense of the highest and lovelist in art.

When we get to heaven God'll give me humans to conduct—you will have angels.

Some few, of course, would never overlook his misdemeanours, but if they were lost forever there were new friends to take their place. Chief among them was Bernard Shaw, who had passed beyond the stage of amused exasperation and now, when not actively assisting Boughton to make a living, was enjoying himself hugely:

2 January 1912

Dear Rutland Boughton,

you really must stop sending me your infernal music. Can't you write anything but Toccatas for the loudest of Collard pianos? Have you never heard of a violin, a human voice, or even a Jew's harp? This ramping, stamping stuff with its strings of chords like a procession of black beetles, its vocal parts that have no sex and no humanity, its bumptious *elans* out of *Die Meistersinger* and *Die Götterdämmerung*, and its arpeggios like Miss Lindsay's with all fifths played a semitone sharp by an bungling player, drive me simply mad. I loathe your music. It isn't music. It is all skeleton in armour, rangle jangle bangle, with nothing but old bones inside. Do you suppose that the impulse to bang a keyboard when you are in high spirits is inspiration, or genius, or anything bearable outside an asylum for the deaf? For Heaven's sake get a professorship at the R.A.M. You will get paid for misleading the young; and you won't have time to compose. This tacking on of the same sort of stuff to any words you can lay your hands on is the mark of a totally inartistic nature. You are worse than a guide to a beauty spot with your eternal Tum Tum Tiddity,

Tum Tum Tiddity, Tum Tum Tid Ditty, Tum Tum. Tum. TUM.

Man: do you suppose if I wanted to play that sort of thing I could not get it from Meyerbeer properly done, and fitted with coherent librettos. Wagner called it Effect without Cause. What would he have said to you? Neither Cause nor Effect.

If you could only hit out an eight bar phrase that people could dance to, you would at least begin to compose. Or a quickstep that a cornet in B flat could play well enough to make a trade union march to a demonstration in Hyde Park. Or even a passable hymn tune. That would be composition.

But no. Tum, tum, tiddity, as thick in the bass as possible, and plenty of G flats and D flats.

I shall warn Lorraine to bar your music. It would kill a cinematograph show.

Happy Czerny! He is dead, and you can turn his *Études de la Velocité* into Songs of Liberty without disturbing him. Of course major sixths are very stimulating; but you can't make music by slinging about stimulating intervals.

And you get publishers to publish these things for you. That is genius. If only Chris knew!

Abscheulicher Musikanter!

What will you do when you find yourself out? Shoot yourself like Haydon? Or begin all over again?

 *etc.*

Send me no more, Rutland, send me no more.

G. Bernard Shaw.

This note represents my opinion of you as a composer.

Even Shaw must have been surprised by an article which appeared in the *Daily Mirror* on 23 June 1912 under the heading: 'Opera for a Holiday. Composer's New Work to be produced in Surrey Woods. Play by Moonlight.' He would then have read:

HINDEHEAD, June 23rd—Do you want a novel and interesting holiday?

If you do—and you need not necessarily be a musician to enjoy it—arrange to come to Hindehead and joint the open-air opera players. Alfresco opera playing is the newest holiday scheme offered to the jaded worker possessed of any sort of liking for music and an open-air 'change'.

Here, in this glorious Surrey countryside, plans have already been made whereby a band of amateur music lovers will be able to spend August in producing a pastoral opera and leading the simple life.

Rutland Boughton, a well-known young singer and composer, who is acting as director, gave me full particulars of the alluring scheme today.

'The idea is to combine an object with a holiday', he said. 'Really we are starting a new sort of summer school for grown-ups.

The scheme is this. We have taken a school here, Mount Arlington, Hindehead, for the month of August. There is accommodation for forty people in the building. 'Term' will begin on August 3 and last until August 31, and during this period we shall rehearse, and play in the woods, an opera.

At present I am busily engaged in putting the final touches to the music, and the opera is not yet named. In a way the work has grown from its surroundings in this lovely country, and the spirit of the free world is its theme.

We shall probably give two complete performances of the opera. It will be performed entirely by those holiday-makers who join the "Summer School of Music". Twenty people, mostly schoolmasters, schoolmistresses, art and music students, have joined the school, and we can enroll twenty more as "boarders" and almost as many more who live as "non-coll" students and performers.

The inclusive charge for from August 3 to 31 at Mount Arlington is only twelve guineas. For a fortnight of the term we are charging seven guineas a head.

We shall probably give two complete performances of the opera. One of them will certainly take place on August 27, by the full light of the harvest moon.

There will be two acts (two scenes in Act I) and we have specially chosen the night of August 27 for playing the first act because it is a full moon then. A night literally elapses between the Acts, Act II being played after breakfast on August 28.

Another novel point about the opera will be that each scene will take place in a different part of the woods. Mr Frederick Jackson of Tarn Moor is giving us the run of his extensive grounds for our "stages", and there will be a mile walk for the players and the audience between scene one and two in the first act.

Performers will be expected to provide their own costumes, which will be of the simplest kind, mostly green in colour.'

This novel summer school for grown-ups is evidently to be run on strictly business-like lines. Here is a specimen day's programme:

9 a.m.	Breakfast
10.30	Opera rehearsal
1 p.m.	Lunch
4	Tea
5	Music and Singing Class
7	Dinner
8.30	Lecture, debate, concert or drama.

Picnics will take place from time to time; tennis, golf and cricket will be played in the afternoons, and there is an open-air swimming bath attached to the school.

It was an odd scheme even for Rutland Boughton, but the events that led up to it were quite ordinary.

The initial impulse had come when Jackson suggested that some of Boughton's singing pupils might like to entertain his friends one week-end. A quartet was invited and spent a couple of happy hours wandering round the house and grounds singing madrigals and partsongs. Boughton was struck by the variety of effects that could be obtained by working in the open air: how in certain spots a quartet could sound like a choir, and then be made to fade into infinity by moving a few steps away. This and the whole atmosphere of the Surrey woods reawakened an idea that had lain dormant since his Birmingham days—to make a setting of Fiona Macleod's Celtic drama *The Immortal Hour*.

He had already begun one version, only to abandon the sketches as being too like Debussy. But now, with the house and grounds as a natural setting, he felt able to make a fresh start. As he took his daily walks through the woods the music seemed to come to him out of the trees. A bird, startled into flight by his footsteps, became for a moment a mysterious symbol, the embodiment of his dreams. So intense was his mood that he half expected the creatures of his fancy to appear before him: it was as if he was a trespasser in the woods that were theirs by right.

Though the work has no direct autobiographical significance, it is none the less part of Boughton's personal experience for it embodies the ideas that Christina brought into his life. Half Irish herself, she was an ardent champion of Celtic revivalism and her enthusiasm kindled his. He was fortunate (inspired, rather) in his choice of libretto. Fiona Macleod's play cries out for music to supply the element of profound mysticism that mere words can scarcely express. With the possible exception of Delius's *A Village Romeo and Juliet* (1901) and Holst's *Savitri* (1908), it is the first British opera libretto of genuine literary merit, and like them it is far removed in spirit and deed from the melodramatic claptrap that had all too often passed for 'opera' among British composers of earlier generations.

In shaping it, however, Boughton had to rely on his own theatrical instincts, for its author, Miss Macleod—otherwise known as William Sharp—had died some years before. Writing under his own name, Sharp (1856–1905) had gained a considerable reputation as a poet, critic, biographer, and man of letters before being seized with the conviction that there was another side of his personality that could only be expressed as Fiona Macleod: a mystic champion of Celtic art living, it was said, in inviolable isolation on some remote Hebridean island. The first literary manifestation of Sharp's new persona, *Pharais*, a romance, appeared in 1894 and created a deep impression among the Celtic revivalists. Thus encouraged, the shadowy Fiona grew palpably in stature, far surpassing Sharp himself in literary fame and threatening to subvert his very existence. So complete was the split in his personality that Sharp was driven to correspond with Fiona, receiving letters in return. Though Fiona, necessarily, never appeared in public (Sharp acting throughout as her 'agent'), she became more real than her creator and it was only with his death that the secret was revealed. There was more than a little of the vampire in the powerful personality Sharp had summoned from his unconscious, and it seems likely that his relatively early death was precipitated by the sheer exhaustion of having to be, quite literally, two people.

Fiona Macleod's 'psychic drama' *The Immortal Hour* was one of two plays from an intended cycle of seven begun in 1899. It first appeared in print in 1900 in the *Fortnightly Review*, and was published in book form posthumously in 1908. Boughton was therefore very quick off the mark in recognizing its potential. Moved by the powerful simplicities of a book of Hokusai sketches that Bantock had given Christina, and impressed by his interest in Celtic literature and music (it may well have been Bantock who first drew his attention to *The Immortal Hour*), Boughton began to study the Hebridean folksongs which Marjory Kennedy Fraser had recently begun to publish. Both elements contributed to a purging of the Wagnerian manner of his early symphonic essays, leaving him free to develop a personal vein of simple lyricism. Many of the melodies that now appear in the work were originally invented as part of his pentatonic studies. During the course of composition they would occur to him afresh and now in the context for which they would seem always to have been intended. *The Immortal Hour* was written in one of those happy moods when all manner of diverse threads suddenly and unexpectedly combine, as if the composer had reached down through layer upon unsuspected layer of his experience to come up with something uniquely personal and convincing.

The *Daily Mirror* article was premature, however. Not only did the idea of a Summer School fail to catch on, but Boughton found himself unable to finish orchestrating *The Immortal Hour*. So far as 1912 was concerned the scheme had to be abandoned. But the ideas set out in *Music Drama of the Future* now caught fire again and the collaborators began to evolve fresh plans. Buckley favoured Edward Carpenter's suggestion that Letchworth would be an appropriate setting for their work—the ethos of the 'garden city' was very much in keeping with his ideals. But Boughton favoured a genuinely rural setting, for he still hoped that the proposed festival might exist as part of a larger plan whereby artists who 'preferred a country life and felt that the means of a livelihood should be other than those of art' might support themselves by running a co-operative farm.

In the end neither of them got their way. The matter was settled abruptly by the intervention of Mr Philip Oyler, who was then running a 'Nature School' at Hindehead at which Boughton's son was undergoing a rather eccentric education.[5] Oyler became aware of their plans, approved of the scheme, and announced that the only possible place where it could be carried out was Glastonbury. The next thing they heard was that he was down there 'testing local reaction'. Local reaction proved favourable, if a trifle bewildered, and Oyler returned with the good news that not only had he obtained an option on a large house known as Chalice Well, but had arranged for a meeting of supporters. At this point he seems to have faded out of the picture, leaving Boughton and Buckley to follow rather meekly in his wake.

Oyler, however, had not been the only person to feel the magnetic pull of Glastonbury. A Mr John Harriman was already occupied with plans for 'Arthurian drama and religious festivals, with choir and craft work, in the town', and had formed an alliance with Miss Alice M. Buckton, author of a singularly sentimental but highly successful 'Christmas Mystery Play', *Eager Heart*, which had been performed at Lincoln's Inn in 1904 and had just reached its eleventh edition. When, in 1913, Oyler's tenuous option on Chalice Well ran out, it was Miss Buckton (1867-1944) who purchased the freehold with a view to securing the so-called 'Holy Well' from 'commercial exploitation' and exploiting it herself as a centre for artistic endeavour and spiritual regeneration. Inevitably Miss Buckton and her friends were drawn into the scheme that Boughton and Buckley were about to unveil before the startled gaze of the more prominent citizens of Glastonbury and Street.

6

1912–1913

The precise timetable of Boughton and Buckley's assault on Glastonbury is somewhat obscure in its early stages. The importunate Mr Philip Oyler had paved the way by making contact with the Reverend Charles Day who had been appointed to Glastonbury's principal church, St John's, in October 1912 and begun his ministry the following January. Though clearly cast in the mould of Muscular Christianity (newspaper reports wax eloquent over his 'clear cut' features, his 'noble nose and pugnacious jaw'), he declared himself fond of a good tune and therefore ready to put his shoulder to the wheel. Through his good offices Boughton came into contact with David Scott, the organist at St John's, who was soon able to announce that the Choral Society he had founded in 1911 in order to perform Elgar's *Coronation Ode* would present its first concert for 1913 on 2 April and that 'the presence of Mr Boughton, one of England's most gifted composers, on the concert platform should secure a crowded house'.

Mr Boughton's was a presence that few could ignore. According to the *Manchester Courier* for 24 February 1912 he had 'discovered—or, at least Madame Marie Brema had discovered it for him—that he possessed a very remarkable baritone voice'. He now proceeded to woo Glastonbury with three Hebridean folksongs, accompanying himself on the Celtic harp, and, more strenuously and provocatively, with a rendition of his setting of Edward Carpenter's poem 'To Freedom!' He appears also to have directed the Choral Society in a performance of *Sir Galahad* which, though an early work, had just been published. Doubtless he took the opportunity to acquaint his audience with the Arthurian delights he had in store for them, and the fact that a 'Preliminary Meeting' of interested parties was to be held in the Guildhall at Wells on 5 June.

On the whole, the meeting was a success and ended with excited forecasts of 'a festival similar to those of Bayreuth and Ober Ammergau'. It was anticipated that £850 would be needed, of which £150 would provide for the soloists and chorus, and £300 for the orchestra. Miss Buckton declared that she would organize a series of plays later in the year, the better to stimulate local interest. Only Canon Scott-Holmes expressed doubt. What, he wondered, did anyone know of the quality of the music about to be foisted upon them? Boughton, however, was ready and promptly produced letters of commendation. First, from Bantock:

4 March 1913

My Dear Boughton,

   I am delighted to know that there is a prospect of your Choral Drama King Arthur being produced under ideal conditions. Not only do I consider the music in every way worthy of its subject, but I consider that such an event would have a really national interest.

   I was much impressed with your work when you played it over to me shortly after you had completed your sketch, and judging from the orchestral excerpts which you conducted for us at one of our Philharmonic Concerts I am led to expect very great results from its performance, both gratifying to you as a creative artist and of some satisfaction to all those who are interested in the progress and development of our native musical art. I shall look forward to the production with the greatest personal interest and hope to be present to add my small share of the congratulations which will attend the success I anticipate for the work. . . .

Even more impressive was the commendation from Elgar, to whom he had gone for a second opinion. After running the icy gauntlet of Lady Elgar's disapproval (Boughton never looked like a 'gentleman') he had played him the whole of *Uther and Igraine* (or *The Birth of Arthur* as it was now called). Elgar, though rather more guarded than Bantock in his response, wrote with sufficient enthusiasm to reassure Glastonbury:

27 April 1913

Dear Boughton,

   I was very glad indeed to make acquaintance with your fine work The Birth of Arthur and am delighted to learn that there seems to be an opportunity to produce it next August in the district most eminently suited by association and tradition for its production, and I cordially hope the scheme may be carried through successfully. The music seems to me to be quite worthy of the great subject you have chosen . . .

To clinch the argument, Dr T. H. Davis, organist of Wells Cathedral, agreed to head a committee to judge the quality of the music from a local point of view. Canon Scott-Holmes was reassured and it was agreed to issue a prospectus, prepared earlier in the year but prudently marked 'Confidential', outlining the joys of a 'Festival Production of THE BIRTH OF ARTHUR—a choral drama by Reginald Buckley and Rutland Boughton, under the stage direction of Alice M. Buckton'. The tone of the announcement is redolent of 'Arts and Crafts' earnestness, Fabian Socialism, healthy bike rides, and Jaeger knickerbockers:

Arrangements are being made to give three performances in a beautiful garden lent for the purpose. The services of the locality will be used as far as possible for the choral dances which will have an important place in the work. In future years it is hoped to develop the local element in the festival; but for the first production some additional help will be welcome, if an earnest spirit is brought to the work. Singers, dancers, instrumentalists, and friends willing to help in preparing the costumes and in other ways can all fill a part in the scheme; but it must be understood that they will be required to be present throughout the month of August.

Beyond the preparation for the choral drama the gathering will have many holiday features such as games, picnics and pilgrimages to the many romantic and historic places in the neighbourhood.

Lectures, Concerts and various merry-makings will be organized during the holyday.

As the spirit of the gathering is to be that of giving, an expression of the joy of creation and recreation, it is not intended to make a profit in any part of the arrangements. The whole holiday, therefore, will not be of an expensive character. Applications should be made at once, stating the kind of accommodation desired, and whether flesh or vegetarian diet is preferred.

The lectures listed included Edward Carpenter on 'Beauty in Everyday Life', Frederic Austin on 'Singing in Relation to Drama', Alice M. Buckton on 'The Festival Element in the Church Year', Reginald Buckley on 'Legend as the Basis of Music Drama', Mary Neal on 'The Dramatic Element in Folk Dances', and Rutland Boughton on 'Music Drama as Divine Worship'. Unspecified lectures were promised from Philip Oyler, Gerald Cumberland, Sidney Grew, and 'perhaps by Dr Ethel Smyth'.

But it was one thing to have drawn up an ambitious prospectus, quite another to put it into practice. Having set his sights on a 1913 festival Boughton had abandoned his various hard-won sources of income and was now decidedly strapped for cash. Though enthusiastic, his Glastonbury supporters were not proving very practical and it was clear that something positive needed to be done if the momentum already generated by the talk and publicity was to be maintained and brought into focus. He turned for help to Bournemouth and the only man who could be relied upon to champion the cause of British music and young composers: Dan Godfrey. Resurrecting the Hindehead scheme, Boughton rented a large house (Gorselands, on the cliffs at Southbourne) and, with the help of Christina and the dancer Margaret Morris, ran a three-week school which culminated in a public performance at the Winter Gardens.

Boughton seems to have met Margaret Morris in 1912 at Grayshott, where she sometimes stayed with friends. Miss Morris (1891–1980) had attracted considerable attention as an actress and dancer with very pronounced views on the art of movement: 'Raymond Duncan [*the brother of the famous Isadora*] had taught me his Greek Positions and I had incorporated them into the new method of training I was in the process of evolving.'[1] That process led to the establishment of a school in Bloomsbury which, in 1912, migrated to Chelsea and a miniature theatre on the corner of King's Road and Flood Street. There, surrounded by her troupe of 'Dancing Children', and encouraged by the novelist John Galsworthy, with whom she was carrying on a frustratingly chaste affair, she was able to demonstrate the validity of her method. Boughton readily succumbed to her charms and talent, supplying her with music for a short ballet, *The Death Dance of Grania*, which she included in a programme at the Court Theatre, Sloane Square, early in November 1912. Thus by 1913 it would have been clear to him that Margaret Morris was someone who shared his ideals and with whom he could co-operate in sympathy and understanding.

The programme which crowned the efforts of the Holiday School at Bournemouth's Winter Gardens on 28 and 29 August 1913 mixed demonstrations by Margaret Morris and her Dancing Children with poetry readings by Mrs Tobias Matthay,[2] and ended with the first performance of the Prelude and Opening Scene from Act 2 of *The Birth of Arthur*. Mrs Matthay sang the part of Igraine, and Arthur Jordan that of Brastias, while the chorus was supplied by the students.

The performances attracted a great deal of attention, partly because advance publicity had held out the promise of 'dancing scenery'. The audience therefore arrived fully expecting to see paint and canvas somehow galvanized into action. What they did see can scarcely have disappointed them, as the *Daily News* for 28 August made clear:

'SCENERY' BY SONG AND DANCE

No Vulgar Realism in a Seaside Drama

EVERYTHING LEFT TO THE IMAGINATION

(From Our Special Correspondent)

There were a good many puzzled heads in Bournemouth this afternoon at the close of a performance in the Winter Gardens of 'A Programme of Dramatic Dances, including Scenes from "The Death of Arthur" [sic]. 'I never saw or heard anything like it', a lady at my elbow remarked as we were coming out.

It was a bold venture to produce anything so unusual for the first time before a seaside holiday audience, but the circumstances were exceptional. For the past month a holiday art school has been at work in a big house in the cliff at Southbourne, near here, the chief 'tutors' being Mr Rutland Boughton, a young musician of great promise from the North, and Miss Margaret Morris, maker of dances.

'We've been camping out in the grounds most of the time, nights and all, and it has been tremendously jolly', Miss Morris confided to me. 'Just see what it's done!' she added, exhibiting a beautifully moulded copper-coloured forearm.

How the Idea Developed

'Gorselands', as this idyllic school is called, was rented furnished, and the pupils are young people on holiday who are interested in music and eager to help Mr Boughton carry out his great musical and dramatic experiment. That experiment is, in his own words, 'dancing scenery'.

The idea developed in this way, Mr Boughton tells me. Mr Reginald Buckley had written a new dramatic version of Malory's Morte d'Arthur, and Mr Boughton was setting it to music for stage production. At first he intended to have a sort of Greek chorus sitting with the orchestra, then the chorus was moved to the stage, and finally came the question—Why should not the chorus dance as well as sing? Only a short step had then to be taken to the conception of 'dancing scenery', i.e., dancers who would before each scene suggest by their rhythmic and symbolic movements, aided by symbolic costume, the atmosphere of what was to follow, so that the imagination of the spectators might create, with far greater vividness than any stage carpenter or scene painter, the right environment for the actors.

Nearly an Artistic Success

'The idea appealed to me at once', Miss Morris confessed. 'Realistic scenery has always seemed absurd, at any rate as a setting for a poetic drama like Mr Buckley's "Arthur". By dancing it is possible to transmit not only an emotion, but a mental picture, and that is what we have been trying to do at the Gorseland classes'.

'To get the best results, the play ought to be performed either in the open air or on a dark indoor stage with limelight, but we have been obliged to make our debut indoors, and in this strong afternoon glare, so I hope everyone will make allowances.'

The performance of the prelude and first scene (there was no time for more) undoubtedly proved Miss Morris to be right and it came very near to being an artistic success, and was from every point of view intensely interesting and suggestive.

This is how the scheme works. The scene required is Tintagel Castle, on the storm-beaten coast of Cornwall, where King Arthur's mother, 'the white Igraine', is persecuted by the stormy love of Uther. Enter four men, in stone-coloured jerkins and caps, from the back, and from the right and left some twenty dancing women in flowing garments of blue and green.

The men form a square and sing how

> Dark and stark and strong
> Tintagel Castle stands

while the women advance and retreat with tossing arms, telling in verse of 'the splash and surge of the sea' on the rocks of Tintagel.

A Strain on the Imagination

To set the scene in this way only occupies a few minutes, and at the close the house should be in a position to furnish Igraine and her noble-minded lover Brastias with the necessary heroic surroundings of roaring sea and steadfast, impregnable castle.

It is doubtful if this afternoon's audience was mentally alert enough to accomplish this feat, but for my own part I found the music and dancing gave just the necessary direction to the imaginative faculty. Whether human scenery should be allowed to remain on the stage during the dialogue is another matter. Four-square Tintagel narrowly escaped the comic, and the flapping of the young lady billows' hands when Igraine mentioned 'the white wave-crests how fair' was a little too near the detested realism.

It was rather disconcerting, also, to find the waves and castle walls turn at the close into 'The Will of the World' when they might very well have been somewhere in the wings.

Tomorrow the piece will be produced at night, with the advantage of stage lighting. It will be a thousand pities if it is not persevered with and brought to one of the more intimate of our London theatres.

Boughton had hoped for a phalanx of at least twenty-five men as his castle and fifty or sixty women as the surging sea. The presentation was, he felt, 'little more than a parody of Miss Walshe's design'.

Whatever the effect the performances may have had on Bournemouth's holiday-makers, they gave a much-needed jolt to Glastonbury. On 26 September the *Central Somerset Gazette* was able to announce that Miss Buckton would make good her promise and give a 'dramatic reading' of *The Birth of Arthur* in the Refectory at Chalice Well on Saturday 11 October at 5.30 o'clock:

The episodes chanted will include the chorus of the wave-beaten walls of Tintagel which was illustrated in such unique and effective fashion in the Winter gardens at Bournemouth last month. The reading will be open to the public free. Those among the audience who may become interested in the scheme of the play and its progress, will have the opportunity of contributing to a collection plate at the doors. Visitors are particularly requested to be in their places 5 minutes before the hour stated, as the doors will be closed during each episode.

Miss Buckton carried the day and the reading had to be repeated on 17 October.

At what point the festival scheme attracted the attention of Roger Clark, his wife Sarah, and his brother John Bright Clark, is not known, but it was exceedingly fortunate that it did. The Clarks were a Quaker family that had moved from farming to tanning and then, by logical stages, into the manufacture of shoes. Their factory at Street was a model of its kind, and their concern for the well-being of their employees—which in effect meant most of the population of Street— was legendary. Roger Clark (1871-1961) took a liking to Rutland Boughton and decided to support him. Soon they had become close friends. Clark's backing—moral and financial—was to prove crucial to Boughton's festival plans. Without it nothing could have been accomplished. And before the year was out their faith in each other was to be tested to the limits.

On Monday, 27 October, Boughton held two public meetings in the Crispin Hall, Street. He spoke eloquently of his festival ideal, read four extracts from *The Birth of Arthur*, and then proceeded to outline the music. Mrs Tobias Matthay again sang the part of Igraine, and Arthur Jordan that of Brastias. Boughton himself sang Merlin's part, summarized the choral passages as best he could, and accompanied the whole thing at the piano. If a letter which Mr T. E. Hodgkinson, owner of the Paper Mills at Wookey, wrote to Roger Clark a few days later is any indication of the general feeling, the effect must have been somewhat less than electric on the majority of the audience:

We have the chance of a lifetime and we are all so stuffy and blind that we shall let it pass without moving. Both my son Ivan and I think the music and the libretto inspiring beyond words. Wasn't Mr Boughton brave to put a sketch like that before people—no one but a man of his personality and genius could have forced a glimmering of his idea on those fat-headed rows— they would have taken the heart out of anyone less inspired, but he took it for granted that they were sympathetic and swept them along—trailing them at his prophetic skirts! If any one of us had been millionaires the thing would have been done.

But the thing had been done. On Tuesday, 28 October 1913, the 'Provisional Committee to arrange for the production of the Arthurian Choral Drama at Glastonbury in 1914' met at the vicarage for the first time, and Reginald Buckley was formally charged with the task of drawing up a 'National Appeal' for approval at the next meeting.

Events now took an awkward turn. Somehow it had become known that the lady whom everyone was calling Mrs Boughton was not exactly his wife. It was a dilemma that called for an explanation. At a second meeting 'Mr Rutland Boughton made a statement of a personal matter to the committee, and then retired while it was fully discussed.'

Beyond this bald report the minutes do not record anything of what was said, or how long it took to say it. It merely states that the Committee was 'unanimous in deciding to go ahead with the production if the necessary guarantee can be raised'. But the first stone had been cast and the waters muddied. Writing from Yardley Boughton offered the Clarks a typically Utopian analysis of the situation:

19 November 1913

My dear Friends,

Christina and the babes duly arrived, full of health and happy and loving memories. But also, I am sorry to say, Christina has the feeling that you are being seriously troubled on account of us. That will never do. Either you must drop us and all our plans, or I must be on the scene to dispel the venom of any scandal that may be abroad. You have only to say the word and I will divert the whole drama scheme to another place.

Otherwise, I must come down with the three older children and live in the district *soon*. I should prepare this move by giving a Concert at which half the admissions would be free—I should have it announced that during the course of the evening I should make a statement in regard to my work as it concerns the people of Glastonbury and Street. I should lightly but frankly touch on the difficulty, saying it must rest with *them* whether I come to live among them or not; *and I should carry the meeting with me*, except perhaps for two or three crystallized minds and two or three music-teachers.

My wife tells me that you kindly enquired as to how we should *live* if I came. Partly by my pen; partly (perhaps) by visiting B'ham and teaching; but chiefly by the gifts of the poorer people of the district who, in return for our work, would keep me as artists used to be kept, still are in Ireland, and shall one day again [be] in England.

This will all seem half mad to you, I fear; but I can do it if it will not inconvenience your relations with your friends.

Fortunately Boughton's faith in the poor did not have to be put to the test. The air cleared sufficiently and the festival plans took a much more ambitious turn. Urged on by Buckley, to whom the idea meant everything, the Committee began to consider the possibility of building a theatre. Though not against the scheme, Boughton was less convinced. His own preference had always been for the self-supporting commune of like-minded artists living in pantisocratic harmony; but, as he ruefully had to admit, 'this part of the project never gained any solid support'. Buckley's dream of a 'Temple Theatre' was approved. Calculations were made and it was decided that land might be purchased and a temporary theatre raised at a cost of £2,000, while the production itself might require a further £1,000. Boughton reminded the Committee of the generous letter he had received from Mr Thomas Beecham on 28 September:

I am most interested in your forthcoming production of 'Arthur' at Glastonbury—of course an ideal background for the work. I shall be willing to lend my orchestra for the three performances suggested. If there is a loss on the undertaking, I will bear my share, according to the proportionate expense of the orchestra. If, on the other hand, the venture pays I shall naturally expect the orchestra to be reimbursed. If I can assist personally I shall be very pleased to do so, provided the period of time chosen does not clash with my other arrangements. In any case you have my very best wishes for the success of the work—a success I have no doubt of if the scheme is well organized.

Beecham, it should be added, had been summarily hijacked by Boughton into taking an interest in *The Birth of Arthur*, pressing the score into his hand as he was about to board a train, and then joining him for the journey.

   With the cost of Beecham's orchestra estimated at about £400, the scheme seemed distinctly possible. By the end of the year approximately £800 had been collected in cash and promises (£225 from the Clarks), and the National Appeal, with a revised estimate of £5,000 for the temporary theatre and the production, was ready for distribution. It was backed by an impressive array of signatures: Lena Ashwell, Edward Carpenter, W. L. Courtney, Gervase Elwes, John Galsworthy, Percy Grainger, Professor Geddes, H. V. Hamilton, T. C. Hedderwick, Ivan Hodgkinson, Josef Holbrooke, Jane Harrison, Edward Hutton, Holbrook Jackson, Mrs Alderman Lees, Neville Lytton, Charles Marson, G. R. S. Mead, Philip Napier Miles, George Moore, Lillah McCarthy, Sir Frederick Pollock, Roger Quilter, George Riseley, Landon Ronald, St Loe Strachey, George Bernard Shaw, Arthur Waugh, and Sir Henry J. Wood. A few weeks later they could have added the names of Sir Hubert Parry, Ralph Vaughan Williams, Charles Kennedy Scott, and a whole galaxy of lesser mortals.

   Equally impressive was the Appeal's preamble:

It has long been thought that England should have its own National Festival Theatre for Religious and choral drama, and that it should not be necessary to go abroad in order to have Music Drama presented under the right conditions as at Bayreuth and elsewhere on the continent.

At last the opportunity has come for producing an English Music Drama upon the Arthurian legend as told by Malory. Glastonbury, the ancient Isle of Avalon, affords the most ideal and appropriate setting for this venture. Glastonbury is, according to well-founded tradition, the site of the first Church in Britain, built by Joseph of Arimathea and his companions. Chalice Hill, Glastonbury, derives its name from the belief that the Holy Grail was buried there. To Glastonbury the mourning queens brought Arthur himself after his last great battle to be healed of his wound, and here he and Guinevere were buried.

Boughton's choral drama, the Appeal continued, was to be produced by Margaret Morris and the composer, the conductor would be Mr Thomas Beecham, and the cast would include Frederic Austin, Robert Maitland, Arthur Jordan, and Mrs Tobias Matthay. The Beecham Symphony Orchestra would be in attendance, and the chorus was to be drawn from local supporters. The building, whether temporary or permanent, would be amphitheatric, having adequate seating for 1,200 people, a suitable stage with modern appliances, and a hidden orchestra. It was, in short, a scheme of great magnificence and startling optimism.

By the end of the year everything promised well for Rutland Boughton. Towards the end of December he was able to write to Roger Clark: 'I have just finished a new music-drama (begun two years ago), Fiona Macleod's *Immortal Hour*. It tells of the coming of the soul to the body, and its going again—I could weep and shout with laughter for the joy and sorrow of it—it is my best work.' And on Christmas Day, Gordon Craig wrote from the 'Arena Goldoni', his studio in Florence, to add his own lofty blessing to Boughton and his dream:

I write to you on the Festival Day of the Birth of Christ, when at least a million pounds are being wasted to do dishonour to the great event.

You are proposing to hold a festival for three days in England to celebrate the Holy Grail, and you have selected Glastonbury as your centre. For at Glastonbury the Holy Grail was buried. You have sent me a letter asking me to sign an appeal for funds for this Arthurian Festival at Glastonbury. You say you want £3000 so that 3000 can hear and witness your Religious and Choral Drama *The Birth of Arthur*.

If you think my voice will carry so far, I have the greatest pleasure in appealing to Glastonbury to find the £3000, and I should say that such a centre under present circumstances with such a cause to do honour to would be jealous lest any other city or centre should be appealed to, now or at any time.

It is not right, or should I say necessary, that you should appeal to the 'whole country', for it is only a small matter of £3000.

If you appeal to the whole country for £3000 or £12,000 I cannot add my name to the appeal. If you appeal to the country for £100,000 I will sign it.

Gordon Craig.

1914

Though encouraged by an enthusiastic article which Robin Legge, music critic of the *Daily Telegraph*, had written outlining their schemes, by the end of January 1914 the 'Provisional Committee for the Glastonbury Festival' had begun to have doubts. At a meeting on 31 January it was decided that if sufficient money had not come in by St Valentine's Day the production of *The Birth of Arthur* and the building of the Temple Theatre would have to be postponed until the following year. In the mean time they noted with gratitude and satisfaction that Professor Adshead of Liverpool University had declared himself willing 'to advise and make sketch plans [for the theatre] and to consider sites . . . and to work with any architect finally chosen, without fee and from interest in the cause'.[1]

But no saint interceded on the festival's behalf. On 16 February the Committee, surveying the contents of their begging bowl, decided that £961. 13s. 4d. was not enough for them to be able to achieve their aims in 1914, but was sufficient to justify their working for a production in the summer of the following year. A suitable site would be acquired and Sir Edward Elgar would be invited to lay the foundation stone at some 'simple and dignified ceremony' to the accompaniment of 'some appropriate musical performance'. The actual sites being canvassed were, according to the *Central Somerset Gazette* (6 February 1914):

A corner site in Street Road, the property of Councillor Chamberlain; a site at the top of the hill next to Mr Jardine's gardens; a site at the top of Bere Lane, also Mr Jardine's, and another property opposite; a site at the north-east corner of Butt's Close, the property of Mr Windmill, and another at the corner of the Recreation Field, where the Mayor understood the directors would have no objection to parting with an acre; and finally, a site in Street Road, next to Mr Applin's house. Personally, the Mayor thought that the site they ought to get if they could was that at the corner of Street Road. It was a question of purchase and it would mean £450.

It all seemed distinctly promising.

For Boughton however, noting perhaps that Mr Gonzague Riviere, a lay-clerk at Wells Cathedral, was even now angling to form a 'Grand Operatic Society' in order to stage *Maritana* in the Crispin Hall, Street,

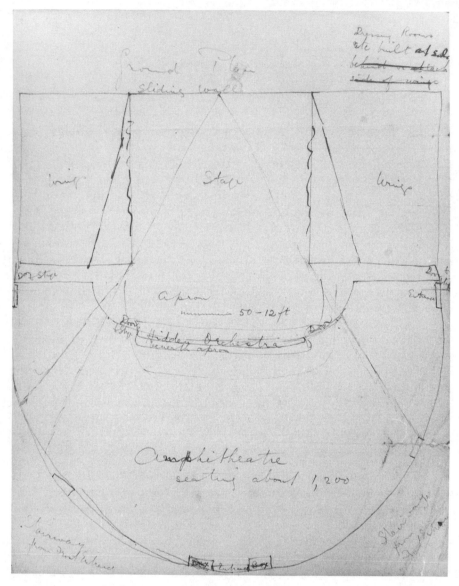

Figure 1   Professor Adshead's preliminary sketch (1915) for the proposed Temple Theatre.

the situation was far from satisfactory. Apart from anything else, he was desperately short of money and had already (17 January) been forced to write to Roger Clark for the loan of £15:

I do not like asking you to help us—not because I do not entirely love and trust you, and can take your money as from a brother—but because I know I am entirely untrustworthy in all affairs of money. This wretched 'artist-life', non-productive on the material side, and dangerously dreamful on the spiritual side, makes a man (at least it has made me) disgracefully unequal to the more serious responsibilities of practical things. For some reasons it would be better that I asked help from no-one, and faced the situation as a friendless man would have to face it—that at least would help to show people that a man working whole-heartedly for a spiritual cause is as big a fool as a man working whole-heartedly for material things.

It was evident that a further postponement would cool such enthusiasm as had been generated, and that it was essential to do something positive in 1914.

It must have been at this point that he decided to hold another Holiday School and that his most recently completed music drama might make an excellent substitute for *The Birth of Arthur*. At the beginning of June the newspapers began to carry announcements of a proposed Summer School under the guidance of Margaret Morris, Charles Kennedy Scott, Edgar Bainton, and Rutland Boughton. Edward Carpenter would lecture, there would be an 'Arthurian Concert' and dramatic readings from Buckley's recently published cycle, *Arthur of Britain*, Mrs Tobias Matthay would present a 'Celtic Recital', and the main work to be studied would be a new music-drama *The Immortal Hour*. The spacious grounds of the Abbott's Kitchen were available ('by kind permission of Mrs Hooper') and it was expected that 'all music and art work would be in the open air where possible', and that expeditions to places of Arthurian interest would be a special feature of the students' activities.

Boughton's head also buzzed with other plans. Earlier in the year he had written to Professor Gilbert Murray at Oxford, and on 26 February Murray replied: 'I should be very glad to consider your plan of making a music-drama out of *The Trojan Women*. I have always felt it might come out magnificently in that form.' Thus encouraged, Boughton submitted his ideas and Murray replied on 4 March:

In general I quite approve of the principle of your cuts, indeed they seem to me very appropriate. In point of detail, however, I would venture to say one thing from my general experience of Greek plays. I have found that cutting out the Gods is usually a great mistake. Everybody begins by wanting to do it,

I among others. But we have always found in the end that the balance and meaning of the play was curiously injured when the supernatural background was taken away. I wonder if you could not keep some fragments of the prologue, just some vision of the conquered gods leaving the city, or the last speech of Poseidon, or the like. It makes such a difference to one's feeling about Hecuba's first scene if you know that the gods have been looking on, and that, so to speak, her redeemer liveth.

Boughton must have agreed, for on 28 April Murray wrote to congratulate him: 'I am delighted that the spirit moves you and that you now feel the value of the supernatural prologue.'[2] But there the matter rested. The idea of a Greek music-drama was dropped and not taken up again until Boughton's own circumstances compelled him to turn to Gilbert Murray's translation of the *Alkestis* to expiate a personal dilemma.

As it was, the problem of earning a living was more than enough to absorb Boughton's energies. This he was forced to do as and when he could, taking whatever came his way. One of the most agreeable opportunities came through Margaret Morris. Each Wednesday, beginning on 3 June, he played for performances at her Chelsea theatre, accompanying the Dancing Children in their displays, and Miss Morris herself when she 'interpreted' the whole of Beethoven's Seventh Symphony. On 17 June, when Miss Morris was indisposed (had Beethoven perhaps proved too much?), he sang, presumably to his own accompaniment, George Butterworth's *Shropshire Lad* cycle and *Six Landscapes* by Josef Holbrooke. At the final recital, on 8 July, he was joined by Mary Neal, Clive Carey, and the Esperance Guild of Morris Dancers. After attending the first of the festival's committee meetings, Miss Neal had proved her allegiance to Glastonbury by lecturing, on 12 June, on the art of Morris dancing—a suitable counter-attraction to the classes in Hellenic dancing, 'according to Raymond Duncan's principles', that were being held at Chalice Well.

The Chalice Well contingent had other plans, too, and were hard at work preparing for a 'Midsummer Festival' at which Miss Buckton's pageant play, *Beauty and the Beast*, was to see the light of day. Though for the moment still linked with Miss Buckton in his Glastonbury schemes, Rutland Boughton may by now have come to the conclusion that they marched to the sound of quite different drums and that sooner or later a break would have to be made. Had not Vaughan Williams referred to the Chalice Well people as 'a sloppy lot' and, while willing to make a financial contribution, had refused to become personally involved in the festival? (Holst, however, would later collaborate with Miss Buckton.)[3] Significantly, Boughton's Summer School plans include no provision for Miss Buckton to produce his work as had originally been intended. Yet even he could scarcely have complained when she linked her own Midsummer Festival with a scheme to 'give a country holiday to some tired worker, or workers, from one of our great cities'—though he may have smiled when, on 19 June, the *Central Somerset Gazette* informed its readers that: 'The Warden [of Chalice Well] has offered a room free of charge for the furnishing of which Beauty and her Court are empowered to collect and receive any simple gifts in coin or kind which will make for the comfort and happiness of the guests', and noted with interest that coins to the tune of £3. 15s. 5½d. were handed over. The same newspaper also recorded the fact that a local labourer, a Mr Henry Coward, had been found hanged in the woods—his last message to an indifferent world, written three weeks before the body was discovered, was simple enough: 'Not wanted in life. This is Christians. Goodbye all.'

In July 1914, regardless of the fact that since the assassination of the heir to the Austrian throne Europe had been toying with the idea of war, the definitive programme for the first Glastonbury Festival was set up in print. Described as 'A Festival of Music, Dance and Mystic Drama', it was to be given 'in connection with Mr Rutland Boughton's Holiday School of Music' and would run from 5 to 29 August, with public performances mainly on Wednesdays and Saturdays. Open air performances were still on the menu: the second act of *The Immortal Hour* was to be performed 'by torchlight'. Realists would have observed with a sigh of relief that 'if the weather is unfavourable [performances] will be given at the Assembly Rooms, Glastonbury, or the Crispin Hall, Street'. The festival would end with Sir Edward Elgar laying the foundation stone of the 'British Playhouse for National Music-Drama'.

The programme on offer was curiously mixed. 'Mystic Drama' was to be represented by a short explanation of the subject by Rutland Boughton, followed by performances of Wilfred W. Gibson's peasant-play *The Night-Shift*, the 'Grail' scene from *Parsifal*, and a new Boughton/Morris ballet, *The Mystic Dance of the Grail*. There were to be children's operas (*Walooki the Bear*, by Reginald Buckley and Edgar Bainton, and *Humpty-Dumpty*, by Lewis Carroll and Walford Davies), performed by the children of Glastonbury and Street, while Margaret Morris and her Dancing Children would perform 'a mystic play of modern life' by Walter Merry entitled *Soul Sight*. Miss Buckton's group would present her pageant play *The Coming of Bride*. An 'Arthurian Concert' linked the 'Grail' scene from *Parsifal* with the closing scene from *Tristan und Isolde*, the 'Tintagel' prelude from *The Birth of*

Arthur, and a chamber cantata, *The Chapel in Lyonesse*, that Boughton had composed in December 1904 to the words of a miniature drama by William Morris. Edward Carpenter would lecture on 'Beauty in Freedom of Life', and Mrs Tobias Matthay would be there to render her 'Celtic Recital' of poems and folksongs. Most important of all, there were to be three performances of Rutland Boughton's new music-drama *The Immortal Hour*, in a production by Margaret Morris and the composer, with 'dresses and dances specially designed by Margaret Morris'. Tickets for this occasion would be priced at 10*s.*, 5*s.*, and 2*s. 6d.*

By 24 July Mr Rutland Boughton had 'taken up residence in Glastonbury in readiness', and promises to attend had been received from Elgar, Bantock, Vaughan Williams, Beecham, Holbrooke, Vincent Thomas, and, perhaps not very surprisingly, Mr Tobias Matthay. The newspapers also made enthusiastic mention of 'Miss Christina Walshe, the designer of the festival poster which, by the way, is probably the first futurist poster to be put on the hoardings in this country, and is attracting a great deal of attention'.

So also was Miss Walshe herself. She had now joined Boughton and the children and stood to remind the righteous of adultery uncondemned. To be on the safe side, Boughton decided that it would be prudent if they lodged in different parts of the town. 'We couldn't offend anyone', he wrote, 'with a hill between us.' And if indeed anyone was offended, Christina's generous, open nature soon dispelled their fears, and when Joy, their baby, fell ill with pneumonia the last barriers melted away.

But as July turned to August new problems arose. War, it seemed, was not to be averted. Artists who had promised their services found themselves unable to meet their commitments. Miss Morris, for example, was with pupils in the South of France; and though, according to her diary for 27 July, she was 'working hard at designs for Rutland', she soon found events were overtaking her: 'I've written to Rutland to ask if the war affects his scheme, but have had no word from him yet!' (7 August): 'Very odd . . . If I hear from Rutland that he still expects me I shall leave here 15th. If I don't I shall stay till September.' Ten days later she had to admit defeat: 'the conditions at the station are impossible, and they can promise nothing more than a train to Marseilles, and then *take your chance* . . .'.[4]

Boughton, however, was not to be deflected. He had said that his Holiday School would commence on 5 August, and the mere Declaration of War on the previous day was not enough to make him change his plans—save that now 'every effort would be made to curtail expenses and a amateur orchestra would be secured for *The Immortal*

Hour. Accordingly, the first concert took place on the rainy Wednesday evening of 5 August in the Glastonbury Assembly Rooms. Only one detail of the programme was changed: Stanford's short choral work *Last Post* was substituted for the promised *Parsifal* scene, in acknowledgement of the cataclysm that was about to unfold. As the trumpet of Mr W. A. Williams declared its melancholy valediction the occasion must have seemed less like a Festival of the Future than a farewell to the nineteenth-century optimism that had brought it into existence.

The Glastonbury Assembly Rooms lie concealed behind the row of shops that line the Abbey side of the High Street. The archway and cobbled yard through which they are approached suggest an old coaching inn—and indeed the Assembly Rooms had been built in 1864 on the site occupied by the stables of the White Hart, once Glastonbury's principal hotel. The building itself, sturdily constructed in grey stone and decorated with a selection of ecclesiastical features, consists of a main floor above an elaborate undercroft. Six steps lead to the entrance, and a further eight to the level of the hall. To the left there was a series of small rooms, to the right the hall itself.

It is not large: 54' in length, 34' across. In Boughton's day there were open fireplaces half-way down on either side, and a stage across the whole width of the back wall to a depth of 14'. Of this, 7' 6" was given over to a back-stage area, leaving 6' 6" for the main acting area, plus a further 5', or thereabouts, for steps which served two-thirds of the stage width, leaving a handy niche (stage right) in which to insert a grand piano. The stage rose some 3' 6" above floor level—a tolerable, though not ideal height from the audience's point of view. Draped curtains slung from the heavy pitch-pine rafters concealed the back-stage area, and when there was not enough natural light from the windows high in the side wall, stage left, the hall was lit by a series of gas chandeliers. There was only one exit: a door, stage left, which led to a staircase outside the main building that was only partially covered. Performers would have to make their way down the staircase, cross the narrow courtyard, and mount the steps to the main door and their dressing-rooms to the left of the entrance. With a little juggling of bent-wood chairs and redundant church pews, the hall could be made to accommodate about 220 people. With all its disadvantages this was to be Rutland Boughton's 'National Playhouse for Music Drama' for as long as the festival remained in existence.

The town itself was pretty enough. Two main streets running at right angles, a handful of lesser streets and alleyways, and, in the background, the magnificent Abbey ruins. Beyond them, dominating everything and surmounted by the pointing finger of an old tower, a strange conical hill, the Glastonbury Tor, rose above the flat marshlands to stir the imagination with dreams of King Arthur and the Holy Grail. Despite its workday appearance as a quiet market town, Glastonbury was sited in a landscape of mystery, heavy with the brooding past, inevitable focal point for dreams and dreamers.

The first Glastonbury Festival departed quite considerably from the programme advertised earlier in the year. Thomas Beecham's orchestra did not appear, neither could a replacement of amateurs be found. The three performances of *The Immortal Hour*, on 26, 27, and 29 August, therefore took place to the accompaniment of a piano, played not by Boughton, who in the absence of Mr Francis Harford was obliged to sing the part of Dalua himself, but by Charles Kennedy Scott. Professional singers, including Irene Lemon, Frederic Austin, and Arthur Jordan, came from Birmingham to sing the principal roles. Miss Morris being still in France, Boughton himself took charge of the production. Costume design was attributed to Margaret Morris, Christina Walshe, and Gerda Giobel—Christina most probably being responsible for the colourful second act. If the example of later productions is anything to go by, she seems to have based her ideas on Stephen Reid's 1909 illustrations for T. W. Rolleston's *Myths and Legends of the Celtic Race*. Miss Morris's contribution to Act I is described by the *Central Somerset Gazette* for 20 August 1915 thus: 'The back and sides of the stage were hung with curtains of a subdued grey-blue shade, and the woods scene in the first act was represented by "choral scenery"—the dancing chorus being attired in robes suitably designed by Miss Morris to resemble trees.' There is no reason to suppose that the 1914 production was any different.

Whatever it was, the public was delighted with what they saw and heard, and the *Central Somerset Gazette* (2 September) recorded their reactions in glowing terms:

On Thursday and Saturday last Mr Rutland Boughton's music-drama The Immortal Hour was repeated with great success. There was an exceptionally large attendance at the last performances, but the audiences bore willingly the discomforts and overcrowding on a warm afternoon, and the spontaneous and enthusiastic applause which broke forth when the curtain was drawn on the closing scene eloquently testified of their appreciation. The principal characters were recalled time after time, and at length, after insistent cries for the composer, Mr Boughton stepped forward and expressed his pleasure that the performance had been so much enjoyed.

On Wednesday, 12 August, Miss Buckton had been able to mount *Beauty and the Beast* in the open air at Chalice Well, but ten days later her pageant play *The Coming of Bride* hit foul weather and had to be given in the Assembly Rooms. Neither of the children's operas was performed—possibly because Glastonbury's inhabitants could not yet be persuaded that a Summer School was a suitable outlet for their off-spring's talents. The children's contributions to the concerts that formed the remainder of the festival were carried out by a posse of Miss Morris's young dancers, already hardened to the lure and dangers of the footlights. There was no foundation-stone ceremony, and Sir Edward Elgar was therefore not called upon to put in an appearance.

The ten concerts were a strange mixture of songs, partsongs, solo and group dances, and short plays, including: W. W. Gibson's *The Night-Shift*, and an 'impromptu scene based on *Alice in Wonderland*', both produced by the poet John Rodker.[5] Lady Gregory's *The Travelling Man*, Walter Merry's *Soul Sight*, and Gerald Warre Cornish's translation of a scene from *The Heracleidae* of Euripides completed the dramatic side of the month's entertainment. With extracts from Wagner (the 'Grail' scene from *Parsifal*, and Siegfried's 'Sword Song'), and arias from Vincent Thomas's Arthurian opera *The Quest of the Grail*, the overall effect was less of a coherently planned 'festival' than the festive celebration of the work achieved by an exploratory Summer School—which indeed was exactly what the first Glastonbury Festival was.

As a trial run, however, it could reasonably be accounted a success. It had cost a little over £200 to mount and had emerged with a deficit of £84. In artistic terms it had been amateurish and enjoyable. *The Immortal Hour* had been a complete triumph, justifying the entire venture. And although it was clear that all thoughts of building a the-atre would have to be abandoned at least for as long as the war might last (but at this stage most people believed it would be over by Christmas), it was evident that the Assembly Rooms could, at a pinch, become a suitable base for work. It was also evident that local talent could not be relied upon to produce soloists in sufficient quan-tity, or quality, but might at least be drilled into an acceptable chorus if regular training could be undertaken over and above the intensive month of the Summer School. Finally, in terms of practical experience if had been invaluable. From this moment Boughton dropped all grandiose Wagnerian ideas and began to deal with the situation as he found it. It was his ability to do this that sets him apart from the average dreamer. If his head was in the clouds, his feet trod a firm reality.

What the experience had meant to those that had taken part in the performances can be judged by the extracts from their letters which Boughton printed in the 1915 brochure:

I have been here [Aldershot] for about a week, drilling Kitchener's recruits, and Glastonbury seems like a delightful dream . . .

With the other performances I was lucky enough to witness it has been to me a delightful revelation of new beauty . . .

I am sending you a little note to thank you for all your hard work with us. I did enjoy the music, and it seemed to grow in interest the more one heard it. I wish we were going to have the last week all over again. You will have heard that the train went out of the station to the sound of the 'laughing' chorus . . .

I thank you for the experience and joy of Glastonbury, and am with you again whenever the gods favour, and you wish it . . .

Even more to the point, the festival had banished any doubts that the Clarks, by far the most important and influential of his supporters, may have felt. They were now wholly on his side, and in October Boughton wrote to Roger Clark:

If details are agreeable to you, I should most gladly accept your offer, on some such basis as the following: that for a number of months (to be specified by you) you guarantee me a weekly income from the neighbourhood. We think Two Pounds a week would cover us, with the few odd amounts we earn in other ways. Then, you promise to take over the problematic profits from The Immortal Hour and The Birth of Arthur, as a friendly means of enabling me to keep my face. In addition to the above I should need immediate help to the tune of about Ten Pounds to get our things from London, as I have earned no money since July.

You don't want us to say how grateful we are, I know. Indeed, I'm not at all sure that we have much capacity for that sort of feeling. But we have a great sense of relief, and can only hope that the work done during the next few months may make you feel that it was worth while.[6]

With this financial assurance Boughton was able to set up regular tuition for the people of Glastonbury and their children. Classes in dancing, voice production, scene painting, and costume design were offered at 10s. a year, and individual lessons were available at special rates. Boughton himself acted as Director of Studies, with Christina and David Scott as his enthusiastic seconds. To supplement the Clarks' subsidy, and incidentally to ensure a steady supply of potential soloists, he resumed his private singing practice in Birmingham.

Not everybody in Glastonbury shared the Clarks' enthusiasm, however. Once the festival was over, Miss Buckton pointedly withdrew her support, and when Boughton announced the setting up of a 'Festival Drama Society' it became clear that the seeds of a less than helpful rivalry had not only been sown but were beginning to sprout. On 23

October Boughton allowed himself the consolations of irony in a letter to the *Central Somerset Gazette*:

Sir—Miss Alice Buckton asks me to make it plain that the Festival Dramatic Society is not founded as a rival to her schemes at Chalice Well. I do this with pleasure; indeed, I do not understand how any rivalry could be imagined by anyone.

The Festival Dramatic Society has been organized as part of the plan outlined nearly two years ago when I first visited Glastonbury to seek support for the National Playhouse idea. This was before Miss Buckton had settled in Glastonbury and indeed before I had the pleasure of her acquaintance.

The scheme for the Playhouse, supported by many of the foremost names in British art and literature, is during these difficult times necessarily in a state of abeyance; but anyhow it is quite distinct from Miss Buckton's work at Chalice Well, and so far from wishing to rival her work I now only point out that our plans for the production of National Music-Drama etc. have no feature in common with the work carried on by Miss Buckton, with the one exception that we both wish so far as possible to use the talent of the locality for our performances.

The object of the Festival Drama Society is to keep alive the National Festival idea during the next few years; its immediate task is the preparation of *The Birth of Arthur* for performance next summer. A committee of townspeople has been formed under the chairmanship of Mr George Alves, with Mr J. A. Gilbert as Hon. Secretary and Mr David Scott, ARCO, as chorus-master, and rehearsals are taking place on Wednesdays at 7.30pm in the Town Hall.

Yours sincerely,
Rutland Boughton.

A postscript plunged the knife in deeper:

It would perhaps have led to less confusion in the public mind if Miss Buckton had not called her society 'The Glastonbury Festival Players'. They are apparently the same ladies and gentlemen whom she described last August as 'Students of Chalice Well'.

It was the first of a series of petty squabbles that were to enliven their relationship for as long as the festivals lasted, matters always being made worse by the irritating fact that it was all too easy to confuse their names.

By the end of 1914 the pattern for future Glastonbury Festivals had taken shape. Classes for the local people would continue throughout the year, with special Holiday Schools at Easter, Whitsun, Summer, and Christmas, each culminating in a festival—the August festival being the most elaborate. Choruses and small parts would be drawn from local talent, while professionals—Boughton's pupils, often, or

young singers at the beginning of their careers—would take the principal roles. A decent piano, played in Boughton's inimitable, full-blooded
way, would serve as an orchestra, and the Assembly Rooms would be
the theatre. If it was not exactly the scheme that Buckley and Boughton
had dreamt of in *Music Drama of the Future*, it was just as exciting.
Above all, it was practical.

1915

The new year began well. Despite the enormous amount of work he had undertaken in 1914, Boughton had found time to compose a three-act ballet, *Snow White and the Seven Little Dwarfs*, which Margaret Morris had commissioned for her Dancing Children. This had enjoyed a highly successful run at her Chelsea theatre, beginning on 30 December, and now the critics were falling over themselves in its praise. 'Mr Boughton's music', wrote the critic of the *Observer* on 3 January 1915, 'is playful, tuneful, quaint and rhythmical, and perfectly fitted to its purpose. The Dwarf music is particularly happy, as are the dwarfs, indeed, from the scenic point of view.' Morale among the dwarfs may not have been as high as it appeared however, as Margaret Morris was later to recall, one of their number had discovered the pleasures of up-staging:

Angela Baddeley gave a really remarkable performance as the First Dwarf. They all wore masks, with big noses, but the expression Angela got into that mask of hers was quite incredible. It seemed always to be changing with the angle of her head and the movements of her hands, and somehow she made that comic little dwarf a most moving and lovable character who quite stole the show.[1]

What her sister Hermione and the infant Elsa Lanchester thought, as mere supporting dwarfs, is not recorded.

From Bournemouth there now came an offer of help from Dan Godfrey, and on 7, 8, and 9 January four performances of *The Immortal Hour* were given at the Winter Gardens. This was the first time it had been heard with orchestra and it was a revelation. Details which could never be reproduced by two hands at a keyboard now took their rightful place in a tapestry of intricate, glowing sound which underlined even more emphatically the Pre-Raphaelite quality of Boughton's inspiration. The cast was largely the same as it had been at Glastonbury, save that Gladys Fisher and Marjorie Ffrangcon-Davies now alternated in the part that Irene Lemon had created, and, instead of Dalua, Boughton sang the less demanding role of Manus. The conductor, once again, was Charles Kennedy Scott. Reporting the event on

9 April *The Times* was judicious in its appreciation of what had been going on:

In our notice yesterday of Mr Rutland Boughton's opera *The Immortal Hour* we spoke of his courage in staking everything upon the musical quality of his work, and there was a certain sporting spirit about the whole production that enlisted sympathy. Instead of wasting time in searching for a wealthy patron to spend thousands of pounds on producing it for one or two nights only at a London theatre, a search which would be likely to be even longer than usual just now, Mr Boughton proceeded to utilize the means at hand.

He got on the right side of Mr Dan Godfrey; that in itself, perhaps, was no great feat, for composers who are worth their salt generally find him a ready friend. Mr Godfrey could only give him a glass-house, known as the Winter Gardens—a place poor enough as a concert-room, but worse as a theatre, with an improvised stage, no scenery, a white lantern-screen for a curtain, and a curtain-boy who always forgot his cue. But Mr Boughton did not want scenery, and was untroubled by the vagaries of the curtain-boy. What he did want was a capable orchestra and a place to perform in which would keep out the rain. He got them both from Mr Godfrey.

He also wanted as a first requirement for his opera a real chorus, not one used to shouting operatic platitudes under cover of brass and drums, but one trained to self-reliance, able to sing in rhythmic counterpoint, and that Mr Kennedy Scott was able to bring as a result of his experience in training the Oriana Madrigal Society in the music of Elizabethan England. The art which Miss Margaret Morris has developed in her dancing school was another contribution to the essential needs of the production, and, in short, Mr Boughton made friends not with the Mammon of unrighteousness, but with artists who are doing distinctive work in their various departments.

This seems to us a thoroughly sound way of laying the foundations of a type of opera which should really belong to this country. Instead of crying out that we have not got things, let us see what things we have, and, bringing them together use them to the best advantage.

So we rather hope that Mr Boughton will not find a wealthy patron, but will make more and more helpful musical friends. Especially we see hope in his reliance upon the chorus as a prime factor in the development of his drama, for choral singing is the very bed-rock of all our national music, and if as a nation we are ever to make a genuine contribution to opera it must be by incorporating our natural means of expression into dramatic action.

As an appreciation of what he was trying to do it could scarcely have been more encouraging had he written it himself. Moreover, it drew attention to his eminently practical solution to the problem that beset almost every would-be operatic composer in Great Britain: where to find an outlet and a proving-ground for their work. Some, like

Rutland Boughton: His Life, 1915 81

Stanford, Delius, and Ethel Smyth, had turned to Germany for help; others, like Holst and Vaughan Williams, were to find support in the London colleges of music. The little encouragement that such companies as Carl Rosa's were able to give was unstructured and intermittent. In short, any composer in this country who wanted to write opera faced nothing but the discouragement of almost impossible odds. Boughton had the foresight and courage to take a step which, even if it had the disadvantage of isolating his development from the mainstream of music, was at least practical. At Glastonbury he could learn his trade and pass its lessons on to others. As the *Musical Times* put it in their issue for February 1915:

The freshness of his ideas bodes well for his future as a leader in that little army which is furthering the establishment of a national musical art. By gathering around him clever coadjutors, imbued with the same sincerity of purpose as himself, it is within his power to erect such a temple of art at Glastonbury as will astonish the musical community.

It was a step that Britten, fifty years later, also found it advisable to take, and for many of the same reasons.

Work at Glastonbury progressed relatively smoothly throughout the early months of 1915, so that a one-day Easter Festival on 7 April was able to draw largely on local talent for its matinée and evening performances of Purcell's *Dido and Aeneas* and the second scene from Act 1 of *The Immortal Hour*. Professional help came only from Gladys Fisher and George Painter, and the loss on the two performances amounted to an easily assimilated £5. Remarkable as Glastonbury was, there were aspects of its productions, as Boughton recalled of the Whitsun Festival (25-6 May) which featured performances of *Dido and Aeneas* and a scene from *The Birth of Arthur*, that created a quite different kind of astonishment:

Our production of *Dido and Aeneas* nearly collapsed because of the masks which Miss Christina Walshe had contrived for the witches to wear. And in fairness to the players it must be admitted than neither the experience of Miss Walshe nor myself was so considerable that we can put the blame for the incident on the players; it was simply one of those things from which one had to learn. The chorus was in open mutiny, declaring they were on the point of suffocation. It was a dress rehearsal, and Mr Edward Dent was unexpectedly a visitor. Now it is a mistake to allow anyone to be present at a rehearsal unless actively engaged on the work in hand; not only may one more easily hurt the feelings of those one has to correct, but, to confess to a personal failing of my own, one is sometimes apt to play up to the audience by showing off or overlooking mistakes. So the only thing to do was to call off the rehearsal for the time being, to enable the players to recover from their asphixiation, while Mr Dent solemnly adjured me not to give way; then, having got rid of our visitor, to assure everybody that art demanded martyrs no less than religion, to beg them to stifle in the cause of a rather effective ugliness, and then get on with the work. But that was not the end of the trouble caused by those masks. At one of the performances the chief witch's face became partly unfastened, and to our horrow slowly descended bar by bar from its rightful position, until by the end of the scene the wicked old lady's head was situated in about the centre of her anatomy. Some of the audience thought it the finest of our stage effects on that occasion.[2]

The presence of so eminent a musician as Edward Dent, Professor of Music at Cambridge University and authority on all matters of English opera, was further proof that Glastonbury was beginning to send out ripples to some purpose. Dent (1876–1957) had been drawn to the festival because of *Dido and Aeneas*, which he had never seen staged. Though sceptical about *The Birth of Arthur*, privately referring to Igraine and Brastias as Migraine and Blasted Ass, he was enthusiastic about *The Immortal Hour*.[3] From that moment he became one of Boughton's staunchest supporters.

    Indeed, to an increasing number of musicians it was becoming obvious that Glastonbury had much to offer. Writing in the *Musical Standard* for 5 June 1915, Charles Kennedy Scott put the case well:

I believe Boughton's work will eventually be recognised as one of the most remarkable achievements in the story of our music. It may not reach its full fruition for many years; but he is advancing ideas (in some tribulation at present) which I am convinced will, sooner or later, come to a very joyful harvest. By reason of his unique endowments he is our national leader in this matter. We have musicians of genius amongst us, often with sensitive literary perceptions, but so far as I know we have no one save Boughton who adds that much rarer gift, a constructive sense of the theatre with all this implies: the feeling for dance, gesture, and (perhaps the most important qualification of all, even in music-drama) scenic treatment. I think it was said of an Elizabethan poet that his words seemed to come to him ready 'cloathed in their note': Boughton's notes come to him already clothed in their gesture. This amazing resourcefulness not only gives unity to Boughton's work, but it means economy in the cost of its realization—a matter of importance in a new movement, which in the nature of things must depend on comparatively limited resources. For every reason, then, Boughton should have our support. Some of us can give money, some can give practical help. Most of us have not the power to initiate or control a movement of this sort. Boughton has. It was Goethe who said that if you cannot yourself be whole, connect yourself with one who can. Let us therefore join with Boughton at Glastonbury and elsewhere in a work that in its inception is undoubtedly vital and complete.

The chance to follow Goethe's advice, and the terms by which 'connection' might best be made, was spelt out in the brochure advertising the 'Glastonbury Festival School (President: Thomas Beecham)' which was to be held during the month of August 1915:

The School will assemble on Saturday, July 31st, for the study of music, dance, opera, and drama, and will continue during the four ensuing weeks. . . . A holiday spirit will prevail as in previous years; but no responsible performing parts will be allotted to members attending for a shorter period then two weeks, and preference will be given to members attending for the month. This rule will not prevent the Committee from engaging professional artists for some of the chief roles. The advantage of the experience to young singers with a leaning to operatic work is sufficiently obvious.

In fact a further reading of the brochure would have made it clear that certain artists had already been engaged. As it turned out, nearly all the principal parts had to be filled by young professionals.

Terms for board and lodging at one of the school's three guest houses ranged from £1 to £1. 10s. per week, with an additional guinea (or 7s. 6d. if paid by the week) for tuition. The works to be performed were *The Immortal Hour*, *Dido and Aeneas*, and Edgar Bainton's one-act opera *Oithona*—written some ten years earlier but as yet unperformed. Tickets for 'central seats, numbered and reserved (with cushions)' were 5s., 'unreserved' half a crown. All performances were to be given in the Assembly Rooms, and assurance was given that 'the hall will be darkened for the afternoon performances, that proper effects of stage lighting may be obtained'.

Unlike the predictions for the 1914 festival, the 1915 programme was carried out to the letter. *Oithona*, together with the second act of *Tristan und Isolde*, received four performances; *Dido and Aeneas* was coupled with scenes from *The Birth of Arthur*; and *The Immortal Hour* was given six times. Students who had successfully competed in a 'Competition in Solo-Singing with Action', adjudicated by Frederic Austin on Thursday 26 August, performed their pieces at an evening concert, and the entire festival ended on 28 August with a concert in the Crispin Hall, Street, which included Wagner extracts (the final scene of *Götterdämmerung*), a scene from *Boris Godunov*, and two of Bantock's heady 'Sappho' songs.

Critics, on the whole, proved to be willing to take the limitations of the Glastonbury Assembly Rooms in their stride, though they may not have been wholly aware of the difficulties the performers had to surmount. The *Evening News* for 12 August, for example, declared: 'One and all were charmed with what they saw and heard, for the artistic spirit had triumphed over the difficulties of the staging'—a sentiment which was echoed by the *Daily Chronicle*: 'The artistic driving force behind it is far greater than in the case of other festivals outwardly more important.' There was admiration, too, for Christina's contribution: 'As for the dresses and general stage settings, they reflect the greatest credit on the artistic sense and sureness of touch of their designer. Christina Walshe deserves almost as much of the glory of the Festival's success as the composer of the works: the collaboration is ideal.' So wrote the critic of the *Music Student* (October 1915). Boughton, however, remembered that some of the backcloths—particularly those for Bainton's *Oithona*—caused something of a sensation because they were considered to be 'very post-impressionistic and "arty" because they were so obviously matters of line upon a flat surface'. Christina had wisely abandoned any attempt at realism and struck out for the symbolic representation of trees and buildings, mountains and lakes, rather in the manner of Roger Fry and his Omega Workshop collaborators. The effect in a small space startled those whose idea of stage scenery had progressed no further than elaborate and wholly unconvincing 'perspectives'.

Even when the critics felt a certain reservation, as *The Times* correspondent did on 12 August when reporting on Bainton's *Oithona*, a genuine admiration still shows through the doubts:

Oithona is the name of the opera and also of its heroine, whose part was skilfully played by Miss Marjorie Ffrangcon-Davies. It is not a promising name. It suggests nothing except what is fact, that the composer has gone to early British legend for his subject, and that type of subject has unfortunately earned a reputation for unreality in opera.

An early British princess, carried away by an early British villain, with an equally impressive name, 'Dunrommath', and avenged by a hero with a simpler name, 'Gaul', does not immediately arouse sympathy, but Mr Bainton has plunged into his subject boldly and succeeded in doing a good deal with it. He has kept the action simple and aimed at making his music express the essential things. Consequently it is manageable in the very simple conditions in which operas are given at Glastonbury.

The Assembly Rooms provide a primitive theatre in which scenery is represented by a broadly-painted back-cloth, and a piano takes the place of an orchestra. A local choir of early British warriors sang with spirit, and with Mr Frank Mullings as the hero and Mr Herbert Langley as the villain, one got a very fair impression of what the opera aims at. Of the two scenes, the second, which begins with the lament of Oithona and ends with her self-sought death in the battle between her two lovers, is the better, because the emotion of the drama is stronger and the music rises to its opportunities.

After the new work the second act of *Tristan* was given with Mme Gleeson White and Mr Frank Mullings in the chief parts, and some local singers in the lesser ones, including Miss Jessie Norman, who showed remarkable intelligence in her singing of Brangaena. Altogether the performance showed that the activities of the school, the object of which is the study of operatic ensemble, away from the conventions of the opera house, are being pursued to good purpose.

Bainton himself was unable to see his work performed, for, having been trapped in Germany by the outbreak of war (he had been visiting Bayreuth), he was in an internment camp at Ruhleben. But it must have been a comfort to know that there was now one place in England where it was possible to stage untried operas both cheaply and effectively. The *Musical Times* (September 1915) commented approvingly on this aspect of Boughton's venture:

One remarkable lesson to be learned from [the productions] is the possibility of staging opera with artistic simplicity which does not require elaborate technical resources. Within the limits of a small hall and a tiny stage Mr Boughton has achieved wonders, in the face of many difficulties. The plan he has adopted of draperies and cleverly designed back-cloths with a simple system of lighting, gives the operas a charmingly effective setting of appropriate atmosphere. It was possible to test this by the performance of a familiar thing—Act II of Tristan—which, staged on the lines just described, was surprisingly good. The striking back-cloth, the work of Miss Walshe, who has been responsible for designing all the scenery and costumes used at the Festival, deserves special mention.

All in all, the festival was judged a success, not because it had achieved anything like perfection, but because it was so clearly developing along fruitful lines. Edward Dent's comment, in a letter to Boughton dated 22 August, makes the point forcefully and judiciously:

It is a wonderful achievement to have accomplished to much with so very small resources, and I was much struck by the way in which you evidently know very clearly what effects you want to get—even when the limited conditions make it impossible to get your effects complete you get near enough to indicate a principle and suggest further possibilities. That was notably the case in The Birth of Arthur. I thought the choral scenery very successful, even as it was. With the additional help of an orchestra, and with a larger stage and subtler effects of lighting, I am sure it would produce very interesting and poetical effects.

I hope some day I may hear your operas again; I don't feel able to form a reasoned judgement on them at first hearing. The Immortal Hour was the one which made the deeper impression on me: I found it clearer in style (and I think clearness is the most important thing in opera) and on the whole more modern in feeling. I was very pleased with the way your singers spoke their words: I hardly lost a line—which is remarkable. I had no knowledge of the poem and heard it for the first time.

And if critics and audience felt pleased, those who had taken part in the performances were delighted. They had been caught up in the adventure of Boughton's dreams and stirred by his burning enthusiasm, had found themselves reaching out beyond the suspected limits of their capacities. For the professionals the festival was a taste of what life might be like, freed of the necessity to compromise with everyday considerations. At Glastonbury it was possible to give of their best for the sheer love of matching a high ideal. 'I can't easily express how much I miss Glastonbury', wrote the young Clarence Raybould, who had come from Birmingham to lend a hand as pianist and conductor. 'It was one of the happiest times of my life, truly an "Immortal Hour", if you will allow "hour" to be rather elastic!' Frederic Austin, writing in the following year, was to be even more emphatic: 'It has been an enormous refreshment to get a glimpse of what an artist's life may be when he is able to live without a thought beyond his work and the things that foster it. I've set my heart on entering the promised land. I feel certain now that with resolution it can be created . . .'.[4]

Precisely how many students had enrolled in the Summer School is not recorded, but a total of £23. 10s. 6d. in fees suggests that there must have been about twenty. The timetable for a typical day's work shows that they got their money's worth:

Monday 9 August	
10.30 am	Dancing Class (seniors)
11.30	Dancers: *Immortal Hour*
	Chorus: *Dido* and *Birth of Arthur*
3.30 pm	Principals: *Tristan Oithona*
8.00	Principals and male chorus: *Oithona*
	Female chorus: *Dido* and *Birth of Arthur*
9.00	Witches: *Dido*

Rambles to Sedgemoor (5 August), and Meare Prehistoric Lake-Dwellings (19 August), plus the inevitable motor-coach trip to Cheddar Gorge (26 August), and each Sunday as a day of complete rest, must have come as a welcome relief from their intense activity. Anyone not involved in rehearsals on 22 August would have been free to acquaint themselves with Miss Buckton's views on 'The Angels of Mons', as, inevitably, she explained their undoubted mystical significance to the gullible.[5]

Boughton's instructions as to behaviour on and off stage give some idea of the cramped conditions at the Assembly Rooms, and the degree of self-discipline the cast needed to carry them through the performances:

There are three dressing rooms at the Assembly Rooms, none of them convenient for the purpose, so a good deal of patience and consideration will be needed. The furthest room will be used by the men, the one opposite the outer door by the women, and the one opposite the inner hall door by both sexes for final make-up.

All performers must be in their places five minutes before the beginning of ordinary rehearsals. At dress rehearsals and performances the chorus and dancers must arrive at least one hour beforehand, and must quit the dressing rooms within thirty minutes to give the principals reasonable accommodation. During performances no-one will be allowed on the stage until they have had their call. They can, however, wait in the stage ante-room. During the performances absolute silence must be maintained everywhere.

No undergarments may be worn which are liable to show. A close-fitting bathing suit with short legs is the best garment for the purpose. Contravention of this rule will cause the performer to be refused admission to the stage.

Members must attend all their rehearsals; the absence of a single performer upsets the stage scheme.

Inevitably the Summer Festival incurred a loss. But, as the detailed accounts show, this was largely offset by donations that spoke eloquently of the goodwill that had been generated:

RECEIPTS:	£	s.	d.
Tickets	189	18	0
Programmes	5	16	11
Donations	93	16	0
Fees	23	10	6

EXPENSES:	£	s.	d.
Artists	123	15	6
Performing Rights	22	1	0
Printing & Publicity	59	11	9
Scenery & Costumes	54	1	6
Staging, carpenter, etc.	17	1	7
Lighting	15	16	1
Hire of halls	10	5	0
Hire of piano	9	9	0
Music	5	6	1

EXPENSES:	£	s.	d.
Postage, telegrams, etc.	5	12	1
Stationary [sic]	2	17	1
Travelling expenses	3	8	6
Sundries	1	15	6
Commission on tickets	1	3	4

Receipts came to £313. 1s. 5d. and expenses to £332. 8s. 0d., leaving a deficit of £19. 6s. 7d., to which had to be added £7. 2s. 9d. carried over from the 1914 productions.

Further donations improved the situation and Boughton felt able to face the rest of the year's activities with confidence. And it was to be a busy year, for the regular 'school' was now in full swing with a Festival Choral Class on Mondays, a Literary and Art Society on Tuesdays, the Festival Dramatic Society on Wednesdays, and the Glastonbury Orchestral Society (conductor Rutland Boughton) on Thursdays. There were lectures on 'Music History and the Appreciation of Music', classes in 'Harmony, Theory, and Form' (ten classes and ten lectures for an inclusive fee of one guinea), and a series of lecture-recitals on 'British Folk-Music'. As part of the Festival Drama Society's activities (they were studying 'Shakespeare', 'Teutonic Legend', 'Greek Drama', and 'The Arts of India'), special 'public lectures' were given on 6 October, when Mr Percy Scholes spoke on 'Wagner's *Ring of the Nibelungs*' with musical illustrations and lantern reproductions of Arthur Rackham's pictures, and on 30 November, when Dr Ananda Coomaraswamy spoke on the subject of 'Indian Art'. The entire season, including the two public lectures, could be enjoyed for half a crown. Boughton, in effect, was running his own Adult Education programme—a one-man venture into Community Arts.

In all this activity Reginald Buckley is only conspicuous for the very minor part he was now able to play. Once his dream had begun to assume a practical shape he found he had increasingly little to offer. 'He was', wrote Boughton, 'always getting to know people, but seldom doing things in any practical way. He would have made a perfect liaison officer—always rushing about being busy, but accomplishing very little, except perhaps to sweeten people'. As the festival began to develop it did so along lines dictated by Boughton and drew further and further away from Buckley's rather nebulous ideas.

Nor did the relations between the two men begin to improve when, immediately after the festival, Boughton began work on the second Arthurian drama: *The Round Table*. In order to make the characters come to life he found it necessary to add new scenes and rewrite others. Buckley fought the alterations, but there was little he could do, having agreed in principle that Boughton should be free to make just such changes if the music needed them. To make matters worse, Buckley had gone ahead and published the entire Arthurian cycle as if it were the finished and accepted libretto—thereby causing considerable bewilderment to later audiences who came armed with what they took to be the 'book of words'.

By the end of 1915 the break had become inevitable, at least in so far as artistic collaboration was concerned, and on 1 December Boughton wrote with characteristic impatience:

Instead of *Arthur of Britain* being the cause of the scheme as you seem to imagine, it has been a millstone round its neck. What progress has been made I have made without you, and sometimes despite you; while you proved in the Letchworth case that you could not move without me. If you won't accept this simple fact I'm afraid it means a rupture between us sooner or later.

The rupture came with the letter that Boughton wrote on 16 December:

Your letter to Gould making proposals behind my back to organize a Festival here on lines other than those which interest me has just come to my notice. This is not the first time you have treated me in a disloyal way, though I have loyally supported you and introduced you wherever I have been able to further things.

In future I decline to have any further business dealings with you, other than those in which I am already involved. I am extremely sorry to have been brought to this conclusion, but it has been forced upon me by our very different ideas of what is meant by working together.

The final stroke was delivered by Shaw during the following Easter Festival when he was overheard to say: 'Buckley will have to go!' Believing Shaw to be capable of almost anything, Buckley began to suspect he was to be the victim of some dreadful assault. Sure enough, a few days later he came indignantly to Boughton with the news that a taxi had deliberately tried to run him down. 'And you know who was inside it, don't you?', he cried. 'Shaw—laughing at me!'

But by that time Shaw had more serious matters to consider. During the course of 1915 Boughton had found himself increasingly drawn to Irene Lemon, the young singer who had created the role of Etain for him. 'I told Christina what had happened and she took it kindly and lightly enough. There was a time, she said, when Langley, one of our singers, had aroused similar feelings in her.' But the feeling persisted and grew stronger when Irene came to sing in the Christmas Festival in December 1915, and stronger still in the summer of 1916:

At the time, and on other occasions later on, I have been sufficiently convinced that without that particular woman I could not do the work I was
intended to do. I have never felt the brake that would have resulted had I
regarded my career impotant as distinct from my work. I had to go all out for
the work whatever happened. That mad and selfish outlook has the only
excuse that it was as honest as it was foolish. I offered to run away with
Irene, but, to the surprise of my conceit, she refused.

Shaw reproved him for being so cruel as to tell Christina, but
Christina, pregnant with their second child, seems to have taken it in
her stride—sensing, perhaps, that the infatuation did not go very deep
and had as its source the typical creative artist's capacity to confuse a
skilled interpreter with his creations. It was not so much Irene he was
in love with as Etain.

   At some point during the autumn of 1915, possibly in reaction to the
differences of opinion he was having with Reginald Buckley, Boughton
turned aside from work on *The Round Table* to concentrate on a completely new idea. Everything about it was designed to take account of
the amateur talents he was hoping to foster. He described the new
work and its origins in an article published in *Somerset and the Drama*
(Somerset Folk Press, London, 1922):

I set the old Coventry Nativity Play with the idea of making a work that
could be performed entirely by local players. It was deliberately composed as
a folk opera, the lyrical quality of the play being increased by the insertion at
suitable moments of Early English Carols, either as choral interludes after the
manner of Greek tragedy, or as part of the play itself—i.e. the two lullabies
for the Virgin Folk tunes were generally used as a basis for the interludes, the
exception being original tunes taken from the Arthurian cycle, for it seemed
worth while to link our various operas not only in the arbitrary way of subject, but also by the interplay of actual musical themes. Besides the entr'acte
carols three other folk tunes were incorporated in the body of the work. I
think *Bethlehem* [*the name finally given to the musical nativity play*] is the most
suited to such amateur societies as enjoy operatic forms. The tearing raging
part of Herod needs a big tenor and we finally had to ask Frank Mullings to
sing it; but the other parts were well within the powers of any body of village
singers. At our first production we also had to engage an old pupil for the
part of the Virgin Mary, owing to the diffidence felt by the family of the
Glastonbury soprano; but any young girl who has not been trained in the airs
of concert room sophistication could easily sing the music.

A similar diffidence obliged Boughton himself to sing the part of the
'Unbeliever', though the redoubtable Irene found no difficulty in

assuming the role of Virgin Mary—only, according to Shaw, to be magnificently upstaged by Christina's Angel Gabriel.

At its first production in the Crispin Hall at Street on 28 December 1915, as the only work presented in the three-day Christmas Festival (28–30 December), the carols which serve as interludes between each scene were sung in unison by the audience which had been invited to rehearse them at the Bear Inn on Boxing Day. In the following year, when the work was revised, they were turned into elaborate choral settings in the style of Boughton's earlier *Choral Variations on English Folksongs*.

Bethlehem was an immediate success. Its simple, subtle melodies which go straight to the heart of the drama, its artless joy in telling the Christmas story, its capacity to tap the depths of communal feeling through the age-old carols it uses, made it instantly and lastingly popular. Boughton's music raised to an even higher level all those touchingly human qualities that make the original play so moving. The spirit in which it was written—the spirit that shines through every page—is summed up in the dedication he placed at the head of the score: 'To my children, and to all children'.

9

1916–1918

Throughout 1916 the Glastonbury Festival continued to expand and grow confident, despite the war which now and again would reach out and snatch away one of the performers. A reprieve was a matter for quiet rejoicing, a signal that the important things in life might continue even when the world ran mad. The Easter Festival, held from 19 to 26 April at a cost of £90. 4s. 8d. and at a loss of £31. 11s. 2d., was particularly significant for Rutland Boughton. Shaw at last 'felt inclined' to pay a visit, did so, and was converted. Recognizing in Boughton's efforts a truthfulness and integrity he could not always find in the professional theatre, he became one of his most helpful and enthusiastic supporters.

The main work at Easter was Gluck's *Iphigenia in Tauris*. It was staged in the more spacious surroundings of the Crispin Hall, Street, and given, marvelled *The Times* on 29 April:

In a very simple manner with piano accompaniment (beautifully played by Mr Clarence Raybould) but with genuine musical and dramatic perception by a company consisting largely of villagers, many of whom have never seen an operatic production of the conventional type. . . . Given the will and energy, a strong guiding hand and a spirit of co-operation, it is possible to get at the heart of a work of art and extract its meaning without the parade of material, the lavish expense, and the fuss which makes opera so often unpractical and inartistic.

Shaw, in an article published in the *Nation* on 6 May, gave a more colourful but no less enthusiastic account of the occasion:

Orestes had been reached at the last moment by the voice of patriotic duty, and had gone to face sterner music in France or Salonica. The result was that the conductor had to take the part; and it may be that some of the freshness and excellence of the performance were due to the fact that there was no conductor.

I do not know what Orestes was like, and so cannot say whether Mr Rutland Boughton resembled him; but he certainly did resemble a well-known portrait of Liszt so strongly that I felt Pylades should have been made up as Wagner; and yet when I looked at Pylades he reminded me so strongly of Mr

Festing Jones that I felt that Orestes should have been made up as Samuel Butler.

Mr Rutland Boughton did astonishingly well under the circumstances. His ability as a composer stood him in good stead; for when his memory gave out he improvised Gluck recitatives with felicitous ease, though his modern freedom of modulation occasionally landed him in keys from which the orchestra (Mr Clarence Raybould at the grand piano) had to retrieve the others as best it could. I do not know whether Mr Boughton's voice is tenor or bass, nor even whether he can be said to have any voice at all for bel canto purposes; but it was all the more instructive to hear how he evaded such questions by attacking the part wholly from the dramatic point of view.

There was fortunately no scenery and no opera house; in short, no nonsense; but there was a Shrine of Diana and sufficient decoration by Miss Walshe's screens and curtains to create much more illusion in the big schoolroom [sic] than I have ever been able to feel in Covent Garden.

Altogether this Easter exploit of the Glastonbury Festival School, as it is called, was very successful and pleasant. Allowances have to be made in judging such performances; and London critics might exaggerate them because, as they are new allowances, they would be more conscious of them than of the prodigious allowances that have to be made in Grand Opera houses in great capitals. But the truth is that there was far less to suffer and far less to excuse and allow for at Glastonbury than at the usual professional performances.

Boughton would have become aware of Gluck's contribution to operatic reform when Bantock staged *Iphigenia in Aulis* in Birmingham during the summer of 1906. Gluck's use of the chorus, his concentration on the spiritual and psychological development of his characters, and, above all, the restraint and purification he brought to operatic conventions must have appealed enormously to Boughton as a further confirmation of his own ideals. The three performances of *Iphigenia* were matched by three of Boughton's Margaret Morris ballet *Snow White*, while on Thursday, 6 April, Boughton and Irene Lemon shared a recital of songs by Stanford and Parry, Holst and Ethel Smyth.

The Whitsun Festival (12–13 June) was an altogether smaller affair: no opera, no ballet, hardly any music at all. Instead the company gave three performances of the morality play *Everyman*, with music borrowed from Walford Davies and Elgar. The parts were taken mainly by local players, but Gwen Ffrangcon-Davies, whom Boughton had known as a child and had taught to play the piano, came down from London to take the part of Everyman. Unfortunately the play proved to have very little appeal for the local audiences. Despite a reprimand from the Vicar of Shapwick, they stayed away. To Boughton, however, this was of small account compared to what his players had gained from the beauty of the language they had to speak.

With the Whitsun Festival safely behind them, plans for an ambitious Summer Festival were put into action. Once again there were difficulties to be surmounted, and Boughton spoke of them at a reunion meeting immediately before the festival began:

The difficulty of getting professional help this year has been very great. We have lost Mr Herbert Langley, who has gone into the army. Mr George Painter is now in the navy, Mr Gerald Cornish is also in the army, and Mr Gilbert, although with us this evening, is also in khaki, and now they are trying to get Mr Scott. . . . I do not suggest that it can be otherwise, for when we read of the German atrocities it makes us feel that we must fight the Germans even if we don't want to. But we believe our work as a Festival School is of national importance in raising the very best in the national spirit. Bad things cannot be expressed by art. Art must be beautiful, and in using the ancient legends and the English language we are evoking out of our own national spirit the very best that is in it. So it is possible to have a month's holiday even in wartime, knowing that we are not idling our time while those we love are being killed at the front. Neville Strutt, who was with us at our first Festival, has been reported 'missing, believed killed', and many other friends are making big sacrifices at the front, so we must feel a solemnity about everything we do this year. During our first year it somehow got about that we were a bad lot, and we have not quite lived that down even now— although many of the people who have come to our performances have given us their support now that they see that art expresses all that is best in us. Still more awkward are those people who have come hoping to find that we really are a bad lot. They have gone away disappointed on finding out what a dull lot we really are. Yet there are still those people who think everything connected with art is wicked.[1]

It is interesting to note that Boughton no longer felt able to include Wagner in his programmes, even though he had vigorously defended his right to do so in previous years. The war had touched nerves that were now far too sensitive to be ignored.

Everyman, *The Immortal Hour*, *Iphigenia in Tauris*, and the ballet *Snow White* were revived for the Summer Festival (8–26 August 1916), and two new works were introduced: Clarence Raybould's setting of a Japanese Noh play, *The Sumida River* (*Sumidagawa*), using a translation by Dr Marie Stopes, and the second part of the Arthurian cycle, *The Round Table*, which Boughton had completed on 3 July. Altogether sixteen performances were given, together with two recitals—one a 'Celtic Recital' by Mrs Tobias Matthay, and the other a mixed programme of songs and piano music by Sylvia and Edwin York Bowen. The festival ended with a singing competition which entitled the winners to three years' free tuition in voice production and dramatic singing.

The Sumida River gave the company a chance to develop still further the methods of stage production that had been tried so successfully in *The Birth of Arthur*. During the course of the action it was necessary to suggest a journey by boat, which, of course, was quite beyond the mechanical resources of the Assembly Rooms' stage. At the same time there were passages of comment and description sung by the chorus. By combining the two and causing the choral singers to carry out a flowing 'water dance' during their account of the journey, a most effective illusion was obtained. The *Central Somerset Gazette* for 18 August reported the occasion in some detail:

The river is represented by eight dancing women who are ranged equidistant at the back of the stage, and their dances, based on the Greek movements, give a vivid impression of the river's flow. Most wonderful of all is the sense of motion given to the boat when supposed to be crossing the river, by the dancers' graceful motions in the opposite direction—each dancer, in turn, having passed off the stage at one end, reappearing at the other end and maintaining an even and continuous forward movement.

Reviewing Alexander Bakshy's *The Path of the Modern Russian Stage* in the same paper on 7 December 1917, Boughton himself looked back with satisfaction to the production:

Against a plain ivory-white back-cloth and the pale-coloured costumes of the chorus in the background, we had in the foreground deeper tones of colour in the costumes of the principals culminating in the black boat near the edge of the stage; and the whole colour was modified by incessant changes of light corresponding to the development of feeling in the drama. It was certainly one of the most satisfactory results we have so far achieved with our poor resources in the matter of light. But however true the colour effects may have been to the music-drama, it was yet subordinate to the dancing of the chorus, the grouping of the principal characters, and the line of their costumes and properties. Form is the vital thing in art.

The similarity of approach between this production and Britten's 1964 version of the same story (the Church Parable *Curlew River*) is altogether striking.

Encouraged by the success of *The Sumida River*, Christina Walshe, whose design and production it had largely been, made clear her dissatisfaction with orthodox stage presentation in an article, 'The Staging of Music-Drama in England', which appeared in the October issue of

Musical Opinion: 'We have hundreds of performances of opera and a good many of music-drama in England every year. But of what kind? . . . Our producers of opera do not seem to realize the necessity for unity in the whole work; although this has been the recognized aim of the dramatic stage for centuries.' After pointing out the effectiveness and integrity of Leon Bakst's contributions to the Russian ballet, she concludes with a spirited defence of Gordon Craig's work:

Russia and Germany . . . have learnt from him. Reinhardt, Germany's best known pupil of Craig, brought a huge production to London a while ago (*The Miracle*, with the comely Lady Diana Cooper as the Virgin Mary). It was watered down and commercialized Craig; but nevertheless it was a huge success financially. And if the patriots who enjoyed the performance so heartily could realize that in supporting Craig's ideas in England they were helping forward a movement far in advance of the best we have yet seen, surely the ideal opera-house would not be long in the building.

Prizes were not offered for guessing where she considered that 'ideal opera-house' might be built, however.

As it happened, Boughton's new work, *The Round Table*, was much nearer to conventional opera, and in only one scene (that of the 'Lake Maidens' in the second act) were there any of the effects that he and Christina believed to be necessary to the proper presentation of music-drama. The critics do not seem to have been disturbed by this slight retreat, noting only, as H. C. Colles did in the September issue of the *Musical Times*, that practical experience had modified Boughton's ideas:

Scenes from *The Birth of Arthur* were performed a year ago, but it has never been given complete, and we gathered not likely to be so given for some time to come. The composer is beset with second thoughts about it. Second thoughts have also modified the course of *The Round Table*, so that the drama is not altogether that which Mr Buckley planned, and it seems by no means certain that the two remaining parts will be finished as they stand.

In short, these years of practical work of training a company which is half amateur and half professional, of studying what can be made expressive on a stage provided only with the simplest accessories, have evidently brought Mr Boughton to a different standpoint from that of three years ago. They have not deflected him from his ideal or chilled his enthusiasm; on the contrary, one had only to visit Glastonbury at the opening of the third summer school to realize how his enthusiasm has radiated to the whole company who gather round him . . .

A short description of what is heard and seen in *The Round Table*, and of how it is heard and seen in the present conditions will best show what is going on. One turns out of the main street of Glastonbury down a little alley to the Assembly Rooms, a building evidently planned with the sole object of looking venerable, and therefore as unsuited to any practical purpose as a building could possibly be. Within, a small stage is framed with curtains. When scenery is essential, a few screens boldly stencilled by Miss Christina Walshe provide the minimum of paraphernalia with the maximum of effect. A grand piano played with extraordinary skill by Mr Clarence Raybould represents the orchestra. Limelights from the back of the hall give us the sun by day and the moon by night. What space is left over in the hall is the auditorium; unfortunately it seems to suffice at present.

But now there is darkness, and the music begins: cold, frosty music (one guesses at the orchestration), and presently the tall, aged figure of Merlin (Mr Percy Heming), his staff of healing in one hand, the magic sword in the other, is dimly outlined. He strikes the sword into the stone of the cathedral porch; his spell decrees that only the King shall draw it out. The legend one feels is perilously near to that of another music-drama, and it may be said that all through the first Act the peril continues, and only the adroitness of the musical treatment save the drama from being a pale reflection of the *Ring*. A sudden change of scene and particularly of light gives us the kitchen of Sir Ector's house on Christmas Eve. The cooks are busy; they bear in the boar's head, the maids roll out the pastry for the Christmas dinner; Arthur (Mr Frederic Austin), the scullion, is blowing up the fire. He sings the carol of 'King Herod and the Cock'. There is fun and fooling in a skilful musical ensemble until Sir Ector (Mr William Bennett) comes in to send his household off to church. On the way thither the knights and people see the sword, and wonder at the legend written around it. There is a big orchestral interlude here, music in which bells are prominent and in which the joy of the festival mingles with the wonder of the people at what they see. Then follows a curious scene in which Arthur first draws the sword and only Kay, Sir Ector's son, sees him do it. Arthur gives the sword to Kay (Mr David Scott), who pretends to have drawn it when his father with the Bishop and all the household come out of church. Sir Dagonet (Mr Arthur Jordan) exposes him by replacing the sword and humorously inviting Kay to repeat the performance. All this seems rather trivial and unsatisfactory. Of course it ends in Arthur drawing the sword a second time, to the mixed wonder and disgust of the people, and the Act is well ended musically, but one feels a certain dissatisfaction with the dramatic scheme. Something no doubt must be put down to the smallness of the stage.

The second Act by the lake of Avalon is more easily made convincing in the conditions of the performance. Miss Christina Walshe with her willow trees and iris leaves makes us see the scene. The light is dim. The Lady of the Lake (Mrs Tobias Matthay) surrounded by her maidens chant the mysteries of life, and hither Nimue (Miss Gwen Ffrangcon-Davies) lures the sage, Merlin, to his death.

The main purpose of this scene is that in it Arthur is charged by Merlin to play the King, to rule his knights, and to establish his Court by his marriage to Guenevere. Some of Merlin's oration tends to prosiness. When he sits down to have the story out one is reminded a little comically of Gurnemanz in the first Act of *Parsifal*. But there is much beautiful and suggestive music, some of it recalling ideas from *The Birth of Arthur*, and the scene reaches a fine climax in Arthur's resolve to rule.

The third Act, much the most consistently interesting in design and character, is at Arthur's Court at Camelot. There is a lovely scene between Guenevere (Miss Irene Lemon) and Lancelot (Mr Percy Snowden), one in which the two characters are sympathetically unfolded and contrasted. Arthur is about to propound to his war-loving knights the peaceful Quest of the Holy Grail, and the whole action turns upon their unwilling acceptance of the quest. It is put forward by the Bishop (Mr William Waite) in a fine song. Arthur urges it with enthusiasm, but the knights only accept it by the decision of Lancelot, who raises his sword at last in response to the appeal of the Queen. The Bishop's oration is naturally and rightly ecclesiastical, Arthur's view seems mainly political, Lancelot's is wholly personal; perhaps it will be the purpose of the third drama to dwell on the deeper, more universal aspect of the quest.

We have mentioned incidentally the artists representing most of the principal characters. One word must suffice to express regard for the sympathy and insight which the singers showed with the meaning of the work. Particularly we would remark on the performances of Miss Lemon as the Queen, Mr Percy Snowden as the fiery Lancelot, Mr Arthur Jordan as the genial Dagonet; but the unity of purpose in the whole presentation was even more noteworthy than any individual success.

For this festival the critics were, almost without exception, enthusiastic. Even Ernest Newman, who had up to that time been rather mealy-mouthed about the whole venture, went so far as to write:

I could not quite believe the experiment was going to be a success. Having seen the whole thing at close quarters I can now testify to the undoubted good it is doing. . . . The best proof of the quality of the work that is being done is that one's interest in the performances grows with the experience of them.

And of Boughton himself he wrote:

It needs a very special combination of gifts to do this kind of thing—original musical ability, a sense of the stage, a talent for organization, a capacity for inspiring belief in others, and unshakeable faith in the value of one's work, and a contempt for the ordinary prizes of life that mean so much for the ordinary man. I know of no musician except Mr Boughton in whom these qualities are blended.[2]

Fortunately the Glastonbury performers were not so earnest in their beliefs as to think themselves beyond the benefit of a little mockery. The festival therefore ended with a performance of a Travesty Play by Frederic Austin. In a delightful mix-up of all the characters from the works being presented in the festival, the play guyed everybody and everything and culminated in a good-humoured but none-the-less heart-felt appeal for money with which to carry out Boughton's dreams:

> Give me the £6,000 and I will build
> A permanent stage with all accessories
> All that to the scene most proper is,
> The best of modern lighting (compliments
> To him who here doth nightly wield the limes
> With handicap severe), Our players house
> In rooms where fittingly they may array
> And eke an auditorium of such a type
> As will withstand the elements for some 6 years.
> For £20,000 we could provide
> A permanent building meet in all respects
> To stand as a memorial to our aims
> And draw our countrymen from far and wide.
> Profits alas! are small within my scheme.
> But think ye of the great gain to the town!

The music for the Travesty Play was, we are told, 'specially dis-arranged by Mr Clarence Raybould, who appeared at the piano in the guise of a charming young lady of the "flapper" type'.

Perhaps the most effective moment, however, was a sudden interruption, stage-managed to look like a genuine protest and written by George Bernard Shaw himself:

MR BOSTOCK [*rising in the audience*]. Stop! I rise to protest against this perfor-mance. I am sorry to interrupt, but I have a public duty to perform. Men of Glastonbury, you all know me. I—

MR AUSTIN [*coming on stage*]. May I ask you to address yourself to me, Mr Bostock? You know we all have the greatest respect for you. We will listen to you with the greatest patience. What do you wish to say?

MR BOSTOCK. I protest against this performance.

MR AUSTIN. In what capacity do you protest, Mr Bostock?

MR BOSTOCK. As a member of the Committee, as a ratepayer—

CHORUS MAN. That's right, Mr Bostock, the rates in this town are something disgraceful—

MR AUSTIN. Order please. Mr Bostock is not protesting against the rates, he is protesting against the performance.

CHORUS MAN. Then, if you ask me, he don't know when he's well off.

MR AUSTIN. I did not ask you, sir; and I don't believe you have paid your rates. Besides, you are only in the chorus; you have no right to assume a principal part. Now Mr Bostock.

MR BOSTOCK. What I say is that classical music should not be burlesqued. I say that the beautiful, tuneful, glorious music of *Iphigenia* should not be burlesqued. I speak with feeling. I say it should not be burlesqued. I go further. I say no music should be burlesqued.

MR AUSTIN. Do I understand you to say that Mr Boughton's work should not be burlesqued?

MR BOSTOCK. I was not speaking of Mr Boughton's work; I was speaking of music.

MR BOUGHTON [*rising*]. What's that you say? Do you mean that what I compose is not music?

MR BOSTOCK. I have the greatest admiration for my friend Boughton. Nobody can appreciate more than I the very stimulating noises he induces our more gifted townspeople to make on these occasions.

MR BOUGHTON. The very stimulating WHAT?

MR BOSTOCK. I said, noise. [*He sits down*]

MR AUSTIN. Shakespeare used the same expression, Mr Boughton. In his *Romeo and Juliet* he speaks of the band as 'sneak's noise'.

MR BOUGHTON. I shall leave Glastonbury tomorrow. You can send for de Souza's band. You can dance ragtimes. You can all go to the movies. Boughton's occupation's gone.

MR BOSTOCK. The immortal composer of *Iphigenia*, the great Gluck—

MR BOUGHTON. Who brought Gluck to Glastonbury? Did you ever hear an opera of Gluck's until I performed it for you?

MR BOSTOCK. Several times.

MR BOUGHTON. What operas of Gluck's—name them.

MR BOSTOCK. *Il trovatore, Maritana, Faust, Carmen, The Bohemian Girl*—

MR AUSTIN. Ahem, ahem! Mr Boughton, I think this altercation has gone far enough. It is very distressing to the audience.

MR BOUGHTON. Not a bit of it; they like it. Besides, who began it? Bostock did. Well, snub Bostock, not me. I won't be suppressed. No man has ever silenced me.

MR BOSTOCK. If you will excuse me saying so, Mr Boughton, you are talking through your hat.

MR BOUGHTON. I haven't a hat. Has any man in Glastonbury ever seen me with a hat? I can't afford a hat. I can't even afford a haircut. I spend all my money on music for the people of Glastonbury. And now because Bostock wants his *Maritana*s and his *Bohemian Girl*s—

MR BOSTOCK. I protest. I cannot allow this. I never wanted a Bohemian Girl. I am a respectable married man. I call upon you, Mr Austin, to protect me from Mr Boughton's scandalous insinuation.

MR AUSTIN. I must request you, Mr Boughton, to keep order.

MR BOUGHTON. I am in order. I am as quiet as a lamb. If you will only listen to me for half and hour or so—Mr Austin. Certainly not. I call upon you to sit down.

MR BOUGHTON. When you call on me, it's my place to ask you to sit down.

MR AUSTIN. Don't quibble, sir. Sit down.

MR BOUGHTON. I won't.

MR AUSTIN. I shall appeal to the Mayor.

MR BOUGHTON. Not even the Mayor of Glastonbury shall muzzle me.

MR AUSTIN. I shall call the police.

MR BOUGHTON. I defy the police. No power in heaven or earth shall prevail against Rutland Boughton—

MRS BOUGHTON. Rutland, you are making a fool of yourself. Sit down.

MR BOUGHTON. [*collapsing abjectly*]. Yes, Dear. [*He sits down*]

MR AUSTIN. Thank you, Mrs Boughton. The hand that rocks the cradle rules the world. Mr Bostock, are you satisfied?

MR BOSTOCK. I will just ask Mr Boughton whether he thinks it fair to burlesque Gluck's music when his own music cannot be burlesqued?

MR BOUGHTON. Why can't it be burlesqued?

MR BOSTOCK. Because it's too funny already.

MR BOUGHTON. (*explosively*) BOSTOCK!

MRS BOUGHTON. (*warningly*). Now, Rutland.

MR BOUGHTON. I am calm, dear, perfectly calm. Mayn't I just tell him why I don't write music like Gluck's?

MR BOSTOCK. Because you can't, Mr Boughton.

MR BOUGHTON. Yes I can; its always easy to so what somebody else has done before. Why is Gluck's music the finest of its kind in the world? Because he didn't imitate foreign composers but wrote the native music of his own country. Well, his country isn't my country, especially just at present. I am writing English music in England for English people; and before I stop, I'll cut it finer still and write Glastonbury music in Glastonbury for Glastonbury people. And I'll make fun of myself all the time if I like, and of Gluck too. So there! [*He sits down*]

MR AUSTIN. I think we may now bring the little episode to a close. [*The limelight man cuts off his light. Austin shouts angrily at him*] Please don't take the light off my face. You are always doing that.

LIMELIGHT MAN. If you could see your face you'd be obliged to me.

MR AUSTIN [*furious*]. What's that you say: [*He jumps down into the auditorium and makes for the limelight man. Boughton and Bostock stop him. All speak together making a fearful row*]

MR AUSTIN. Let me go. He insulted me. I'll let him know. I'll put a face on him. I'll teach him his place. Let go, will you?

MR BOSTOCK. Order. Order. Remember where you are, Mr Austin. The man meant no harm. Calm yourself. This is most unseemly.

MR BOUGHTON. Steady, steady, Austin. Never mind, old chap. Don't kill him: we can't get along without him. Easy, easy.

LIMELIGHT MAN. Come on. I'm ready for you. It's me that makes this show fit to be seen. I'm fed up with your complaints. Come on the lot of you.

At this excitable moment a crash of thunder on the piano signalled the appearance of the goddess Diana who poured oil on the troubled waters.[3]

As soon as the festival was over Shaw had the pleasure of starting off another comedy on hearing that Boughton had received his army call-up papers. Understandably, Boughton was reluctant to hurry to the defence of a country that had so far contrived to do well enough without him. He had offered his services in 1914 but had been rejected on medical grounds. Now, that much older and deeply involved in work that he at least considered important to the nation, he lodged an appeal for exemption. It was supported by Sir Frederick Pollock, Sir Edward Elgar, Granville Bantock, John Masefield, and George Bernard Shaw. With the perversity that only the military mind seems capable of, the hearing of his case had been called for the last day of the festival. Feeling obliged to put his work before the claims of patriotism, Boughton begged the Wells Military Tribunal for a few day's grace. This was at first refused in a welter of sarcasm about the work at Glastonbury being 'all very artistic and interesting, no doubt', and vindictive hopes that the case would be 'reported fully, and read by all the soldiers in the trenches, that this gentleman appeals and then asks for adjournment because he is engaged in a *festival*'. But he was eventually allowed to tie up the loose ends on condition that he report for service immediately afterwards.

The idea that Boughton was about to be skewered by a disapproving military tribunal proved too much for Shaw, who, on 27 August, leapt to his defence with a lengthy and scathing letter to the national press. 'Are we really being held from Bapaume', he demanded to know, 'because the British Expeditionary Force is paralysed by the absence of a solitary man of 38, no bigger physically than Beethoven or Wagner, who can handle a grand piano or a conductor's baton like a master, but who does not know the right end of a gun from the wrong one?' A storm of letters descended to tell Shaw exactly what was thought of him, and Boughton, and art generally, in the retired military purlieus of Cheltenham and Leamington Spa. It was clear that Boughton's country needed him desperately. The situation was made even more farcical when it was revealed that 'a young musician of genius' aged 19, who had had a waltz published and wanted to sit an exam, had been granted just the kind of temporary exemption that Boughton was seeking.

And so, in the middle of September 1916, the daily papers had the pleasure of informing their readers, in surprisingly large letters, that Rutland Boughton had, after all, gone to do his duty. No mention was made of his age, or his low medical category (C3), or of his previous attempt to enlist, but it was remarked with some satisfaction that the Glastonbury Festival School would close for the duration.

The appeal and the subsequent newspaper fuss was not calculated to make his first few weeks in the army any easier, and Boughton's initial reaction to a new and uncongenial way of life was to fall sick. The army suspected malingering, but eventually realized this his illness was genuine and put him in hospital. This, if anything, was rather worse then being in barracks, but it did not last long. 'I am just out of hospital', he wrote to Charles Kennedy Scott, 'turned out because they wanted to strike the hospital tent, so they declared the patients well. However, I think one's life is safer outside a field hospital than in it. On the last night I slept with a pool of water beneath the bed and a shower bath above.'

His army career began in the Cambridgeshire Regiment, first at Killinghall ('Kill-em-all') Camp near Harrogate, and later at Doncaster. In the middle of a rather uneasy attempt to transform himself into an NCO (he disagreed with the RSM over methods of voice production) he was transferred to the Sherwood Foresters, informed that he was now a bandmaster and told to produce a band from the various regiments in the brigade. By dint of a great deal of bluff and a certain amount of surreptitious homework he managed to form and train a band and soon found himself leading the brigade on route marches and playing hymns for church parade. He then set about teaching himself to play the cornet and trombone, and began to make arrangements to suit the musical limitations of his outfit. Within the band, and indeed the regiment, he achieved a certain amount of respect over and above the claims of rank when it was announced that his music-drama *The Immortal Hour* was to be published under the terms of the first 'Carnegie Collection of British Music'.

In March 1918 the battalion was sent to France. The band was broken up and Boughton was transferred to the King's Royal Rifles at Meance Barracks, Colchester. But on hearing that Walford Davies had become the first Director of Music of what was then the Royal Flying Corps, he applied for a further transfer and was soon able to join him at Blandford. The distinction of being the first bandmaster in the Air Force carried with it little more than the melancholy duty of playing 'The Dead March' from *Saul* as counterpoint to the flu epidemic then raging. Eventually this came to an end, by popular demand, and the epidemic was allowed to complete its work in silence. Boughton joined the rest of the band in listening to Walford Davies's lectures on musical appreciation, delivered in the same avuncular tones that were later to prove so successful with less cynical, though equally captive, audiences of schoolchildren.

At last the war was over. For Boughton service life had been a waste of time. The little it had taught him hardly compensated for the energy dissipated in carrying out the mysterious routines of army life. He had, however, written a short dramatic piece, *Agincourt*, as a morale-booster for some mutinous troops who eventually turned out not to need a Shakespearian corrective (it is a setting of words from *Henry V*); in April 1917, a Piano Trio (the *Celtic Prelude*) for two chamber-music friends he had met in Doncaster; and finally, in June 1918, a powerful anti-war ballet innocently entitled *The Death of Columbine*. The only other work was a first draft of the libretto for the third part of the Arthurian cycle. He called it *The Lily Maid*, and it had no prototype in Buckley's scheme.

The decision to write his own libretto, and therefore cut loose from the strait-jacket of Buckley's conception, seems to have been encouraged by Shaw who, on 1 February 1918, acknowledged Boughton's effort in typical fashion:

Dear Rutland Boughton,

I read The Maid, but have not had time to write to you since. It is an extraordinary production, full of that contempt for all the decencies of literature that musicians acquire by reading translations of foreign libretti, and mangling our native poetry. Yet a lot of it is quite presentable, obviously by accident; and it is all good enough to hang music on: it has the sort of feeling you want.

It is, however, incomplete. I don't see how you are to work a Parsifal panorama at Glastonbury; but clearly the dead girl passing down the river in her boat with the ancient man next to her finally arriving at Lancelot's feet, is indispensable to a musical picture of the whole tragedy. You can't stop Romeo and Juliet at the apothecary's. A preface is all nonsense: what on earth does it want with a preface? Write an overture. One must not do senseless things for the same of tempting a publisher.

Have you learnt anything from the band?

Ever,
G. Bernard Shaw.

Boughton, exasperated, turned Shaw's letter over and sketched his reply on the back:

Still the same stimulating and unhelpful person. First you give a chap a shove and send him spinning down a track he has not taken before—then you gibe at his ungainly appearance. I'm not going to let you off like that. I like my Maid better than any verse except Chaucer's, Blake's and William Morris's. So let me know, pray, in what way does she outrage 'the decencies of literature'? Tell me that without more ado—especially generalized ado. As for your idea that it should end with the barge scene at Camelot—Lordy me! you must have been reading Tennyson. My play is the drama of Elaine and the barge scene is the first part of the next drama, of Lancelot's quarrel with Guinevere.

He was, evidently, keen to get back to work.

Fortunately Christina had kept things going at Glastonbury. On 24 January 1917 she presented W. B. Yeats's play *The Land of Heart's Desire*, with incidental music by Rutland Boughton (which he later reworked as the *Celtic Prelude*), and Miles Malleson's *Paddly Pools*, in which Malleson himself was joined by Gwen Ffrangcon-Davies. She repeated the programme at Street on 20 February, and in April launched a three-day festival consisting of the Yeats play and, as a last-minute substitute for the *Alkestis* of Euripides, a play by George Calderon, *The Maharani of Arakan*, based on a poem by Sir Rabindranath Tagore. It was produced, presumably with some degree of authority, by K. N. Das Gupta, honorary organizer of the 'Union of the East and the West'.

On 24 April it was announced that the 'Festival Reading Circle' would continue to meet on Wednesday and Saturday evenings until 17 July. Plays by Shaw, Wilde, Sheridan, Shakespeare, and Ben Jonson were to be studied, but there appear to have been no further public performances until 11–14 September 1918 when a second miniature festival was presented. This time there were five short plays on offer: Tagore's *Sacrifice* (with music by Boughton), Shaw's *The Dark Lady of the Sonnets* (for which Boughton obtained leave from the army and 'made a fool of [himself] as Shakespeare'), two plays by Miles Malleson, *Young Heaven*, and *Paddly Pools* (the latter with music by Adele Maddison), and what the *Central Somerset Gazette* described as a 'quaint little composition' by one of Boughton's most loyal supporters, John Bostock, entitled *The Robin, the Mouse, and the Sausage*. This, in the event, turned out to be a parable for children, performed mainly by children, in which a valiant dog makes short work of a hapless sausage—symbol of Prussian militarism. Boughton supplied music for this piece of patriotic inanity which even the *Central Somerset Gazette* (20 September 1918) had to admit 'threatened at times to become boring'. In addition to the plays and a recital by Arthur Jordan and Gwen Ffrangcon-Davies, three performances were given of the hut scene (Act 1, Scene 2) of *The Immortal Hour*—Gwen Ffrangcon-Davies making her first appearance in the role that would soon carry her name throughout the length and breadth of the kingdom.

A few days after the armistice the Glastonbury Festival Committee, together with Shaw and Elgar, met in the Board Room of Novello's in Wardour Street to discuss the future. Boughton himself was anxious to forward his dream of a self-supporting community of artist-farmers, but got little support from the others. It was John and Roger Clark who came up with the most workable idea in a letter written on 19 November, immediately after the meeting, and it was their suggestions that formed the basis for the future work at Glastonbury. Roger Clark wrote:

John Bright Clark and I came away from the meeting in London with the feeling that what was really urgent and vital to the success of the scheme had received little or no attention. I mean the arrangements for financing the performances year by year, including such provision for you as would, along with your other work, allow you to live in Glastonbury in security and reasonable comfort. The building of a theatre, or in our opinion the providing of a few acres of land for you to work, in no way meets this need. The theatre in fact will need a considerable extra sum for upkeep.

The public interest and enthusiasm that was aroused by your productions at Glastonbury was most encouraging, and though naturally during the two years you have been in the army it has been somewhat diverted we are sure it can be reawakened by carrying on performances as before, until a real live thing is again in being. Then the demand for a theatre will become irresistible and will appeal to a far wider circle of interested people.

We would, therefore, suggest that the Committee and known sympathizers should be asked to join in guaranteeing, say for three or five years, a sum sufficient to cover deficits on performances to, say, the amount of £250–£300 a year, and a retaining fee income for yourself for the same period, the amount to be discussed with you. We believe that to be the only sound foundation and that once secured the site could be chosen and bought at a favourable moment and the Theatre built to house the living creation of your Art and Work.

10

1919

Boughton returned to civilian life in January 1919, appropriately on the eve of his forty-first birthday. Preparations for the year's festivals began at once and proved easier to organize than might have been expected. Christina had worked to such good effect that a nucleus of keen interest was already in place and eager to start work. By the middle of February the local papers were able to announce that students were meeting three times a week to study singing, dancing, diction and elocution, stage work, and Elizabethan music. Opera classes were scheduled to begin after Easter, together with classes in chamber music, sight-singing, ear-training, and eurhythmics. The works being studied included *A Midsummer Night's Dream*, and J. M. Barrie's *Quality Street*. Most of this activity was to take place in Glastonbury's Avalon Hall, on which John Bright Clark had taken a lease as his contribution to the festival. When neither teaching or planning, Boughton occupied himself in making music for a Marie Stopes translation of a Japanese Noh play, *The Robe of Feathers*, which he had come across in 1912 when the composer Dalhousie Young had used it as a dance scenario for one of Margaret Morris's Royal Court productions. Boughton renamed his version *The Moon Maiden*. Scored for chamber orchestra and requiring only two soloists and a female chorus, this delicate dance-drama explores, in symbolic terms, the nature of virginity. The Moon Maiden loses her robe of feathers to a fisherman, who restores it only when she promises to dance for him. In writing it Boughton had in mind his eldest daughter, Ruby, who was to create the role in the Easter Festival. It was his way of reassuring her (she was 14) during the difficult period of her adolescence.

Before the festival could begin, however, the Temple Theatre scheme lost its chief supporter. On 21 March 1919, after a long struggle with ill health, Reginald Ramsden Buckley died in a London hospital. He was 36 years old. In an obituary tribute Boughton described him as 'one of the sweetest, gentlest men who ever lived', and commented on his burning enthusiasm for the theatre project. 'Had he lived', he said, 'he would sooner or later have pushed it through.' Without him, and as time went on, the idea came to be quietly dropped. Privately Boughton must have been somewhat relieved to be in sole charge, particularly where the libretto for the Arthurian cycle was concerned. From this moment he abandoned any attempt to follow Buckley's plan as laid out in *Arthur of Britain*. What eventually emerged was to be entirely his own.

Glastonbury's first post-war festival began on 13 April 1919. 'Festival' is not perhaps the right word, for it was more a series of musical events spread over the Easter period: demonstrations of the students' achievements in their study of 'British Music, Drama, and the Allied Arts'. Thus, two performances were given on Wednesday, 23 April, each consisting of a *Midsummer Night's Dream* ballet to ' Mendelssohn's overture; a group of madrigals and songs; an amusing 'Puritan Ballet', *The Wickedness of Dancing*, arranged to William Byrd's variations on *O Mistress Mine*; a group of dances to music from the Fitzwilliam Book, in which 'little Joy Boughton' (as yet unacquainted with the delights of the oboe) 'simply took the house by storm'; and the first performance of *The Moon Maiden*. The *Central Somerset Gazette* (2 May) waxed lyrical at the thought that the Assembly Rooms had been transformed into 'the land of the fairies and gnomes and the regions where nymphs dwell'. But behind all the journalistic hyperbole it can be seen that the Festival School was drawing good work from its students, children and adults alike:

They dance for the sheer love of dancing, and the apparent accuracy of every movement of limb or body is done so naturally, so gently, so gracefully as to enrapture the hearts of those present. Their dancing is simply delicious; and the credit is reflected back to Mr Boughton and his helpers, Miss Florence Jolley and Miss Marjorie Gilmour, particularly.

Further proof of the school's serious intentions came on Thursday 24 April, when Professor Edward J. Dent arrived from Cambridge to deliver a lecture on 'The Foundations of English Opera', which was rounded off by a performance of a scene from his edition of Matthew Locke's *Cupid and Death*. It was the first performance in modern times and went a long way towards underlining the truth of the graceful tribute with which he concluded his lecture when, after commenting that English opera had died with the death of Henry Purcell, he turned to Boughton and said: 'And now you have to revive it here in Glastonbury.'[1]

Friday, 25 April being Good Friday, Boughton took over the Parish Church for a performance of the *St John Passion* (described by a rather ingenuous *Central Somerset Gazette* as having been written by 'Bach, the eminent 18th century German composer'). For this occasion he gathered together a choir of some sixty voices, a small orchestra, and Gwen Ffrangcon-Davies and Percy Snowden as soloists. It seems that the work was given in a somewhat truncated form, for the redoubtable Miss Buckton is credited with having read certain narrations in a voice that was 'full of expression and sympathy'. Wednesday's programme was repeated on Saturday, with the addition of the scene from *Cupid and Death*; and on Sunday, as part of a chamber-music recital that included sonatas and trio sonatas by Purcell, Henry Eccles, and William Boyce, Boughton gave a talk on the nature and importance of the madrigal. It was on this occasion that he was able to announce a new development: the festival had been offered £100 towards the formation of its own permanent string quartet.

Although the Easter festival ended on Monday, 28 April, with a recital of songs, partsongs, madrigals, and anthems (with a violin sonata by James Lates thrown in for good measure), a further performance of the *Midsummer Night's Dream* ballet and Boughton's *The Wickedness of Dancing* was given on 2 May, presumably as an advertisement for the school's eight-week Summer Term that was to commence on Monday 5 May. The weekly timetable was carefully arranged so as not to conflict with the adult students' daily work and the children's obligations to their schools:

Mondays:	6.30pm:	Theory of Music.
	7.45:	Chamber Music
Thursdays:	6.30:	Hellenic Dancing (children)
	8.00:	Dramatic Class
Wednesdays:	5.15:	Madrigal Choir (ladies)
	6.00:	Madrigal Choir (mixed)
	6.45:	Voice Production and Sight Singing
	8.00:	Opera Class
Thursdays:	6.30:	Hellenic Dancing (children)
	8.00:	Hellenic Dancing (adults)
Fridays:	6.00	Eurythmics (children)
	7.00:	Eurythmics (advanced children)
	8.00:	Eurythmics (adults)
Saturdays:	2.00:	Drawing from Life and Nature

Most of the classes were of 45 minutes duration, and all were geared to the performances being planned for the Whitsuntide and Summer Festivals. Additional classes were to be given immediately before the festivals: 'All who realize the value of a thorough musical training, and the importance of physical development and graceful carriage, will wish to take advantage of these special courses . . .'. The cost of a term's tuition was set at 10s. 6d. for the adults and 2s. 6d. for the children.

However reasonable these terms may seem—and despite low average wages they were reasonable—the local middle class (the fees would

have been too high for the average labourer) proved increasingly unwilling to expose its children to the benefits of a 'graceful carriage'. As time
went on they began to remove them from a sphere of influence they,
rightly or wrongly, felt to be morally doubtful. It had been hard enough
to accept that in the eyes of the Church, and for that matter the Law,
Boughton could not call Christina his wife. But when it came evident
that he could become enchanted by a pretty face—particularly if it was
complemented by a useful voice and a convincing stage presence—then
it was clearly time to think again. Though his infatuation with Irene
Lemon had passed almost unnoticed, it had still given rise to rumour.
And when rumour was allied to the relatively free and easy ways of
Boughton's professional friends, rumour was all set to have a field day.

Such doubts do not seem to have affected the Easter Festival, and
there was no shortage of children to impersonate Shakespeare's fairies
and dance to the music of the Fitzwilliam Book. The Whitunstide
Festival (9–16 June) also passed off without difficulty, being largely a
series of recitals (including one by the poet Henry Bryan Binns). It was
only during and after the Summer Festival (17–30 August) that a change
of emphasis became necessary. It was, in any case, an altogether more
ambitious affair, and as such required the help of young professional
singers more than anything that had so far been attempted. To ensure
the success of what must have been the first complete performance of
Cupid and Death since 1653, Edward Dent turned to some of his former
Cambridge students: Steuart Wilson, Clive Carey, and Harry Sheerman
Hand—all survivors of the war. Two of them (Wilson and Hand) were
to become Glastonbury regulars and of crucial importance to Boughton
in the way the festivals were to develop.

For the revivals of *The Immortal Hour, The Sumida River, The
Moon Maiden,* and *The Round Table,* Boughton brought in a series of
singers quite new to Glastonbury—William Johnstone Douglas,
Winifred Lawson, Hamilton Harriss, Phyllis Jewson, Seymour Dosser,
Constance Lermit, and Lewis Godfrey among them—to reinforce such
pre-war stalwarts as Gwen Ffrangcon-Davies, Arthur Jordan, William
Bennett, and Percy Snowden. Reginald Paul replaced Clarence
Raybould (tangentially a casualty of the Irene Lemon affair) as festival
pianist. Of the Glastonbury locals, apart from the children, only Agnes
Thomas, Percy Holley, and David Scott remained to take any
significant part in the productions. One of the troubles, Boughton
recalled, was that the festival was 'beginning to bring visitors to the
town, and when we most needed our people for the performances they
felt obliged to stay away and look to their businesses. If they did not,
the greater material benefits would be reaped by the very people who
were opposed to the organization'.

Nevertheless, the festival was a success and a marked advance on all previous occasions. Preceded by the usual Holiday School, it began on 2 August. Three performances of each of the music-dramas (*The Immortal Hour, The Round Table, The Moon Maiden, The Sumida River*), and three performances of an entirely new production: *A Pot o' Broth* by W. B. Yeats, were interspersed with a series of recitals. On 20 August Miss Marie Hall, accompanied variously by Philip Napier Miles and Reginald Paul, gave a programme which included John Ireland's D minor Violin Sonata and Parry's Suite in D major. Even more impressive was the contribution made by the festival's own String Quartet (Harold Batten, Marjorie Forbes, Gladys Home, and Dorothy Forbes). Their first concert (17 August), in which they were joined by Reginald Paul, included John Ireland's *Phantasie Trio*, Boughton's chamber cantata *The Chapel in Lyonesse*, and, in what must surely have been one of the earliest performances since the Brodsky Quartet introduced it to the world on 1 May 1919, Elgar's String Quartet in E minor. It was repeated at a concert on 24 August, this time in the company of songs by Clive Carey and Vaughan Williams's cycle *On Wenlock Edge*. The soloist on this occasion was Steuart Wilson, one of the cycle's most admired interpreters. Both concerts had been given on a Sunday evening, rather to the disquiet of certain locals—though the *Central Somerset Gazette* (29 August) decided that the innovation had added 'lustre to the day of rest' and observed that 'in an age of rigid conformity to an even more rigid (and frigid) conventionality these concerts would have been considered daring to the last degree. But we are now entering the portals of a new era in which unconventionality can run without fear of contradiction dogging its footsteps.' Besides, Boughton had taken precautions. The concerts were free.

The presence of the Quartet added lustre to some of the other performances also. Augmented to about eight players it was used to flesh out both *Cupid and Death* and *The Sumida River*, and to provide an accompaniment to the children's dance programme which largely repeated what had been done at the Easter Festival. Nothing, other than the voices of Gwen Ffrangcon-Davies, Gladys Ward, Steuart Wilson, and Clive Carey, was needed to illustrate the comprehensive lecture on 'Madrigals and the Elizabethan Composer' delivered by Dr Edmund Fellowes on 27 August. Deeply immersed in the preparation of the thirty-six volumes of *The English Madrigal School*, Fellowes was the leading authority on Elizabethan music, sacred and secular, and, as Dent's had done at the Easter Festival, his presence must have added considerable weight to the visible evidence of Boughton's seriousness of purpose. By the end of the 1919 Summer Festival it must have been clear that, however idealistic and quixotic he may have seemed,

Rutland Boughton meant business. One young man certainly felt so. Gerald Finzi attended performances of *The Round Table* and *Cupid and Death*, and heard Fellowes lecture. He returned the following year, this time with a note from Vaughan Williams to introduce him to Boughton. Again he was impressed and decided that it would be an excellent thing if every small town in England became a Glastonbury.[2]

Seriousness of purpose, as we have seen, was not the only element in the Glastonbury Festivals, and it is reassuring to note than when the Committee turned the Victoria Hall into a place where the artists and audiences might obtain meals they called it the 'Cramalot Restaurant'. Cheerfulness broke in also at the end of the Festival when Mr Bernard Shaw stepped on to the Assembly Rooms' stage and bade Miss Christina Walshe sit in a replica of the 'Glastonbury Chair'. He then turned to Boughton and said:

My dear Rutland, when you have made a great many affectionate friends, and are doing a distinguished public service, there is one dreadful thing that is bound to happen: that is, someone is bound to make you a presentation. I don't want to embarrass you. It is a very useful present. As a matter of fact your wife is sitting on it. It is, I believe, what is known as the 'Glastonbury Chair'. What is the significance of offering a man a chair? It is because you do not want him to go. [*loud appluuse*] I do not know whether you are conscious of the fact, but we are, that a man of your talent and genius is not very often found out in this country until after he is dead. There are places in this country where you would have better chances than at Glastonbury, but I do not think there are any places where you would be better appreciated, and loved. [*applause*] I hope you will stay long enough so that we may add all the rest of the furniture, and also a house, including a garden, with greenhouse and theatre—one of those little extravagances which you want so badly.

Then, pulling out a wallet and eyeing it curiously, he continued: 'Here are a few Treasury Notes. I will not mention the amount, but leave it to the imagination of the audience. The reason we are putting in this necessary thing is that when you go home the children may want to know where they come in. But I think on consideration I will give it to Miss Walshe. [*applause*][3] Rutland Boughton was overwhelmed and, for once, seemed almost lost for words. The chair—a heavy, oak affair, liberally carved in an ecclesiastical manner—became one of the few items of furniture he treasured. From that moment it found an honoured place in every home he was to inhabit.

As soon as the festival was over, Boughton and his friends boarded the train to Bournemouth to give, once more with Dan Godfrey's orchestra, two performances of *The Immortal Hour*, timed to coincide with a major public appeal for funds with which to build the theatre. The appeal, endorsed by the Right Honourable J. R. Clynes MP, Mr John Drinkwater, Sir Edward Elgar, Mr Thomas Hardy, Sir Frederick Pollock, Mr G. Bernard Shaw, and Sir Henry Wood, asked for 'at least £10,000' to cover the cost of providing 'a site, with grounds and hostel, permanent stage with modern lighting, and a temporary auditorium'. Philip Napier Miles, of King's Weston, Bristol (a wealthy amateur composer whose music, diplomatically perhaps, had already figured in the festivals), had promised a cheque for £500.[4] Lord Dysart had offered £200, and Lady Radnor £100, while Mrs Kennerly Rumford (better known as Madam Clara Butt) had pledged £10, and Mr Adrian Boult, well down among the also-rans, a useful two guineas. Altogether over £1,000 has been promised, so there was every reason to feel optimistic.

In addition to the theatre appeal, Roger, Sarah, and John Bright Clark set about establishing a Guarantee Fund to underwrite the work of the Festival School for three years, 1919-21. It also became necessary to reorganize the Festival Committee, dividing it into an Executive Committee at Glastonbury itself, and a London (presumably advisory) Committee that included Sir Edward Elgar, Clive Carey, Percy Scholes, Charles Kennedy Scott, and Dr Richard Runciman Terry, with H. C. Colles as its Honorary Secretary. This done, the new term commenced on 29 September.

In fact Philip Napier Miles had done more than simply promise a handsome cheque, he had secured the lease on a large house: Mount Avalon. This was to be the school's headquarters on the understanding that a fair rent would be paid if and when the venture made a profit. There was a large garden, various glass-houses, and a three-acre field which might become the nucleus of a self-supporting commune. But Boughton's mind was so occupied with the musical side of the festival that he was obliged to let the fields to a neighbouring farmer and employ a gardener to get what results he could from the kitchen garden.

Mount Avalon added greatly to the attractions of the Festival School. Students no longer had to search out lodgings and make shift with erratic meals. Soon some twenty had enrolled, among them an attractive 19 year old who had attended the Summer School and now, at the suggestion of her father, hoped to further her general musical education. Her name was Kathleen Davis, and Boughton had last seen her as a 9 year old in Birmingham. She had developed a pleasing soprano voice, and already showed a little talent for composition. More to the point, she was beautiful and radiated a warm sensuality and showed every sign of worshipping the ground he trod. By the end of the Autumn Term he found he could not get her out of his mind.

The exact progress of their relationship is not easy to trace, though much may be surmised from the evidence of Kathleen Beer who, with Dorothy D'Orsay and Desirée Ames, was also a resident student.[5] She remembered Dorothy D'Orsay saying that Kathleen Davis must have been sleep-walking as she had somehow ended up in Rutland Boughton's study one night. When questioned about it Boughton said: 'No, she was not sleep-walking. I was giving her extra tuition in harmony.' Being very young at the time, neither girl had thought any more about it. If Mr Boughton considered it reasonable to give harmony lessons after lights out, who were they to imagine any impropriety?

In the mean time the term went its way. Each Sunday the students presented a free concert to anyone who would listen. Most of these took place in the Crispin Hall, Street—much to the dismay of the local incumbent, the Reverend A. C. Schofield, who drummed up the ready support of the Reverends G. Sutton Read and Ernest H. Lawrence and launched a protest against this unseemly profanation of the Sabbath. Roger Clark leapt to Boughton's defence, and as he was the town's principal employer the matter quickly assumed a rational perspective. The three clerics backed down and the concerts continued their peaceful ways.[6] By 14 December Boughton and his friends were able to celebrate their 100th performance with a concert of extracts from *Bethlehem* and W. B. Yeats's one-act play *The Hour Glass*.

The final event in a very eventful year was the Winter Festival, 26-8 December. This was preceded by a tour which took *Bethlehem* to Bournemouth for five performances. Boughton's singers were again accompanied by Dan Godfrey's orchestra, conducted on this occasion by Glastonbury's new acquisition W. H. Kerridge—described by an awestruck *Central Somerset Gazette* as a 'former assistant conductor of the Municipal Theatre, Zurich'. The actual festival began in Glastonbury's Victoria Rooms with a matinée performance of *Snow White*, followed by an evening performance of *Bethlehem* in the Assembly Rooms. Both were repeated on the following day, and then, after a Sunday 'free concert' of Christmas music, repeated on 29 December at the Crispin Hall, Street. The Glastonbury performances were on offer to subscribers only (one guinea per annum), but those at Street were open to the general public: tickets, 5s. 9d., 3s. 6d. reserved, and 1s. 3d. The programmes for these performances carried details of the new term which was to begin on 26 January. Listed among the teaching staff were: Gwen Ffrangcon-Davies, Grace McLearn (certified teacher of Eurythmics from the Dalcroze Institute, Geneva), Christina Walshe, Laura Wilson (formerly Mademoiselle Olkhina of the Diaghilev Russian Ballet), Desirée Ames, Rutland Boughton, William

H. Kerridge, BA, Mus.Bac., and David Scott, ARCO. In addition to the
Sunday free concerts, stage performances of *Deirdre* and *The Hour
Glass* by W. B. Yeats, John Blow's *Venus and Adonis, Everyman*, and
Laurence Housman's *Nazareth* were promised, with an Easter Holiday
School and Festival from 25 March to 7 April.

And for those whom it concerned, sceptics and enthusiasts alike, the
year's accounts were laid out for examination:

RECEIPTS:	£	*s.*	*d.*
From the Guarantee Fund	535	7	4
Donations to General Work	51	9	0
Subscriptions for tickets	172	15	0
School fees	99	9	3
Easter and Whitsun ticket sales	76	10	0
Sunday and other Concerts ticket sales	51	1	11
Collection for Buckley Memorial Library	2	13	7
Rent Avalon Hall	5	0	0
Lettings Victoria Rooms	7	11	0
Christmas Shows, ticket sales	58	14	0
August Shows—net takings	449	1	11
Bournemouth (August) share of profits	2	12	11
Bournemouth, December takings	81	0	3
Bristol, Christmas takings	123	4	1
TOTAL RECEIPTS:	1716	11	1
PAYMENTS:			
Balance to Treasurer from 1916	48	9	0
Salaries: Direction	320	16	8
Clerical	41	13	9
Artists: Fees and Board	449	19	10
Purchase of typewriter	21	2	6
Purchase of Harmonium	57	4	0
Hire Piano	20	8	0
Music and Copying	27	18	3
Printing, Stationary, Sundries	80	8	9
Advertising	67	2	7
Postage	16	10	10
Stage and other work	107	6	10
Various Concert expenses	14	5	0
Hire of Assembly and other Rooms	19	19	3
Tax Stamps: Easter and Whitsun	20	14	3
Costumes	73	9	11
Avalon Hall expenses	8	16	8
Victoria Rooms, rent and expenses	36	15	1

PAYMENTS:	£	s.	d.
Wardrobe, purchase	3	0	0
Performance fees	62	9	6
Director's Travelling Expenses	20	12	8
Bournemouth, Christmas Expenses	99	2	0
Bristol, Christmas Expenses	31	18	6
BALANCE: cash £23. 15s. 2d.			
bank £42. 12s. 1d.	66	7	3

The cash balance might have been healthier it is true, but at least it was on the credit side. All in all, the year 1919 had been a success and Boughton felt able to face 1920 with complete confidence.

11

1920–1921

During the tour of *Bethlehem* Rutland Boughton came to understand that he had fallen in love with Kathleen Davis completely and utterly. The little girl he had once delighted in as part of a delightful family was now an enchanting young lady. He could not get her out of his head. Christina was not amused. 'Fool!' she said. 'It's always a younger girl. It'll be an infant next!'

Others, less personally involved, took it upon themselves to be even more outraged. William Kerridge was particularly upset and told the Festival Committee that their plans would not succeed 'with Rutland Boughton at the head and under these conditions'. A meeting was held to discuss the matter, but no action was taken. Everyone thought it was no more than an infatuation that would pass as others had passed. Christina seemed to be unconcerned, why should not they? Kerridge, however, would have none of it and promptly withdrew his services. He rather spoilt the effect of his gesture by, according to Boughton, attempting to set up a rival operatic scheme in the area. Though this came to nothing, the mere fact of his obvious disaffection added fuel to the rumours that were beginning to accumulate around Boughton and his doings. He was soon being credited with a reputation that would have been the envy of the most dedicated philanderer, and was even thought by some to be the prototype of Sanger in *The Constant Nymph*.[1] He was not a natural Don Juan. He was, if anything, rather puritanical in his attitudes. But having fled from a marriage in which neither partner had the slightest thing in common into a relationship that nourished his soul and allowed him to find himself as a creative artist, he had now become aware that the sensual aspects of his nature required attention if he was to find any sort of inner peace. Kathleen, it seemed, had it in her power to answer all his needs.

He believed, moreover, that it was necessary for the artist to take such steps for the sake of his art. If he was to continue to grow and renew himself, then Wagnerian decisions might have to be faced. Since the birth of their second child, Maire, in July 1916, his relations with Christina had grown weaker, until by the end of 1917 they were on no more than companionable terms. Though this suited Christina, it did not suit him. And there were other factors that he might have been less willing to admit. Photographs suggest that by 1920 Christina, at 32, had grown to look rather haggard. The comparison with Kathleen's 19-year-old nubile promise was bound to put her at a disadvantage. At 42, and a natural a prey to all the doubts of middle age, Boughton can only have found Kathleen's attentions flattering and reassuring.

But much water was to pass beneath the bridge before any irrevocable step had been taken, and in the mean time he seems genuinely to have tried to put her out of his mind. The new term commenced, as planned, on 26 January and followed the routine established during the previous year. The Sunday free concerts were again a regular feature. In an 'all-Beethoven' recital on 15 February Boughton, doubtless to the amusement of his more cynical friends, sang 'An die ferne Geliebte' in a programme that included the B flat Trio, Op. 97, and the F major Violin Sonata. More Beethoven (the Pathétique Sonata) was included in the next concert, together with Bach's E minor Violin Sonata, Coleridge Taylor's *In Memoriam* song-cycle, and a talk by Boughton on 'Religion and the Arts'. On 14 March Frederick Goudge lectured on 'Folksong', and on the following Sunday there was a cello recital (Beethoven and Grieg, mainly). On 28 March the Easter Festival began.

The 'new' work on this occasion was John Blow's *Venus and Adonis* (part opera, part masque) with Clive Carey as Adonis and Gladys Moger as Venus. *Everyman* was revived, again with Gwen Ffrangcon-Davies in the title role, and there were recitals of Elizabethan music and an all-Purcell concert on Easter Sunday. Once again anticipating modern preferences, and perhaps encouraged by Edward Dent's opinion, Boughton employed early instruments for Blow and Purcell: a Kirkmann harpsichord, dated 1789; a viola d'amore by the nineteeth-century enthusiast George Saint George; and a viola da gamba by Barak Norman (1718). Together with a violin and cello of unspecified date, they were played by Mabel, Kate, and Nellie Chaplin, and their presence, like the Dent and Fellowes lecture-recitals of the previous year, must have helped to raise the status of Boughton's endeavours in the eyes of serious musicians, even if the subtleties passed over the heads of such Glastonbury music-lovers as were prepared to brave the wrath of the Sabbatarians.

Venus and Adonis made a considerable impression. Writing in the May issue of the *London Mercury*, Edward Dent declared:

In many ways the opera showed to advantage by being presented in such an intimate manner. Blow makes his effects with the fewest possible notes, and at such close quarters the protagonists, Miss Gladys Moger and Mr Clive Carey, were able to make every little phrase tell with an unerring sense of dramatic effect. The stage pictures were charming. There was no scenery, merely a background of dark curtains, against which the group of shepherds and shepherdesses all in white attended upon Venus and the Cupids in pink. In the third act they appeared clad in purple and green, giving at once the note of tragedy to ancitipate the death of Adonis. . . . The whole company have greatly profited by the teaching of Miss Laura Wilson, formerly a member of the Russian Ballet, who has now established herself in Glastonbury as a teacher of dancing. She has not only taught them a lithe and supple sense of movement and attitude, but has given the whole production a certainty and clarity of design and grouping which was rather noticeably lacking in earlier productions.

Glastonbury, he concluded, was all set to become 'a leading centre of the revival of our native classics', and there was even talk of a festival devoted entirely to the music of Henry Purcell.

Though the Sunday concerts continued throughout April and May, there was no Whitsun Festival. Instead, Boughton and his team removed themselves to London and from 31 May to 12 June presented a selection of their Glastonbury achievements at the Old Vic. Their programme consisted of *The Immortal Hour*, *Venus and Adonis*, extracts from *The Birth of Arthur* (including the notorious 'dancing scenery'), *The Moon Maiden*, and two new ballets: *Music Comes*, by Philip Napier Miles, and *The Children of Lir*, by Adele Maddison. The Boughton-Byrd entertainment *The Wickedness of Dancing* was also included, as was a short ballet for solo dancer and soprano to Ethel Smyth's song 'La Danse', suitably anglicized.

   The critics, on the whole, were impressed. The standard of the dancing was judged to be superior to the singing. The orchestra, conducted by Charles Kennedy Scott, got off to a shaky start but improved as the week went on. The music of Adele Maddison and Philip Napier Miles was summarily dismissed, one critic commenting on *Music Comes* was driven to writing 'I wish it did!'—a verdict with which it would be hard to disagree. *Venus and Adonis* was praised, but the 'hit' of the season was *The Immortal Hour*: the 'one solid justification' of the whole venture, as the *Musical Times* (July 1920) put it. Something of the impression it made can be judged from the letter Geoffrey Shaw wrote on 8 June:[2]

Dear Rutland Boughton,
   I must write to you.
   I heard *The Immortal Hour* for the first time last night. You have made some wonderful music, and I want you not to think me impertinent in writing to say how fine I think it is, and how deeply I sympathize with your splendid work.

How glorious it is to hear English opera that is not Wagnerian, or Russian, nor anything else but just its true self. The whole of it struck us all as being sincere and clean. The beauty of your music is quite extraordinary.

I brought a party of ten last night, including an underwriter, a curate, and a bank clerk. I was delighted to find that you had captured them completely! One of them is a regular Covent Gardener, and came under compulsion. He said, afterwards, it was a revelation to him. I certainly think your work is popular in the best sense of the word—genuine people's music, which is what the world wants so much now. One of the ultimate tests of a thing's worth is, I believe, its power of direct appeal to the sense of beauty that is latent (nowadays almost dormant) in the mind and soul of poor Everyman. You have lit your beacon, and you ought to be able to die happy!—though I hope you will live to give us many other lovely things and to show us how a courageous facing of difficulties can win through to achievement. Your people seem inspired with your spirit.

I should like to write many more things—but I've been overlong already. If you hate answering letters as I do, you won't want to answer this. Don't. I only wrote because I felt I had to—and it doesn't want an answer.

Yours,
Geoffrey Shaw

At the end of July, after facing up to the bitter truth that, however encouraging the artistic impression of the Old Vic season had been, it had not been able to make ends meet by some £276. 5s. 2d., the London Committee of the Glastonbury Festival School made a last desperate public appeal:

The visit of the Glastonbury Festival School to London has created two dominant impressions: firstly, that the Glastonbury movement contains the seeds of a healthy musical and dramatic development in the life of the country, and secondly, that Mr Boughton's music-drama *The Immortal Hour* is a work of genius; but, unfortunately, unless the movement can receive further financial support, it must cease with this year. The local resources are too slender to support even its present modest outlay of expenditure. To carry the work on requires an additional income of £750 a year to pay a small living wage to the musicians who take part in the School's activities, and to defray the expenses of the periodic performances. So long as these have to be given in the small Assembly Rooms at Glastonbury they cannot pay their way, because large audiences cannot be accommodated.

It then suggests that if 150 people can be found to pledge £5 for each of three years the work can go on. An additional £1,800 must be found if the £3,000 for the site for a permanent theatre is to be secured. The appeal, signed by Sir Edward Elgar and Sir William Hadow ends:

We believe that Glastonbury has in it the possible development of a movement of the greatest importance, both for British Music and for the regeneration of the life of the countryside. It will form a musical and dramatic centre for the study of British Music from the Elizabethan period down to the present time. Its methods will be largely experimental. We believe that at present Mr Boughton's work is unique, but we hope that it will also be an example and that centres may be started in other parts of the country on the same lines.

Boughton and his friends returned to Somerset to begin the new term and continue with the regular Sunday free concerts. Intensive preparations for the Summer Festival began on 2 August, with special classes in Chamber Music (Ethel Robb), Orchestra (Edgar Bainton), and Dance (Laura Wilson). The festival itself began on 15 August.

It was by far the most ambitious programme Boughton had attempted. Lasting nearly a month (15 August to 11 September), it was built around six performances each of *The Birth of Arthur* (heard complete for the first time), *The Round Table*, and *Dido and Aeneas*, with seven performances of *The Immortal Hour*. There were in addition no fewer than ten chamber concerts devoted largely to British music. The first (15 August) was billed as a 'Parry Memorial Concert' and included his Quartet in A flat, a group of songs, and the *Lady Radnor Suite*. Later concerts included Edgar Bainton's Quartet in A major, Elgar's String Quartet and Piano Quintet (neither more than two years old), Vaughan Williams's *On Wenlock Edge*, and music by Percy Grainger. Napier Miles's choral ballet *Music Comes* and Adele Maddison's *The Children of Lir* each received six performances.

Although most of the professionals had appeared on previous Glastonbury occasions, Boughton continued his policy of providing young singers with a chance to find their operatic feet away from the cut-throat pressures of the London stage. Among the new recruits were Edith Finch, Arthur Jaques, Colin Ashdown, and, most talented of all, the baritone John Goss. Young composers also had their chance, as at the Chamber Concert on 29 August when Michael Head, then a student at the Royal Academy of Music, appeared as accompanist to a group of his own songs. The 'resident string quartet' had changed its personnel and now consisted of Harriet Solly, Desirée Ames (plucked by Boughton from the boring routines of a Glastonbury bank), Gladys Home, and Dora Petherick. Reginald Paul remained the festival's ubiquitous pianist: orchestra, accompanist, soloist, and chamber musician— slipping effortlessly from one role to another, splendid in each. And Edgar Bainton sacrificed his vacation from Newcastle University to conduct all the staged performances, save that of *The Immortal Hour* which Boughton kept for himself. It was, in short, a team that admirably mixed experience with youthful enthusiasm. The results were not lost on the press: 'ALL PREVIOUS SUCCESSES ECLIPSED!' exclaimed the *Central Somerset Gazette*, and the London critics were inclined to agree.

Commenting on the sheer improbability of the Glastonbury Festivals, *The Times* (28 August 1920) decided that 'the performances we have seen this week have gone a very long way to dispelling this scepticism'. It praised the two parts of the Arthurian cycle, noting that Boughton's music 'does not merely furbish up old and picturesque legends; it strikes to the human emotions and impulses from which the legends took form'. Particular interest centred on the complete production of *The Birth of Arthur*. The *Bristol Times* (18 August) was lyrical in its praise:

The story hinges upon the prophecies of Merlin the Magician and the love of Queen Igraine and King Uther, from which bond springs the wonder-child, the fabled Arthur.

The most notable features of the drama are the phenomenal effects which are obtained from the chorus by their triple use as singers, as dancers, and as scenery. There was no scenery in the ordinary sense—merely a draped stage— but the ladies of the chorus, clad in graceful robes of artistic hue, indicated the scenes by a series of rhythmic evolutions, often singing as they moved. The artistic value of the method was much enhanced by the excellent taste of the lighting—red and blue and yellow limes being employed with great skill. This new method of creating atmosphere by means of living scenery carries a wonderful appeal and, to quote a hackneyed phrase, it 'must be seen to be appreciated'.

Truth (8 September), as might reasonably have been expected, saw the funny side of Boughton's innovations but still had to admit that the thing worked:

The mass effects of the choral scenes were wonderfully impressive. Having no scenery in the ordinary sense of the term, Mr Boughton suggests the pictorial background by means of his chorus and ballet. It may seem ludicrous to represent Tintagel Castle by four men who hold their arms up, and the sea by a dozen females in green who sit on the floor and wave bits of chiffon, but in actual performance the romantic atmosphere is obtained with complete success.

Having gone this far the critic than rather spoilt the effect by remembering that the chorus had 'hard work to do in the storm scene, for many of them were singing complicated music as well as rushing round and round the "castle"; they finally subsided on the floor and lay there panting loudly'. But, by and large, the comments were favourable and most critics would have agreed with Edward Dent when he wrote, in *The Athenaeum* (10 September):

No one, however learned or experienced, can go through the Glastonbury rehearsals without learning something new and valuable. For Glastonbury has no convention and no routine. It is perpetually trying experiments; it faces everything as a new problem, and faces the problem for its own artistic interest, and not with a view to popular or commercial success. People learn to throw themselves into an opera for the sake of the work of art, without any care for individual distinction. It is a school of idealism, and Mr Boughton sets the example by being always ready to welcome new ideas. Glastonbury is a place where new ideas are welcomed and have every chance of being carried out with sympathy and understanding.

On the last day of the festival a meeting was held in the Assembly Rooms to discuss the financial situation. The London Committee's appeal had brought in only £375, and with additional promises (including a handsome £500 from Napier Miles) the Theatre Fund stood at £1,295, leaving an additional £1,300 to be raised by the end of September if the building site was to be secured. The prospect was far from rosy, and when account was taken of a loss of nearly £500 on the year's work it could only be described as grim. It seemed likely that the school would have to close and the festivals be wound up. Speaking in his capacity as Inspector of Music to the Board of Education, Dr Arthur Somervell painted an apocalyptic picture of children denied the benefits of music after they had left school and being thrown back on the public house and cinema for their entertainment. 'The Kinema', he thought, was the 'most dangerous influence for evil that we have in this country', and he therefore moved that every effort should be made to preserve the benign example of the festival.[3]

A glance at the local papers would have told him that there were additional temptations in the very heart of Glastonbury. Even as he spoke the *Central Somerset Gazette* was advertising the Berkshire Aviation Company's flights at 10*s.* 6*d.* a time and from 10.30 a.m. until dark, with 'Thrilling Exhibitions' on early-closing day, Saturday, and Sunday of 'Looping, Spinning, and Walking on the Wings in Mid-Air'. Though perhaps he would have been reassured by their boast that, so far, 34,000 passengers had been carried 'without mishap'. Elsewhere in the paper his worst fears would have been confirmed by an inquest on an 11-year-old boy who had died from chronic nicotine poisoning. 'He was', we are told, 'a heavy smoker, beginning before breakfast and continuing all day', though no mention is made of how he was able to finance his hobby, or what his family had done to discourage him.

Boughton, had he been aware of local aeronautical delights, might have reflected that his own position was almost as hazardous. And he cannot have been particularly pleased to read (*Central Somerset Gazette*, 3 September) that on 28 August the foundation stone had been laid at Chalice Well for 'Britain's First Orchard Theatre'. Miss Buckton, it seemed, had gathered her cohorts for 'the building of an open-air playhouse, with good floors and four uprights, careful protection for large choirs and orchestra, with fine acoustic properties, amid beautiful surroundings suitable for pageant work and allegorical work of all descriptions'. It was irritating enough that her group had adopted a name ('The Guild of Glastonbury and Street Festival Players') that frequently became confused with his own Festival Players; now, it seemed, she was setting up in direct rivalry in a theatre that, though likely to be draughty, might be easier to realize than his. The author of *Eager Heart* (known to some Glastonbury observers as 'Eager Tooth') was becoming a distinct thorn in the flesh.

The rivalry between the two groups undoubtedly added fuel to the extraordinary sitaution which arose when Glastonbury decided to remember the war by building a new town hall. Philip Napier Miles immediately offered a substantial sum towards the cost on condition that the council would agree to include a reasonably equipped stage and orchestra pit. The town council expressed considerable interest but, as a democratic body, necessarily had to put the matter to the vote. As they entered the council chamber they received a sharp reminder of their civic responsibilities when one of Glastonbury's more prominent citizens whispered to each in turn: 'A vote for Boughton is a vote for adultery.' Plans for the new town hall were duly approved— without any allowance made for the needs of the festival.

So desperate did the situation become that Boughton seriously began to think of moving to a different part of the country altogether. On 23 November 1920 the *Scarborough Post* reprinted a letter he had written to the town council asking if they knew of any residents who would 'take pride in securing the national movement for your town, and who would provide the necessary financial guarantees (which are not large)'? But Yorkshire proved even less willing than Somerset to venture its capital—though the threat of such a removal may have sharpened the edge of his Glastonbury supporters' desire to keep him for themselves.

There was nothing for it but to begin yet another bout of fund-raising. As a start, he and Kathleen joined forces in providing 'Musical Teas' as part of 'Ye Olde Yuletide Fayre' held on 17 December in the Victoria Rooms, Benedict Street. Autographed books and scores had been presented for auction by John Drinkwater (*Poems*), W. H. Davies

(*The Autobiography of a Super Tramp*), Elgar (a sumptuous edition of *Sea Pictures*), and Ernest Newman (*Music's Motley*). Sir Henry Wood contributed a signed photograph. How much was raised is not known, but the Elgar volume became part of Boughton's own library.

In the end, of course, it was the generosity of the Clarks, and the weight of their reputation, that saved the day. A New Year Festival (2-5 January) was mounted for 1921, followed by a tour (6-15 January) of Bath, Burnham-on-Sea, and Bristol. *Bethlehem* was the main item on offer, though there were recitals at Glastonbury on 3 and 5 January (Bach and Handel, mainly), and Vaughan Williams's *Fantasia on Christmas Carols* at Bath on 7 January. Among the soloists were two new recruits: Dorothy Silk and Frederick Woodhouse. Both were to become important additions to the Glastonbury Players. For some of the performances Kathleen Davis had graduated to the role of the Virgin Mary. Less charitable observers may have found this ironic.

No sooner had Boughton's team vacated the Assembly Rooms than Miss Buckton leapt in with a production of *Eager Heart*, complete with what she over-enthusiastically described as 'Mr Gustav Holst's famous string band'. Holst had already set some of Miss Buckton's poems to music, but, unlike his friend Vaughan Williams, had shown very little interest in Boughton's activities. For his part Boughton found Holst 'rather a cold fish', but their relations improved considerably when Holst expressed a genuine pleasure in the London triumphs of *The Immortal Hour*. In the mean time, all Boughton could do was grit his teeth and hope that no one could confuse Miss Buckton's efforts with his own and divert much-needed funds in her direction.

Something of the confusion and consequent thrashing around for ideas that might solve the festival's financial problems becomes apparent in a series of letters Boughton wrote at the beginning of 1921. The first, dated 11 January, was to Philip Napier Miles:

My dear Miles,
    I think we have at last found a way of saving the work here: a way that depends entirely on the help we can secure from other professional musicians (chiefly singers), but judging by the reception the proposal has had from Dorothy Silk, Steuart Wilson and others of our present company, I think there is no doubt it could be carried through. It is to form a professional Touring Company on a co-operative basis, self-controlled but affiliated to the Festival School, Glastonbury. The Tour I am arranging for Easter will be the first experiment on these lines; and it remains for the Festival School Committee to decide whether this touring body shall become an additional source of propaganda and saving of expenditure so far as the local Festivals are concerned, or whether it shall be left to drift away as a separate organization.

I have not seen Roger Clark's figures lately, but we cannot be far short of £500 in the money we needed for the purchase of Mount Avalon. Surely it would be worth the Committee's while (if you have not already disposed of the property) to take it over on a mortgage, sub-let the fields and gardens, and if necessary even the house—although the gardens and especially the house are desperately needed at Festival time. For the present, I hoped that the Committee, if they bought it, would let me have part of it rent free until my family could move into another house in Glastonbury. There is no other house-room available, and when we do leave I shall be obliged to ask for something by way of an increase in salary at the reconstruction (if it takes place).

I am putting these matters before you for your personal consideration before the Committee meeting and hope during our stay in Bristol to be able to have a thorough talk with you.

The Bath visit was a very great success, although the money taken was not quite enough to meet my personal needs as well as pay the Company; but I hope to make up at Bristol. Anyway, Hatton of the Pump Rooms offered me a return visit in the Spring on more favourable terms. He also called an influential committee too discuss the possibility of transferring the whole work to Bath if Glastonbury fails; so you see there is plenty to talk about. Could we spend Saturday morning together? On Sunday I shall probably go to Town.

Yours ever,

R.B.

The idea he had discussed with John Hatton, manager of the Pump Rooms, envisaged an all-Purcell festival at Easter—an idea doubtless fuelled by Dent after the success of the *Dido* production in 1915 and its subsequent revivals.

A week later (17 January) Boughton was canvassing alternative ideas in an atmosphere that seems somewhat more hopeful:

My dear Vaughan Williams,

there are signs that even Glastonbury may not come to an end; but if it should we are certain to find a home somewhere else, so I want to be prepared with my programme for the Summer Festival.

You very kindly said in 1919 that you might be willing to give us the honour of the first production of your opera. Would it be convenient for you to lend me the vocal score for a short time so that I may be able to consider it as a possibility for this year when I am weighing up the problems of cast and items of expense generally?

Yours ever,

R.B.

In the event, either the expense, or disinclination on the part of Vaughan Williams, put paid to the idea and *Hugh the Drover* (completed in 1914) had to wait until July 1924 and the greater resources of the Royal College of Music for its first presentation.

A letter written on the same day to Gwen Ffrangcon-Davies lifts the lid on the kind of problems that inevitably arise when even a group of 'idealistic' performers have to work under difficult conditions. However genuine Boughton's concern may have been, the touch of paranoia in his expression of it, not to mention the clear indication that he was inclined to drive himself and everybody else too hard, cannot have made their working relations any easier:

Dear Gwen,

We have just now had a very pleasant little tour of Bethlehem and are now considering the formation of a professional company for the festivals here, with tours at Easter, in the late Autumn, and after Christmas. This company is to be founded on co-operative lines and we should be glad to have you as one of the Advisory Council, and as one of those to join in the actual work when mutually convenient.

I trust you will remember the talk Chris and I had with you about twelve months ago when you were concerned with the little group of people who had made work rather hard. You agreed then to hold yourself free from the disaffected person; but a similar thing happened in the Summer and your lodgings were the centre of the trouble as before. At the time I put it down to Brenda, but I had a conversation with her last September which satisfied me that her loyalty was as real as your own; for both Chris and I have never doubted your personal loyalty, and admire you too much as an artist, and care for you too much as a friend to wish to break our working connection. But just as you can do without us professionally, so we can do without you—although we would much rather have your help.

The problem is them for you to let us know what is your chief objection to us when we are working together. Our chief objection to you is an incessant grumbling and magnifying of petty troubles until it is possible for an outsider like John Goss to say that 'he came down having heard of the wonderful loyalty we all had for our cause and found instead all kinds of objectionable things' which he then proceeded to enumerate. Now, my dear, you've to make up your mind once and for all whether you'll work with us or whether you won't. If you come on this scheme you must try to avoid working your nerves to rags with unnecessary late parties, as all our energies are needed for the work itself. If you feel that a little court is more necessary to you, who have so little energy to spare, please be frank with me so that we can remain friends even if we cannot work together.

Yours ever,
Rutland.

He had, of course, known her since childhood, so the heavy hand was perhaps to be expected. Even so, he was lucky that, after a brief spasm of annoyance, she allowed the matter to pass and not affect either their work or their friendship.

On 26 January Boughton was able to write to Sir Edward Elgar that 'we are very nearly within sight of a secure basis for the continuance of our work. An offer has come from Bath which, with your help, should make it possible to realize complete security for us.' He then suggested that if Elgar would take the chair at a meeting on 9 April, during which Hadow and Shaw would speak on behalf of the festivals, the future might be secured. By the beginning of February, however, his optimism had turned almost to despair:

My dear Miles,

the meeting last night was a wash-out. Of the two proposals laid before them the Committee ruled out Mount Avalon entirely, as a Company or anything else. Nor would they take on themselves the raising of £600 per annum in ticket subscriptions that would enable us to accept the co-operative proposal. Their counter-proposal was to form an Advisory Committee to sit with the Executive to consider how the whole Festival Scheme might be reformed. They would not listen to the idea of an Easter Festival. As I personally cannot wait till the Summer, the work is likely now to collapse.

Mount Avalon: please send me your instructions regarding the gardeners, and whether you wish seeds to be purchased and the ground cropped. Shall I ask the gardeners to make a list for you to order? I shall, of course, make arrangements to quit the place at the earliest possible date.

It is a sorry end to all the labours you and the others have made. Some day in a saner world it may be realized that the things of the spirit yield better dividends in happiness than the things of the material world.

Christina left yesterday for London, and proceeds to Paris on Thursday. She at any rate is out of the rottenness of this place.

However casually Boughton might refer to Christina's absence from the Glastonbury scene, the fact that she had decided to leave even temporarily was bound to add fuel to the speculation that surrounded their relationship. For one motive or another, people were beginning to talk. Writing to Tom Goodey on 8 March Boughton complained:

You are being quoted by Kerridge as a person who 'knows' that Christina has left me because of my loose or immoral conduct. This will be news to Chris as it is to me, and if you in any way believe such an absurdity I will report it to Chris and ask her to write and tell you why she is in Paris for a while.

On the same day he wrote to Arthur Jordan: 'Kerridge, who is trying to form a co-operative Company, is putting about the most preposterous tales of us. So do help us to show the world which is the really artistic and going thing.' For all his denials, there can be no doubt that the sudden possession of a small legacy had enabled Christina to get away from the triangular situation that, despite all Boughton's efforts, still existed between Kathleen and themselves. She needed time to think, time to analyse her feelings and decide upon the next step. Paris was a reasonable enough excuse, for she had arranged to work and study with Raymond Duncan.

She stayed for several months, and seems to have felt no qualms at leaving Kathleen to look after the Boughton children as a kind of nursemaid-cum-governess. The fact that they were to live with Miss Agnes Thomas, with Boughton lodging elsewhere, was not much of an insurance, for the arrangements left the door wide open for his interest in her to blossom unchallenged. Nor did everything in Paris go well for Christina. Once there she fell in with a group of Russian *émigrés* intent on experiment with a new system of stage lighting—powered, it seems, chiefly by other people's money. Her work with Raymond Duncan also led to problems, and on 22 May Boughton was forced to write to him:

Sir,

I fail to see why the fact that Miss Christina Walshe is my wife entitles you to refuse to pay her for work she did for you. Either there is some amazing misunderstanding, or your methods are even more infamous than at first appeared.

I am, sir, yours indignantly,
Rutland Boughton.

Duncan paid up, but only after Boughton was obliged to go to Paris and sort the situation out in person.

Against this background of rumour and denial, it can be understood that the idea of transferring the festival either to Bath, or Bristol, or anywhere other than Glastonbury, had much to commend it. A larger community might provide a larger audience, and it would certainly provide a degree of anonymity. In a city Boughton and his marital problems might hope to pass unnoticed and unchallenged.

Even so, his Glastonbury supporters were not entirely put off. They rallied round sufficiently for the presentation of a small-scale festival at Easter, 1921, which turned out to be surprisingly successful. It consisted of two programmes: *The Immortal Hour* and *Dido and Aeneas*, together with the choral ballets *Music Comes* and *The Moon Maiden*. This (31 March–2 April) was followed by an extensive tour of the West Country designed to capitalize on the support that previous occasions had proved to be available. First came Bournemouth (6–8 April), then Bath (11–16), and finally Bristol (19–23), various recitals being given in each place in addition to the main works.

The tour began well enough, and everything seemed set for a financial success. But disaster struck, in the shape of a miners' dispute. On 15 February Lloyd George's government had announced that all wartime controls over the mining industry were to be ended. The mine-owners immediately declared they would in future only negotiate wage claims on a district basis. The intention, clearly, was to break the miners' national solidarity. The miners refused to accept these conditions, and on 31 March the mine-owners declared a lock-out. The miners at once appealed to their allies of the 1914 Triple Alliance, the railway-men and transport workers, and a strike was called for 12 April. The government proclaimed a state of emergency and then, in a moment of rare commonsense, agreed to talk with the miners. The strike was postponed until 15 April. But the negotiations came to nothing. The Triple Alliance failed to hold, and on 15 April ('Black Friday' in trade union annals) the miners struck alone, and after two months, were driven back to work on the owners' terms.

The effect on the tour was catastrophic. Visitors to Bath, unnerved, simply stayed at home. Writing to Gerald Cumberland (7 May), Boughton was driven to a pathetic offer to write a book for Grant Richards: 'the strike synchronized with our Easter Tour and I have come a cropper and shall only be able to avert a sell-up (including my books and music) if I realize some money on the future'. On the same day he wrote to Kenneth Curwen in an attempt to renew his interest in publishing the cycle of five *Symbol Songs* he had composed to poems by Mary Richardson, Curwen having expressed doubts on account of certain 'inartistic' words they included—'drunken' seems to have been the main culprit! And finally he wrote in abject Mozartian terms to John Bright Clark:

Dear John,

Many thanks for the cheque, which carries my salary to the end of June. I quite see your point about paying me so far in advance, nor would I have asked for it but John Bostock advised me that in the present state of things I should do all I possibly could to postpone the inevitable sale of my goods. It is such crises as these that make a position like mine not only impossible but unendurable. This continual begging for money lowers our self-respect, in spite of the fact that we know we have worked for it about twice as hard as the average person. In normal times I could work half the time and get twice the pay as Musical Director of a Picture-House; and although I know what we made in pocket would be taken out in intellectual torture, the sheer commercial fact is not altered. You, Roger and J[ohn] B[ostock] have sacrificed a good deal for the sake of this work, and I quite understand that circuses must follow a long way after bread. I would not have inflicted this yarn on you but for what seemed to me an undercurrent of reproach in your letter . . .

A week later (14 May) he was to write to Philip Napier Miles:

Of course I was sorry that you could not lend us Mount Avalon as headquarters for the Summer School, because that seems our only chance of tiding over the present national crisis; but it did not occur to me from your letter that you were in any way disaffected until comment was made by Sheerman Hand to the following effect: he had been informed that you—and especially Mrs Miles—had made up your minds that the rumours arising from Christina's stay in Paris have basis in fact, and has further been told that Colin Ashdown has done us a lot of harm with you and Mrs Miles. As I am sick of these stabs in the back, in every case so far from people who are either under an obligation to me, or on the other hand have, like Colin Ashdown, had to suffer a rebuke for misbehaviour, I want to find out exactly who are my friends and who are not, and I want to know exactly how I stand with you.

Miles evidently did his best to reassure his friend, but it must have seemed to Boughton as if everything he had worked for was about to fall in ruins at his feet.

12

1921–1922

In fact the worst was over. On 2 June 1921 Rutland Boughton was able to write to Philip Napier Miles:

It is very gratifying to get your letter. It comes at a moment when the omens seem to predict the passing of the mid-night hour. A most successful meeting was held at Bristol last Tuesday. It overflowed beyond the doors of the Folk House—132 applications received for class membership, and guarantees beginning to come in, including £50 from one of the Wills's.[1] The people promoting this know of our special difficulties so that the project cannot be injured by rumour, and of course we shall not be living there. I am quite amazed to find the strength of the idea in Bristol peoples' minds.

The meeting had been held on 31 May, Boughton enthusiastically describing the benefits of a school for 'the collective and individual study of music'. Students, he said, would be able to join the Festival Players when they toured, thus putting their studies in dramatic and choral singing, dancing, gesture, and eurythmics to good, practical use. It was likely, he thought, that a Summer Festival would continue to be held each year at Glastonbury, but that for the remainder of the year work would be transferred to Bristol and Bath, where a similar scheme was being canvassed.

In the end only the Bristol Festival School got off the ground, its first term commencing on 3 October with some 200 students: juniors at 2s. 6d. a term, seniors at 10s. (music extra). The works to be studied were: *Bethlehem*, *Everyman*, and Dr Ethel Smyth's *The Boatswain's Mate*. But even before work commenced Boughton had received more good news. Appleby Matthews, the music director of Barry Jackson's Birmingham Repertory Theatre, had attended one of the Glastonbury performances and had been impressed. He persuaded Jackson to come to Glastonbury and see for himself. He too liked what he found—most particularly *The Immortal Hour* and Miss Gwen Ffrangcon-Davies. He decided to present the opera at Birmingham in a manner based fairly closely on what he had seen at Glastonbury.[2] The opening performances (beginning on 23 June) were not well attended, but gradually the numbers picked up and soon the Repertory Theatre was packed. Jackson extended the run, and then extended it again. What had started out as a possible half-dozen performances ended up as an unprecedented (unprecedented for serious opera) run of forty-two. Audiences were enchanted by the music and Paul Shelving's evocative designs. And as for Gwen Ffrangcon-Davies, she was safely launched on a dazzling career.

Thus by the middle of the year things were looking up for Rutland Boughton. Though the Bath Festival School was not to take root, the school at Bristol was well established. A little money was coming in from Birmingham, and the fact that Barry Jackson had taken his work seriously helped to bolster the confidence of his remaining Glastonbury supporters. Some sort of Summer Festival was now a distinct possibility. He could even be reconciled to Shaw's refusal to entertain the wild proposal he had put to him on 26 April:[3]

I want to make a comic opera version of your Dark Lady. Am I to make the adaptation myself, or will you? I suggest Queen Bess and the Bard for the title and the following for the modified scenario:
1. Song of the Beefeater
2. Scene: Beefeater and Shakespeare
3. Song of the Snapper up
4. Sleepwalking scene: Parody of Grand Opera
5. Dark Lady, Queen and Shakespeare on the Immortality of Art
6. Scene: Queen, Shakespeare and Dark Lady
7. Song: from a Poet to a Prince
8. Quartet: Drama of the Future
This means, of course, that the Dark Lady must be kept on for the sake of a conventional ensemble, giving four points of view to the said drama: the points of view of the Official Person, the Artist, the Smart Set, and the Crowd.

This comic opera is in a way our forlorn hope. We have just run a month's tour on a co-operative basis, and the co-operators, instead of drawing money, have to pay their share of the mine-owners' tribute. We have now to go under entirely, or carry on through the dark days with as few people as possible.

Chris has been in Paris since January, studying Cubism and other kinds of shape-making. She will be home in August.

What finally emerged was a festival on a very small scale. The Summer School opened on 25 August and the festival commenced the following week. It lasted from 29 August to 3 September and consisted of eight recitals, each a mixture of plays, ballets, and chamber music. The new productions were three short plays: *The Fairy* by Laurence Housman, for which Kathleen Davis wrote incidental music; Lady Gregory's comedy *Spreading the News*; and *All Fools Day* by a Bristol author, Josephine Baretti, with music by Clive Carey. Boughton's *The Moon Maiden* and *The Chapel in Lyonesse* were revived, and the anti-war ballet, *The Death of Columbine*, which he had written in collaboration with John Bostock in June 1918, was given its belated first performance. At a chamber concert on 30 August Boughton joined Kathleen Davis in the first performance of his *Symbol Songs*, and accompanied Desirée Ames in the first performance of a three-movement Violin Sonata.

The Summer School had opened with Laurence Housman giving a reading of several of his short plays, including two of the 'St Francis' cycle that were to dominate the work at Glastonbury in its final stages. During the festival itself, Edward Dent lectured on 'The Growth of the Masque', and on 26 August John Drinkwater arrived to give a reading of his new play *Oliver Cromwell*. Unfortunately, as he explained, something had gone wrong with his packing and he had arrived in Glastonbury with entirely the wrong script—a talk on 'The Nature of Art'. This he proceeded to deliver. But rather than disappoint an audience that had come hoping to hear something quite new, he decided, on the spur of the moment, to round off his recital with two of his latest poems. It was then that the trouble began. Although the poems dealt, as the *Observer* (4 September) delicately put it, with 'spiritual matters', one of them, 'Lake Winter', had rather more earthy undertones. This proved altogether too much. Up rose a clergyman: Vernon Herford, a bishop of the Nestorian Church of India no less. 'In the name of the Lord Jesus Christ', he thundered, 'I protest against this justification of fornication!' It was a moment more dramatic than anything *Oliver Cromwell* could have vouchsafed.

Drinkwater survived the outburst and dealt with it efficiently and with tact. But to Boughton, smarting under the lash of rumour and more than a little guilty, it was only what was to be expected. On 5 September he wrote to the offending poet:

There was a short report in yesterday's Observer and a longer one in our own local Central Somerset Gazette of your lecture-reading and the interruption. It would be good if the matter could be pursued further, as the time cannot be far distant when we have got to force upon the non-thinkers a new code of morality very much more definite than their disgusting mixture of wedded sexuality and don't-do-it-again forgiveness. Should you deal with the matter in any way involving our own local aspect of it, please feel perfectly free so far as I am concerned; but be gentle with the reporter if he has in any way misrepresented you. He came here a foolish boy, with nothing but a stock of lurid adjectives; and it seems to me that he has succeeded in reporting your reading and lecture rather well.

With many thanks for the courageous and dignified stand you took and for the help you have given. It is still making itself felt.

But Drinkwater, more discreet than ever Boughton could find it in himself to be, wisely let the matter rest.

The precise state of Boughton's marital mix-up is difficult now to gauge. In taking herself off to Paris, Christina had left the field open to her rival. But after the Summer Festival Kathleen also took herself off, to continue her studies at the Royal College of Music. The relations between the two women, judging by an undated letter evidently written soon after the festival, were remarkably warm and civilized:

Dearest Kath,

I tried to get you to myself for a moment, but it was hopeless. I wanted to thank you for the way you behaved all the time—it was simply splendid, and made things so much easier for R and for me too. I had been in an agony of dread before the Festival as to the possible consequences; but you were so good and so sweet that I find myself looking forward to our next meeting with joy instead of fear. How wonderfully you have developed in all sympathetic and understanding ways. I do hope you won't have too bad a time with your people—they will have to remain 'disappointed' with us I'm afraid. It is a shame that such dear, good people should be so hurt—don't go against them any more than you can help, my dear, will you, because it is no use: they can never understand, and they are right too, from their point of view!

It has been a wonderful month from every point of view—now I am only wishing one thing, to hear from you that you think so too.

Ever your loving friend,
Chris.

Both, it would seem, were doing their best to cope with an impossible situation in ways that would cause the least possible distress. As for Boughton, he was addressing the emotional tangle in the best way he knew—by composing an opera.

He had begun it towards the end of 1920, retiring for weeks at a time to an isolated cottage in Shapwick, a village some 6 or 7 miles south of Glastonbury. Though referred to by the postal service as Vicarage Cottage, and presumably on loan from the local incumbent, the Reverend Charles Seamer, it was little more than a two-room shack. But it was wonderfully quiet and he could work there in peace. The subject he had chosen was pitifully appropriate: *Alkestis*, the story of a wife who volunteers her own life to save that of her husband. It was, in essence, what he would soon be asking Christina to do.

For his libretto Boughton had turned to Gilbert Murray's translation of the Euripides, thus making good the abortive approach he had made in 1914 when *The Trojan Women* filled his mind. 'The choral form of Greek drama', he wrote, 'seemed finer than the more primitive Noh Play and very much akin to the form planned for *The Birth of Arthur*.' As the music progressed it began to turn into a statement of artistic faith: a deliberate protest against the indiscriminate use of discord that seemed to him to have taken over the music of his contemporaries. Discord there would be in *Alkestis*, but only where the dramatic situation demanded it. Using an orchestral palette that laid particular emphasis on flutes and harps, the opera was to vindicate the power of simple, concordant harmony and clear, unsensationalized textures. It had been conceived in the spirit of Gluck and Winckelmann, the very paradigm of 'noble simplicity and calm greatness'.

To a large extent Boughton succeeded in his aim. Of all his works, *Alkestis* is the most transparent. Clear and lucid in all its aspects, it wonderfully evokes the sense of classical restraint that yet embodies deep emotion. Its chief glory is its magnificent choruses, but its musical delineation of character is also impressive. Alkestis and Admetus stand out as real, suffering human beings; Apollo and Thanatos as remote, impartial gods. Only Herakles seems rather crudely drawn—with themes (borrowed, in fact, from the early 'Oliver Cromwell' Symphony) that are somewhat less than sensitive. But since Herakles is a boisterous, crude fellow, even this slight lowering of the inspirational temperature had its positive side. Having once again found a theme with which he could identify on a personal level, Boughton was able to produce music that was unlike anything being written by anyone, anywhere. Like all his finest work, *Alkestis* flew in the face of everything contemporary and yet achieved an individuality and sincerity that make it timeless.

During its composition (inevitably a lengthy process) Boughton found the time and energy to complete two other substantial works. The cycle of five *Symbol Songs* to words by Mary Richardson were composed in June and December 1920 and scored for medium voice and string quartet. He dedicated them 'To Kathleen', and though on the surface they can be read as an exploration of religious ecstasy they are clearly a personal declaration of love.

Even more substantial is the three-movement Sonata in D major for Violin and Piano completed in May 1921 and dedicated to Desirée Ames who gave the first performance. A virtuoso work, it does not quite live up to the promise of its thematic material—a splendidly powerful slow movement, for example, dissipates its strength in a trivial folk-dance ending. But on the whole it is an impressive work, somewhat in the manner of César Franck. The critic of the *Daily Telegraph*, commenting on the first London performance (given on 7 February 1922 by Marie Hall and the composer) described it as 'a work of vigour and vision' that 'never once gave the feeling of straining after effect; the music seeming rather to flow straight from the words which the composer has taken as his motto'. Those words were from Nietzsche's *Also sprach Zarathustra*.

The works of Friedrich Wilhelm Nietzsche were, in fact, very much on Rutland Boughton's mind at this period. On 26 July 1921 he wrote to his friend the photographer Herbert Lambert:

I want the remaining use of my life to be more concentrated than ever; but I am just beginning to realize that ideas which move me have a force which is so destructive to what good folk call law and order that I either have to forego my soul, so to speak, or sometimes offend my friends. Don't let this sound too terrible to you; it only means that I intend much more boldly to declare in my work, and so far as possible in my life, the ideas promulgated in Nietzsche's *Zarathustra* and Shaw's *Methusala*, but I have only been conscious of where I stood in the last few days.

By which he seems to have been warning his friends that his music, and most particularly his music-dramas, might in future touch on matters that could offend their political beliefs, and that, even more dangerously, he intended to allow his love for Kathleen lead him where it would.

Obliquely, the quotations attached to each of the Violin Sonata's three movements bear this out. The first ('Quick and Passionate') movement is headed:

Most honestly and purely the healthy body speaketh, the perfect and rectangular; it speaketh of the significance of Earth.

The second ('Slow and Severe') and third ('Fairly Quick') bear the following mottoes, respectively:

Once having passions thou callest them evil. Now, however, thou hast nothing but they passions. Thou laidest thy highest goal upon these passions: then they become thy virtues and delights.

I am the advocate of God in the presence of the Devil. But he is the spirit of gravity. How could I, ye light ones, be an enemy unto divine dances? or unto the feet of girls with beautiful ankles?

Puritan inhibition, it seems, was to be a thing of the past—though what Christina had too say about 'girls with beautiful ankles' must surely have been even more uninhibited.

That matters between them were growing more difficult by the hour was obvious to them both. On 30 July 1921 Boughton wrote to Shaw for advice: 'Chris is in France on her own job (though she came back for a holiday a fortnight ago); but things are very difficult and you are the only person I know who would understand them sufficiently well to help.' But he must have known that Shaw would only smile his Mephistophelean smile and repeat what he had said when he first realized his friend's dilemma: 'It's a pity you can't live with Kathleen. I believe she would hold you.'

Fortunately, as we have seen, Boughton was able to detach himself from the worry of his obsession with Kathleen and the difficulties of organizing the Bristol Festival School, not to mention the restructuring of his Glastonbury schemes. In this he was greatly helped by the presence of Harry Sheerman Hand, someone he at last felt able to trust. He had always found it difficult, indeed almost impossible, to delegate, but Hand (1893–1922) was different. Since arriving in Glastonbury in 1919 he had become part of the family and was soon engaged to Boughton's eldest daughter, Ruby. Now, as Boughton's official second-in-command, he took over the day-to-day administration of the Bristol School, Boughton confining his contribution to a series of Thursday classes in music appreciation.

With Sir Edward Elgar as its President, Philip Napier Miles as Vice-President, Boughton as Director of Studies, Sheerman Hand as Assistant Director, Margaret Morris in charge of Dance Studies, and Christina Walshe directing Scenic and Costume Design, the Bristol Folk Festival School (as it was renamed) began to settle in well at The Folk House, College Green. The work achieved was impressive. Visiting lecturers included Edmund Fellowes (on 'Early English Lute Songs' and 'Madrigals'), Margaret Morris (on 'Hellenic Dancing and Modern Art'), John Drinkwater (at last able to read *Oliver Cromwell*), Barry Jackson (on the work of the Birmingham Repertory Theatre), and Laurence Housman (reading several of his St Francis plays). There were regular recitals, and even a symposium on 'Modern Music', with illustrations from Scriabin, Ravel, and Boughton's pet hate, Stravinsky. On 2 January 1922 the term opened with a New Year Festival: three performances of *Bethlehem* in Bristol (2–4 January), and three in Bath (5–7), together with two 'Children's Concerts' which included performances of a play by Gwen John (not the painter) called *The Goblin and the Huckster*, and a ballet, *The Enchanted Princess*, arranged to Frank Bridge's *A Fairy Tale Suite*.

The performances of *Bethlehem* saw Kathleen again in the role of the Virgin Mary, and when, by 22 January, she had returned to her studies at the Royal College, Boughton, alone with Miss Brownlee's photographs of the production, wrote to her in terms that left no doubt as to the seriousness of their relationship:

My precious wife, fibre and heart of me: I have come in here this morning that I may be alone with you in spirit, as we had hoped to be alone together in the flesh today. For me, darling, the separateness is a little relieved by the wonderful Mary portraits which I now have ranged along the shelf in front of my table. First, there is the reproachful one because of Joseph's distrust of her, then an enlargement of the bust of the same. Next, the sweet and tender young mother bending over the manger. Then, two mysterious clairvoyant poses for the second lullaby. Then, the full, quietly joyous, motherly one standing with cloak thrown resolutely round her, ready to face the journey into Egypt. And these, my very wife, are symbols for us. You have reproached me when I misunderstood you, comforted me, looked forward to the coming of our baby, and will finally hold yourself ready with that soul of courage which is so wonderfully yours to go into Egyptian darkness with me if necessary. For today I live with these, my Maries. Tomorrow I lend them to Miss Brownlee to show in her case for a week; and then they return to me for all my life-days. They are beautifully mounted and I will have them well framed, together with the two lovely ones you promised to send me. And O my sweet Kathleen I shall grudge the loss of every *one* that I don't have. Forgive me for that; for I know that the freer we leave one another the more completely we belong to each other. You are growing in knowledge and loveliness. You will become one of the noblest women of your time, as you are not only the most beautiful, but the best, truest, and sweetest of wives who ever lived—you enrich my life past all my hopes or understanding. God bring me to a fair appreciation and worship of you, and give me such music as will be worth your singing.[4]

Your husband,
Rutland.

Thus it would seem that, even though the 'baby' was as yet a matter of wishful thinking, by January 1922 the die had been cast.

As a preliminary to a short Easter Festival (17–19 April) consisting of three performances of *Venus and Adonis* (plus music by Byrd, Weelkes, and Gibbons), Boughton and his pupils presented a tour of *Everyman*. Because the performers were now being drawn from all over the West Country, they dropped the name 'Glastonbury Festival Players' in favour of 'The Western Players', and as such performed in Bristol, Burnham-on-Sea, Winscombe, Bath, and Bristol again: a total of eight performances, beginning on 18 March and ending on the last day of the month. The printed programme shows a commendable and rather touching caution in its assurance that 'the part of the Deity will not be personified, but sung to a kind of plainsong by a small chorus'. The fact that Boughton himself had composed the 'kind of plainsong' may have alarmed the truly susceptible, but no one seems to have complained.

On 2 April Boughton's pupils joined forces with the Bristol Philharmonic Society in a performance of *The Immortal Hour* at the Colston Hall, Kathleen Davis singing the part of Etain for the first time. Fears that the enormous hall would be half empty proved groundless. It was packed, and the performance was a triumph. Buoyed up by the success and apparent stability of the Bristol Folk Festival School, Boughton was now able to fulfil the complementary part of his overall scheme by returning to Glastonbury for a full-scale Summer Festival.

Advertised as a 'Festival of Greek Drama', the 'School' commenced on Saturday, 29 July, in a 'tremendous downpour of rain which fell for a couple of hours'. But enthusiasm was not drowned out, and on Monday, 31 July, the students began their daily routine:

    10 am: Hellenic Dancing
    11.15:   Rehearsal of Greek Play
    3 pm: Junior Hellenic Dancing
    5.00:   Oral Classes in the Elements of the Greek Language
    6.00:   Senior Hellenic Dancing
    7.30:   Lecture, Dramatic Reading, or Recital

The plays chosen for study were the *Oedipus* and *Antigone* of Sophocles, *The Cyclops* and *Bacchae* of Euripides, and *The Frogs* and *The Birds* of Aristophanes—all, of course, in translation. The lecturers were Professor Gilbert Murray on 'The Religious Basis of Greek Drama' (2 August), annd R. W. Livingstone, Fellow of Corpus Christi College, Oxford, on 'The Genius of Sophocles' (9 August). The fee for the fortnight, including a 'Ramble and Picnic' on Saturday, 5 August, came to three guineas.

A similar fee, or a reduced 'inclusive' fee of five guineas, gave access to the 'School of Dance and Drama' which ran from 14 to 25 August. This followed a similar daily pattern, except that the works to be studied were Blow's *Venus and Adonis* and Boughton's new music-drama *Alkestis*. In addition to Murray and Livingstone, the lecturers were: F.W. Cornford, Fellow of Trinity College, Cambridge, on 'The Psychology of Greek Tragedy and Comedy' (16 August); and Professor Gilbert Norwood, on 'Ancient and Modern Methods of Play-making' (23 August). Boughton himself was advertised as lecturing on 'Mozart' (14 August), but in the event seems not to have done so. Much more to the point was a rumoured lecture by Shaw on 'The

Figure 2   Festival programme 1922, designed by Christina Walshe; black on white

Evolution of the Drama' which, to everyone's delight, actually took place on the last day of the month. The attendance figures speak volumes: 39 for Gilbert Murray, 10 for Livingstone, 25 for Cornford, 18 for Norwood, and 212—as many people as the Assembly Rooms could squeeze in—for George Bernard Shaw, who, if the *Central Somerset Gazette*'s verbatim report of 8 September is to be trusted, regaled his audience with a rambling discourse that was clearly more improvised than planned. It ended, however, in a welcome, if somewhat provocative, tribute:

Here in Glastonbury [you] have achieved a musical festival which is, though [you] may not know it, the most important thing in England at the moment. All other festivals—like the Three Choirs Festival—are only marking time. There are no signs of any development in them. They have done a certain amount of mischief in causing people to write noxious music which had better not been written. Here you have a development of music-drama and many other things which is very much greater than you imagine.

Shaw, it should be added, had probably been frightened into giving his lecture by a letter Boughton had written on 21 November the previous year in which he declared, rather airily and with a certain cheek, 'Please remember I want you to play the part of Euripides in *The Frogs* [next August]. Gilbert Murray says he mustn't break the Oxford tradition by appearing as an actor, but John Drinkwater will play the Aeschylus part.' The scheme, however, progressed no further, and Glastonbury was denied an experience that would have crowded the Assembly Rooms even more satisfactorily.

Alkestis was given six performances, beginning 26 August. Two important new singers were brought in: Astra Desmond, to play the part of Alkestis herself, and Frederick Woodhouse for Thanatos. Arthur Jordan sang Apollo, Steuart Wilson was Admetus, and Clive Carey took the part of Herakles. In *Venus and Adonis* Kathleen Davis found herself in the character of Venus—which must have given Christina occasion for a wry smile, especially as she herself had been cast as Deianeira, the betrayed wife in *The Trachiniae* of Sophocles, performed on 22 August in Plumtre's translation under the title *The Death of Herakles*. The musical direction for these performances was in the hands of Boughton and Julius Harrison, a new recruit to the cause. Bantock himself came down to accompany Astra Desmond's performance of his *Sappho Songs*—a wonderful occasion, according to the *Central Somerset Gazette*, and much admired by the audience.

All three productions were well received—the *Sunday Times* (13 August) innocently praising Christina's presentation of a woman 'depressed by the obsession of coming evil' as 'a fine bit of intuitive characterization'. Of Boughton's new music-drama, the music critic of *The Times*, having studied the orchestral score beforehand, said:

It is a work which adds an important new resource to the operatic repertory, and it is much to be hoped that it will not be long before it gets the full presentation it deserves. . . . The chief power of Boughton's music is one which very few of his contemporaries share with him. One comes away from a first hearing of his works, and from this one as much as any, with certain phrases ringing in the head—obvious phrases, possibly, which anyone might have written, but which no one else did—and they are phrases which mean all the world to him, and therefore mean much to anyone with ears to hear.

The *Sunday Times* (27 August), equally enthusiastic, gave its readers a description of the performance:

The decor is a simple platform with apron stage and steps leading from the auditorium; curtains of purple and white are suspended from the ceiling round a darkly draped doorway, denoting the entrance to Admetus's house. Against these the black and white garments of the chorus strike a note of simplicity and strength. The auditorium is also pressed into service, the funeral procession of Alkestis leaving and returning through the audience with singularly striking effect.

Altogether Boughton had every reason to be pleased with the festival. Not only had it been an artistic success, it had actually made ends meet. Without counting the many concerts and lectures, Glastonbury could now look back on some 290 staged performances, including 47 of *The Immortal Hour* and 28 of *Bethlehem*. There was every reason to suppose that 1923 would see an Easter Festival of 'Greek Music and Drama', and a 'Celtic Festival' in the summer, for which Boughton was rumoured to be writing a new music-drama and Bantock had pledged his Hebridean opera *The Seal Woman*. Not even the fact that Glastonbury was agog with the shooting of a 'Pageant Film' by the Steadfast Film Company, with Miss Buckton and her Chalice Well adjuvants prominent among the leading players, could mar Boughton's sense of well-being. And there was, he knew, more to come. On Friday, 13 October, Barry Jackson would present *The Immortal Hour* in London!

Even his private life seemed to be moving toward some sort of resolution. After the festival, Christina, possibly gambling on the hope that a period of uninhibited proximity might finally exhaust their obsession, suggested that the lovers should go away together. Taking her at her word, they went to the Isles of Scilly—and returned more in love than ever, for Kathleen had the happy knack of never making him feel ashamed of his physical needs:

The union between Christina and myself had always been effective on the higher levels; but she generally made me feel ashamed of my physical appetite. We could always work together in happy accord; we could even be gentle and tender in our relations with each other and our children—Flo's children as well as her own—but of those deeps where the courtesan, the mother, and the fellow-labourer are fused we had never been able to learn. Our work was warmed by affection but not fired by passion . . .

It was not that the creative artist needed a sexual stimulus: it was rather that mental work cannot be effective unless the sexual appetite is satisfied and so relegated to its own sphere in the background of life. Kathleen was as sensual as I was, and she was single-hearted; so that for the first time I learned how sensuality could be combined with innocence . . .

For Christina there was nothing for it but to accept the situation. She left for London, leaving Boughton to cope as best he could in Glastonbury.

Although it had now become clear that the work in Bristol would have to be modified—reduced, in fact, to classes in opera and drama, with performances in the spring and autumn—there was every reason to suppose that it would continue. Plans were laid accordingly. But on 26 September disaster struck. Sheerman Hand committed suicide by jumping from the Leicester–London Express as it passed through Pinner Station. Railways officials reported that he 'fell in front of a slow down-train. He was carrying his shoes in his hand as he leaped.' He was 29 years old.

No one ever found out why he did it, though it was said that he had been very upset by the melancholy part (Herakles) he had to play in *The Trachiniae*. It was also said that he had a morbid fear of cancer, from which his father had died. Boughton was greatly affected. Not only had he lost a trusted assistant, but a future son-in-law; and he had the additional bitterness of wondering whether the fact that he had taken it upon himself to reprimand him for letting his attentions wander momentarily away from Ruby might have contributed to his sense of unworthiness. One thing was certain: the work of the Bristol Folk Festival School was now in jeopardy.

But life had to go on. *The Immortal Hour* duly opened at the Regent Theatre, and on 26 October Shaw wrote in jocular mood:

Dear Rutland,

I am getting out a new edition of The Perfect Wagnerite. On turning to Who's Who to find out where you were born, and write you up a little, I find with impatience that you have not taken the trouble to supply that invaluable book of reference with any usesful information whatever. Don't let this go on for another year, if you can help it (you may be too late; for I corrected my Who's Who proof some time ago). Make out a dossier at once—thus:

Boughton, Rutland; musical composer, b.1830, e.s. of Telemachus Boughton of Boughton Park, Hull, and Cornelia Tompkins; (you must omit the marriage, I fear); educ. Eton, Oxford (or private schools), Royal College of Music (Wooden Spoon for Oratorio); founded the Glastonbury Music Festival 190?; Compositions: The Skeleton Horseman 1842, Leaves of Grass 1844, Towards Democracy 1845, Sweeney Todd, an opera, 1846, Paradise Lost 1847, King Arthur, a heptarology 1848-96, The Immortal Hour 1898, Alkestis (not the Gluck one) 1922. Recreation: being seduced by his pupils. Clubs: The Taproom at The George, Glastonbury. Address: Acacia Villa, Glastonbury. Telephone: Glastonbury 1. Telegraphic address: Arimathea, Glastonbury.

Send this to the editor of Who's Who c/o Messrs A. & C. Black, Soho Square, London W.1., and apologize for not having sent it before on the ground that you are heavily preoccupied with your artistic work and did not realize the importance of Who's Who to you until your business friends remonstrated.

Also read the enclosed proof and see whether it is alright. The sooner you can let me have it back the better.

in haste, ever,
G.B.S.

By the end of the year there would be no doubt in anyone's mind who Rutland Boughton was.

13

1922–1923

To begin with it did not seem as if *The Immortal Hour* could possibly be a success in London. The Regent at King's Cross was well outside the charmed circle of West End theatres. Designed by Wylsen and Long it had begun life in June 1900 as the Euston Palace and as such was largely devoted to music hall, until it was taken over by Sir Nigel Playfair and reopened in August 1922 under its new name and with Arnold Bennett's short-lived play *Body and Soul*. Moreover, there was no reason to suppose that a serious opera could enjoy, or even endure, a run of any length. The precedent of Sullivan's *Ivanhoe* was not particularly encouraging. The first weeks justified the pessimists. But Jackson held on, for the box office had noticed that people were returning for a second and third time. Gradually the tide turned. London woke up to the idea that it was being offered an unusual masterpiece and responded accordingly.

The first run of 216 performances, from 13 October 1922 to April the following year broke the record for any serious English opera. A revival of some 160 performances began on 17 November 1923—broken into by 36 performances of *Bethlehem* over the Christmas period. Two major revivals followed: at the Kingsway Theatre in 1926, beginning on 30 January, and then at the Queen's Theatre (9 February 1932). In supporting the initial run Barry Jackson (1879–1961) lost a considerable amount of money. Reports vary, but it seems to have been in the region of £8,000—though whether he made up the loss during the subsequent revivals is not clear. In so doing he earned the gratitude of his audiences, who soon came to regard the work as their private property. They returned to see it time and again. Gossip columns had a merry time recording the visits of the Princesses Marie Louise and Helena Victoria, devoting to them some of the fervour normally accorded to batting averages. In the end it was Princess Marie Louise who came out on top, with 52 performances to her credit. Yet even this heroic effort pales into insignificance beside the claims of a certain Miss Parker to a record of 133 performances. 'She has seen it', remarked one paper, 'from every corner of the house.'

At the end of the first run a presentation was made to Barry Jackson at which Sir Denison Ross referred to the subscribers in terms that summed up the attitude of all the work's admirers: 'They are all people, I think, who like me have grown to look upon a visit to *The Immortal Hour* as almost a religious service.' Hannen Swaffer added his comment: 'That was the extraordinary effect *The Immortal Hour* had on many of us. There seemed something reverential about it; the last act of *Parsifal* is the only thing like it I can remember.'[1] Boughton's fellow composers were equally glad to acknowledge his achievement. Arnold Bax, himself devoted to a more opapque and gargantuan celebration of the Celtic Twilight, wrote: 'I have just heard *The Immortal Hour* again and feel I must tell you that I think this work is undoubtedly the best opera written by an English composer. And the *human* love motif in Scene 2 and its harrowing allusions later on will always haunt me.'[2] Gustav Holst forgot his reservations and, on 25 April 1923, wrote:

I am off to the USA tomorrow for about two months and before I go I want to tell you:

(a) It is a fine thing to reverse the old order of things and turn a Music Hall into an Opera House.

(b) It is also a fine thing to establish a record run of English Opera.

(c) But it is not only a fine thing, it is also a beautiful thing that the work to accomplish (a) and (b) should be The Immortal Hour.

Many congratulations! Many thanks!! More power to your elbow!!![3]

There is a similarity in temperament—the same feeling for mysticism—to link both these composers with Boughton, even if in practice it led to very different musical results. But there is little to connect his work with that of Dame Ethyl Smyth; yet her praise (on 17 November 1922), though characteristically unconventional, was just as wholehearted:

I came away yesterday with a great, great 'impression on my chest' from the Regent. When I tell you I'm not fond of mysticism you can imagine what I think about your music. I hope to hear it again next Thursday. I know you dislike the performance and I know how the ideal one has in one's mind is brutally treated by most renderings. But I don't think we makers know how many facets our handiwork has and what it can stand. All I can tell you is that the performance enchants me in every way . . . the whole thing gripped me.[4]

Later (16 February 1923) she was to add: '*The Immortal Hour* is I think doing more good for English music life than anything that has happened in my lifetime.' Even Eugene Goossens, scarcely a Boughton fan as his autobiography shows, wrote on 6 March with all the enthusiasm of a typical 1923 devotee:

May I take this opportunity of saying how sincerely delighted I am with your well-deserved success at the Regent. I must say that the work has given me the utmost pleasure every time I listened to it, and though my sympathies are suppose to lie in other channels, I still have sufficient balance left to appreciate the subtle beauties of *The Immortal Hour*.[5]

Among Boughton's contemporaries only Kaikhosru Sorabji seemed incensed by the opera's success. On 19 July 1924 he felt it necessary to inform the world, through the columns of *Musical Opinion*, that it was 'a tissue of abject commonplaces and banality, utterly without distinction of idea, style, or expression'. At the other extreme, Lord Alfred Douglas who, according to John Betjeman, 'saw *The Immortal Hour* at least fifty times', declared that it was 'the best thing that had happened in England since Shakespeare'.[6] Doubtless the truth lay somewhere in between.

What gave it its popularity? The quality of the music is beyond question. For this shadow play of dreams tested against reality, Boughton found the perfect musical expression. Yet material and method were the very antithesis of post-war attitudes to life. It was not the music of disillusioned, 'bright young things' intent on a hedonistic escape from a new and harsher world; nor was it an expressionist scream of anguish, an atonal retreat from emotion, or a Stravinskian exploration of the primitive. It was the voice of an authentic dreamer from an earlier, saner age, and as such it had a powerful appeal. A war-weary generation was eager to escape into this dream world. It had done so already with J. M. Barrie's *Mary Rose* and Frederic Austin's sanitized version of *The Beggar's Opera*, and would do so again when Delius added music to James Elroy Flecker's *Hassan*. In the early 1920s *The Immortal Hour* was the perfect mirror of society's needs.

Unfortunately society did not look closely enough. As Boughton feared, it eagerly missed the real point of the story. Behind the simple tale of a love that is found, enjoyed, and finally lost forever, there is the subtle parable of the transience of beauty—the bitter knowledge that perfection cannot last. That these audiences missed the true significance of the work may, in part at least, be due to Gwen Ffrancgon-Davies's performance as Etain. Her small, pure voice, the incomparable grace of her movements, her strange fey charm, drew all the attention and all the sympathy. Etain won back to the Land of Youth seemed only an affirmation of joy; the darker side, the destruction of Eochaidh's dreams, was easy to overlook. And soon everything was swamped in the cult of popularity that sprang up. Young ladies began to look wan and droopy, palely loitering in the hope of hearing strange, far-off voices. Pictures of Gwen in various somnambulistic poses peered soulfully from the pages of the illustrated magazine, and, for a time, the 'Faery Song' displaced Mendelssohn at society weddings.

Boughton was horrified. He had not wanted the production to go to London, but had given in way to the entreaties of his singers. They had served him for a pittance, and London would be the chance of a lifetime. As he read the daily papers his worst fears were confirmed:

The Marchioness of Londonderry is said to have seen it 25 times, Princess Louise of Schleswig-Holstein many times, and Lady Patricia Ramsay (Princess 'Pat') several times . . .

I saw Lord and Lady Ridley in the foyer during the interval, Lady Gervase Elwes and Lord Clifton were in the audience, and, in a box, I caught sight of Lord Beaverbrook . . .

The other night I saw Lady Maud Warrender, herself the owner of a fine voice, in the audience entertaining some friends. Lady Cunard, too, was playing hostess to a big party, so was Lady Falmouth to a group that included Mr and Mrs Lionel Fox Pitt . . .

The Duchess of Sutherland, wearing a black marocain gown with a magnificent sable wrap, was present with the Duke and her sister, Lady Betty Butler . . .

Lady Audrey Chetwynd-Talbot and her young bridegroom, Lord Stanley, have chosen the 'Faery Song' to be played at their wedding . . .

It was a nightmare. A work into which he had poured his very soul had been taken up by the rich and idle as a fashionable plaything: 'During the final performance at the Regent Theatre I went up into the gallery to study the audience. The hysterical response to the last curtain, when only silence will prove that the effect made has been a true one, sickened me.' The work was tainted and he could scarcely be persuaded to acknowledge it any more. He refused to conduct it for a suggested visit of the King and Queen, and is said to have turned down the offer of a knighthood. As far as he was concerned the opera had 'gone on the streets and could earn its own living', which, of course, was precisely what it was doing.

Critical reaction was almost universally favourable. In its review of 14 October 1922, the *Daily Telegraph* put the case well:

Mr Rutland Boughton has always followed an independent line of thought. Worldly wisdom and worldly advantages have never had much weight with him. A less scrupulous or less uncompromising attitude might perhaps have reached sooner the contact with the great public that is a necessity of the artist. But he has also escaped marvellously the contagion of other composers and other schools. The Immortal Hour last night left two very clear impressions—the first was the fitness and mastery which bound together words and music; the second was the no less astonishing freedom of the music from any taint of imitation. The best pages of the opera have a kind of vestal simplicity and perfection which are all the more remarkable for the complexity of some of the best and of all the worst moderns. It is a British opera, not merely because it has been written by an Englishman, but because it possesses characteristics that are not found in German, Italian, or French opera. The text or story is as alien to the emotional plots of Puccini as to the Heroic Wagnerian stories or the phantastic medleys of some Straussian works. But it is familiar in the main outline to all students of Irish poetry and drama. . . . It is just the kind of story that gives music the right opportunity to add its voice, heightening the effect of the supernatural elements, making the passions and sorrows of men more poignant. Mr Boughton has long been attracted by the poetry of Fiona Macleod, and in this tried sympathy lies one of the secrets of his wonderful felicity in handling the story. His music is as novel as Mr Yeats's and Lady Gregory's plays were in the field of drama when they first came out. There is no straining at new rhythms, no hankering after crashing discord. Mr Boughton can tell a simple story in a way that is both simple and beautiful. Some of his melodies have a haunting charm that, once heard, cannot be forgotten. But the great merit of the opera is not only in its lovely melodies. The fine harmony of the whole structure, the sureness of touch with which Mr Boughton deals with the most delicate as with the most robust aspects of the play are equally striking and convincing.

In the face of such sympathetic appreciation even Boughton must have felt that his efforts had not been entirely wasted. But he had never liked every aspect of the production. Paul Shelving's design for Act I was marvellous—a glade of slender trees and a moonlit pool, not unlike the woods he had roamed in at Grayshott when he first sought music for Fiona Macleod's words.[7] But the second Act was altogether too fussy and exotic. It was as if Celtic Ireland had suddenly been invaded by a Mexican fiesta. Even the work's admirers had their reservations, as photographer Herbert Lambert made clear in a letter written on 14 October 1922, immediately after the first night:

It really went very well and there was an appreciative audience. Everybody who mattered seemed to be there. In my immediate vicinity I found a lot of people I knew. Holst, Clive Carey, Goossens, Adrian Boult, Mark Hambourg, Arnold Bennett, and lots of others. Barry Jackson made a little speech afterwards and said how much he regretted you were not there.

As for the performance it was really nothing to be ashamed of. Of course on a first night there are bound to be things which could be improved with regard to ensemble, but the thing came alive and you could feel the audience were being held by it.

The setting you have seen of course. The 1st scene satisfied me entirely—it was just right and was one of the most beautiful things I have ever seen on the stage.

The Hut Scene I do not like. There is too much colour, or rather too many colours, and the actual painting of the scenery is poor—a sort of compromise between Realism and Convention that worried me. There was a beam of green light which came from the side of the stage on to Etain at the end—you could have cut it with a knife. It would have fairly made your hair stand on end I know!

The last Act struck me as a good design but overleaded with detail and again badly painted. When the light lowered at the end it looked very well. This scene badly wanted a larger stage. The crowd was very tight and the dresses and headgears, and chin-gears, struck me as being a bit overdone. I have always been terrified of this sort of thing since I saw Chu Chin Chow!

Now that's enough of grumbles. On the whole it was a very fine production and even spoiled London was impressed.

Gwen was simply marvellous and it was nice to feel that some of your original tradition was being carried on so perfectly.

All of which was echoed in his own way by the young John Gielgud, annotating his programme on 13 February 1923:

Very charming indeed. Imaginative and exquisite. Miss Davies perfectly lovely throughout: both acting, posing and singing. Well sung on the whole; Dalua good, Midir and Yeodaidh [*sic*] alone might have acted better. Colouring excellent all through, save the last act, and beautiful lighting throughout. Last act dull beginning, and scene and dresses not a success in this scene—lacking in design and breadth. A beautiful story and the music beautiful, particularly the choruses. Altogether most poetic and idealistic, a most interesting and enjoyable evening.

While of the first revival he wrote (25 April 1924): 'Enjoyed it still more, seeing it again. The new set in the last act and different lighting improved the general decor and there are other minor advantageous alterations. Gwen Davies still leading brilliantly a very adequate cast.[8]

The Christmas production of *Bethlehem*, which began at the Regent on 19 December 1923, was equally well received, though the work did not become a cult. It 'shows Mr Rutland Boughton quite at his best, and proves him to be almost the only young English composer who at present can write successfully for the stage', declared the *Daily Chronicle* on 20 December. Once again Gwen Ffrangcon-Davies charmed and delighted her audience, and Paul Shelving's set ('a proscenium suggesting the facade of a church, through the windows of which appear the chorus, represented as angelic beings') created a touchingly medieval atmosphere. If the audience had anything to complain about it was the fact that, at Boughton's insistence, they had been asked not to applaud. A few enthusiasts tried, but were quickly hushed into an appropriately respectful silence.

Shaw, however, apparently under the impression that Christina had designed the costumes, had reservations:

21 December 1923

Dear Rutland Boughton,

I saw *Bethlehem* for the first time last night; and it was a complete success as far as I was concerned. That says everything; and now for a few criticisms.

The angel is impossible. Chris has scored a magnificent success with the three Kings and the Herod ballet on top of her reds and browns and greens for Joseph and the ordinary folk. But the angel spoils it all with his beefeater's crimson, his absurd peruke and his halo that has no halation. And somehow you have put the final touch to him with your harp and cornet accompaniment, and made him a public house angel—the angel at Islington. Barry Jackson assured me it was a trumpet; but for melodic purposes it might just as well have been a cornet: the manufacturers aim at cornet tone always: and the only trumpet that sounds like one is the straight Bach trumpet playing florid divisions and fanfares and not sentimental melodies. Chris must redesign the angel; and you must rescore him for something strange, say the ocarina. Or, if you must have a popular and obvious angel, let him have a harmonium and sing the Lost Chord as an encore.

When the 3 Kings come before Herod the band raps out a phrase from the very hackneyed scherzo of Tchaikovsky's Pathetic Symphony. The resemblance is too great; and it is really painful to be reminded of such stuff in *Bethlehem*. Couldn't you turn its nose the other way?[9]

That scarlet old woman did not get a single word over the footlights: you should insist on at least *some* articulation. And Ruby and the other angel must learn to move like angels, and not just drop in to tea. The glockenspiel was a failure because the player had no touch: it would be all right if it were purposefully played and not roughly clicked. The choral variations will be splendid when the chorus gets quite steady and confident. That's all the grumbling I have room for.

G.B.S.

The success of *The Immortal Hour* and *Bethlehem*, combined with his own rapidly increasing domestic difficulties, made it impossible for Rutland Boughton to contemplate any work at Glastonbury in 1923. The final event of 1922 had been a Christmas tour of *Bethlehem* by the Festival Players. This had begun on 6 December, lasted two weeks, and involved a dozen performances: Bristol, Weston-super-Mare, Bath, and

Bournemouth being visited. It was successful enough, save for a disastrous 'charity matinee' on behalf of the Bristol Rotarians. After the tour all activities ceased and the Bristol Folk Festival School simply faded out of existence.

The exact sequence of events that led to the final collapse of his relationship with Christina is not easy to follow, for all the more personal letters that passed between the three protagonists were left undated, and the correspondence is in any case incomplete. But it seems that by the end of the year the situation had become impossible. Writing to the Clarks on 28 January 1923, Boughton tried to explain:

Another attempt to come to terms with Chris has failed, and it seems perfectly clear that she means the break to be permanent. So I wish you to know that her complaints of my unfaithfulness have foundation in fact. I would like to tell you all the facts, but the one friend to whom I have done so seems to regard the mere statement of them as an additional wrong-doing on my part. They were already known to Tommy, Ruby and G.B.S.—the last of course being the only one to understand them. I would gladly have told you all long ago, but I knew you would have to withdraw your friendship, though not, I trust, your love. And once you had been told, the Festivals must come to an end. They don't matter to Chris now that she has got work in London; my own future actions depend mostly on the children's wishes and welfare. Many thanks to you all, my dear good people, your part in it at least will be beautiful—not that I am ashamed of mine, but I am sorry Chris is hurt for I shall always love her.

Christina, for her part, seems to have fought a rearguard action, desperately trying to be fair to everyone:

Darling Kath,

So glad to find you still love me, anyhow! Now I want to reply a little to your letter, although I am not at all fit and am feeling very wobbly inside (these storms do affect one's internal apparatus terribly, you know of course).

Evidently R did not tell you what led up to our letters. I didn't tell you because it was the most horrible experience of my life so far, and I was trying to forget it. I suppose I must let you know that following upon my saying 'I had hoped she would have found she would be able to do without that now' R said something so cruel and caddish that my heart simply crashed—we had the most frightful quarrel in front of Tommy who we kept there, I for protection, he for a witness. The outward results of that storm were that my little silver ring is lost forever, and R's gold ring is broken. Also something has happened in my body which will never be mended.

The inward results were—oh K, how it hurts me to write about it!—anyway, I was determined to leave him for good and all. But Tommy begged us to wait, and remembering our agreement to wait 3 days we waited. On the second day R came with a proposition: that we should continue to live together for the sake of the children; that he should see and write to you no more—also for the sake of the children; and that I was not to expect him to act as a husband to me anymore.

I was minded to refuse it, and thought it over for a day; but Tommy and Ruby urged me to accept, and I came to think it would be best in the long run. So I write to you on the 4th day.

That is roughly the outline, you can fill it in according to your heart's knowledge of our characters.

And don't forget dear, when you blame me for what you think is my change-ableness, that the idea of your marrying somebody to make things right seemed fatal to me and I rightly objected—don't you see that in such a case you would be putting yourself under a curse for life—but in this way you will be cleared absolutely and can marry whom you wish to without taking him as a refuge.

But in objecting to that and agreeing to your continued freedom I was trusting you to return generosity for generosity and to see that you could not go on as you have been without driving me mad, even if I denied it to myself. I daresay you would have seen it like that if there had been time, but this came upon us before anything had time to work.

In any case I believe that if you had gone on with it, it would have all been discovered by your people—and your's and R's attitude must seem to them so suspicious that they would surely have watched you very carefully.

Whatever you may say or think my dear, I have done my best for you and if I have failed it is because I am not strong enough to bear it. Even now it may kill me. R can easily do so by acting in a certain way, and it will not take much to make me feel that life is not worth going on with.

Of course we all only know things from one side.

I have to come up on Wed. (tomorrow) for the opening of the exhibition, though it is the only money we've got, R thinks it is my duty to. But I don't think I'd better see you dear. I am so weak and not really fit enough just now. I should love to, though.

My love to you, if it is worth having.

Ever yours,
Chris.[10]

It would seem that Kathleen responded to her obvious misery and veiled hints of suicide by proposing a solution that brought an immediate protest from her lover:

Precious life and wife of mine,

Your sweet little note was brought to me by Stelle. I had been lying awake most of the night with a bad throat, longing for you to be there and massage it. So I fell asleep (in Arthur's room) about 5 o'clock, and then overslept. And, funny and contrary person that I am, now that you write and say you will steel yourself to my giving myself again to Chris, I am almost horror-stricken lest it means you love me less. For I want no-one but you; and though I am glad you have given me my entire freedom of action, it is only because I want to keep myself for you only for the rest of my life. I am an animal, my darling, and if you did not love me I could go wandering sexually as I used to do; but you have made me feel the whole of that side of life as the deepest kind of sacrament leading to a union that is mystical—the gate of heaven itself. You must never feel bound to me; but the fact remains that we are most strangely united, and I do not think we need consider the breaking of that union now Chris is so good. She has not even hinted at my sleeping with her again. If she changes I will let you know. In the meantime, you have made her understand that my withholding myself is my own free action, and that is all to the good . . .

I am aching to be with you my darling—oh, how I love you!

Yours, wholly and entirely,

Rutland.[11]

As the situation became more and more painful, Boughton, characteristically, sought ammunition in printed authority:

Dearest Love of my Life, my Kathleen,

By this post I am sending on to you Havelock Ellis's *Essays on Love and Virtue*—and I want you to read at once all of them except, perhaps, the last which doesn't so much matter for our immediate purpose. And then, in the light of our love, our conditions, and what you know of life, make some more definite decision for the future. We both find that whatever ideas of love and renunciation we get when away from each other we only have to meet for them to seem absurd and impossible. But things cannot go on as they are, my darling. Chris cannot bear it, and I cannot bear it. She was very sweet on seeing me again, but says she must either leave me (so as not to be tortured by any personal proximity) or have me again sometimes as her husband; and now I have learned of her sufferings through the Havelock Ellis book, I feel that I must give way.

But the book touches not only on her sufferings in being suddenly and for so long a time cut off from one she loves and yet lives with, it also shows us how deep and holy is the relationship which has been set up between you and me. And now that you and I are wife and husband it is every bit as cruel and wrong for us to be separated, and unless we can be often together we must both get deeper and deeper into the torture of it.

Now you sometimes say you will never live with me, and at others you hold out hopes which you defer—generally for my sake. And I quite realize that it is the beautiful and self-sacrificing side of your nature which hesitates to take a step so serious for many people; but as you will find in Ellis, the self also has its rights, and by ignoring these rights we may *all* live to regret it. I never come away from a time of tenderness and intimacy with you without feeling renewed in power; and it was the time in Scilly which made *Alkestis* possible. Chris realizes this and does not want to stop it now. But unless her love for me can have physical satisfaction she must leave me. And unless our intimacy (yours and mine) can continue we must part, for I cannot live near you and not possess you. I cannot even write to you and work with you without knowing you are mine.

If then you will not go right away with me, will you share me with Chris, either by living with us both in London, or she remaining here, and I coming up much more often, and as soon as possible sharing a little home with you? If you can do neither of these things, my wife, you must let me go and not see me for a long time, till I can bear it. Please do not be angry with me. And know that with you or without you in the body, my feelings, my heart, my thoughts, my music, are only yours; and whatever I can do to serve you at a distance that I will always do. It shall be the chief joy of my life. But if you decide on the hardest course, you must make it easy for me by having nothing to do with me for a year or two. I hope you will understand this, my darling. The thought that you may feel unable to take a course which will make it possible to share a good deal of life together makes this letter the hardest thing I've ever had to do. So forgive anything that may seem cruel in it, and with that sweet and wonderful understanding of yours see through all I have expressed badly, and know me for all my life-days in all ways, your friend, your love, your husband,

Rutland.

And remember, dear one, if what this book says is true, I am your husband as I have never been to any woman.[12]

The idea of a *ménage à trois* seems also to have been canvassed, in some desperation, by Christina: 'If he is happy in loving us both, and can give me back the old comradeship and freedom of body and mind, then we shall all be happy, for I shall be happy in the knowledge that he loves you devotedly too, and shall realize that it is necessary for you both to spend time together.' But it was not to be. By the end of February 1923 Kathleen had written to her parents to announce ('this will come as something of a shock to you') that she intended to live with Rutland as from the autumn—presumably having brought her studies at the Royal College to a neat end.

The letters expressing shock and horror that descended upon her can have come as no surprise. Her mother presented a picture of her lover as an unprincipled lecher, untruthful, and liable to become 'horribly angry' and 'resent any refusal on your part to submit to him in any way'. As for the future, she saw only that 'after a time he will tire of you and throw you over as he has done Christina . . . and end by becoming degenerate'. Her aunt Edith was equally appalled: 'The *awful waste* of all your youth and hope, and the ruin of your career and everything that makes life worth living . . . how can you think of giving up so much for the sake of a man who you must know in your heart to be absolutely worthless!' And then, remembering that Kathleen had borrowed money from her to help with her studies, she aimed a final shot: 'Never think for an instant I could touch a penny of anything you may earn while you are living with this man. I should feel it would bring a curse with it.'[13]

Other people tried other arguments. Steuart Wilson, writing from Petersfield on 10 March 1923, appealed, rather illogically, on behalf of art:

Don't think that any of us are judging you for what you have done. None of us can throw stones. But do let me implore you, before it is too late, to consider again that if you go off with Rutland and stand, in any degree however small, in the way of a reconciliation between him and Chris, you are ruining his chances in England. Let me tell you frankly that people care too much about music to tolerate this sort of irregularity in life from its musicians. You think, I am sure, that you are giving everything *for* Rutland. My dear Kathleen, I wish I could tell you how deeply I feel that it will be the finish of him in England. None of us want to see that for his sake, for your sake, and for the sake of music.

We are not and never will be your enemies, but your friends and we *implore* you to think over again what you are doing, and how it is bound to affect Rutland.

Inevitably the Church was wheeled in—Father James Adderley, now translated from socialist Birmingham to St Paul's, Covent Garden, asking what would happen 'if everyone acted on mere selfish impulse', pointing out that 'the world is in a mess enough as it is', and wondering 'what it would be if pure marriage and a happy family life were eliminated?'[14] Which only brought from Boughton an enraged:

I am amazed at the Pharisaical drivel of the parson who, being more or less bloodless, slops about his mother to excuse the impertinence of his interference with you. I daresay he did love his mother. So do you; but that doesn't prevent your blood being mixed with fire and his with water. Fire burns its way to heaven. Water can only get to heaven by the goodwill of fire, and then only as steam to weep its way back to earth sooner or later. Heavens, what children they all are![15]

Kathleen, of course, was required to leave the Royal College of Music as soon as the seriousness of their intentions became known.

Besides, she was now pregnant and could not expect to conceal her condition indefinitely. With money coming in unaccustomed abundance from *The Immortal Hour* the lovers were able to set up house in Notting Hill (19 Arundel Gardens), and in August they invited Christina and the children to join them on a holiday in Wales. Although it began well, matters soon became tense. Kathleen took to the mountains in quite the wrong way, and in attempting to climb Snowdon brought on a miscarriage. To make matters worse, Christina began to talk of suicide again and was only persuaded not to carry out her threats by Boughton agreeing, at Kathleen's insistence, to a reconciliation. They returned to London—Kathleen to join the chorus of *The Immortal Hour*, and the others to the inevitably futile attempt to recreate their former life together. By the end of the year Christina had to accept failure. She returned to Glastonbury with the children and abandoned her claims forever.

Muddled and sordid as such entanglements always are, it has to be said that the three protagonists behaved decently and honourably towards each other as soon as the dust had settled. Christina continued to collaborate with Boughton on all artistic matters, she and Kathleen remained friends, and none of the children (there were to be eight in all) were ever made to feel at a disadvantage. Boughton continued to pay Flo, his only legitimate wife in the eyes of the law, the agreed proportion of his income (he never failed in this, however hard times became), and Christina—a much more self-sufficient and intelligent woman—found ways of supporting herself and leading an independent existence. But she never quite got over the break, and never quite escaped his spell—few people did. The life she carved out for herself (joining the Workers' Theatre movement, and later marrying a fellow socialist, Tom Richardson, and going to live in France) was broken-backed and unrewarding in comparison with what they had achieved together. And whatever Boughton may have gained from Kathleen in terms of personal solace, he paid handsomely for it in terms of artistic stimulation. Without Christina's cool appraisal to temper his enthusiasm, the chances of his making a fool of himself were now to multiply out of all proportion.

14

1923–1924

Relieved of the need to arrange festival performances in 1923, and buoyed up by the comfortable thought that money was no longer a problem now that *The Immortal Hour* was earning him a living, Boughton was able to devote more time to composition. This he managed to do in spite of his domestic problems, finding detachment enough even during the worst crises to complete two String Quartets: a 'Greek' Quartet in A, and a 'Welsh' Quartet in F. The 'Greek' Quartet was composed in London during the early summer of 1923, benefiting from the unaccompanied choruses based on Greek folksong he had written for the 1922 production of *The Trachiniae*. He dedicated it 'To the memory of Harry Sheerman Hand'. The 'Welsh' Quartet, composed on holiday in Beddgelert during the August of 1923, also has folkish overtones, but is more particularly a programmatic reaction to the Welsh landscape—quickened by his love for Kathleen.

Though perhaps not Boughton at his most convincing, both are works of considerable power and expertise. At their first performance, as part of a series of three chamber concerts in London's Aeolian Hall on 12, 19, and 26 October 1923, he provoked critical reaction by announcing:

> These concerts are NOT FOR HIGH-BROWS but for the
> general musical public who still believe in the common-chord and
> an occasional tune.
> NO FREE TICKETS even for 'the Profession'.

The printed programme for the first concerts offered an explanation of his headline-grabbing challenge:

From the mere musicians point of view, chamber works form much the most important section of musical art. But owing to a general impression that such pieces are only written for musical high-brows, there is but a limited audience for them (on these occasions the audience will probably be extra-limited as I don't intend to admit even my dearest friends unless they pay for their tickets and so help to make a stand against the farce of the London Free-list System). But I cannot help believing that although the primary appeal of Chamber Music must and should be to that small body of professional musicians and amateurs who make music in their private lives, yet remembering the Saturday and Monday 'Pops' and the crowded St James' Hall of my boyhood, I feel that once the right direction is found the subtler appeal of these smaller art-forms may find a more general response than at present, so that finally a greater number of young people may be induced to learn stringed instruments and play such music in their homes. No country can be called musical in which concerted music is not constantly played and sung in the private lives of the people. For chamber music is the natural corrective to the vulgar and dissolute elements in the larger and more popular art-forms. We can bluff a crowd of music lovers into a belief in our musicianship by means of the colours of a large orchestra; we can switch off their musical intelligence from time to time by means of the even more popular accompaniments of opera. Indeed, it is because I am chiefly interested in the dramatic values of music, and to some extent aware of the dangerous ease of stage-effect, that I have hoped from time to time to check a tendency to vulgarity (obvious and in one sense unashamed) by trying to make pieces of music which should be endurable with a minimum of colour and dynamic effect, and quite unsupported by the attractive accompaniments of colour and movement.

He goes on to explain that if the new quartets are liked it may induce his publishers to lose money by printing them, and wonders if it is not high time that a Society be formed for 'the dissemination of such Chamber Music as presents the likelihood of a general appeal'. Why not, he asked, a 'Society for the propagation of Tunes and Common Chords'?

The programme note for the second concert is headed 'A Note for High-Brows and Dead-heads' and discusses the necessity for all human beings to be encouraged to enjoy the arts and not be put off by notions of its exclusivity: 'I confidently look forward to the time when the lives of human beings are so leisurely that the arts fill a very much greater part in the world. In that spirit I weave notes; and therefore object to be classed with the high-brows. I do not think it will happen again'. He then attacks the habit of giving away free tickets to paper the house:

Very few people are taken in by it. You get the sort of audience that listens as to the neigh of a gift horse. . . . The first step in a musical revival should be to put everything on a self-supporting basis. It is not dead heads that will achieve such a revival, but live hearts. And so I hope that before long my offended professional friends will realize that I have not insulted them but, on the contrary, helped them.

Under the heading 'A Study in Musical Criticism' the third concert's programme note tackles the problem of critics head on. Noting that they had been less than kind about the first two concerts, especially where the 'Greek' Quartet was concerned, Boughton explains that it is to be repeated in the final concert in order to give the audience a chance 'to decide the matter for themselves'. Acknowledging the difficulty of making a useful judgement on one hearing (he had been a working music critic himself) he finds it hard to balance the critics' unanimous dislike 'after one hearing' with the judgement of the four artists who had rehearsed the piece 'and still like it'. He ends the case for the defence:

Now let me say how I think the Quartet made so bad a first impression. In the earlier part of the programme I made the very serious mistake of following unhappy song by unhappy song, and further depressing the audience by turning the lights down; so that by the time the Quartet was reached everybody was in a state of the most comical gloom. When listening to music I like to get as near to a condition of twilight as possible; but there is evidently a limit to the aesthetic melancholy the human mind can enjoy. So here's to a happier performance tonight, and if the Quartet is really bad, may it have a still worse trouncing in tomorrow's newspapers. But I shall still believe in it.

And a trouncing was what it got.

By the end of the year Boughton was too absorbed in composition to be much concerned with critical disapproval. The subject that now gripped his attention was singularly relevant to his own domestic situation and the conflict of loyalties he had been grappling with. Probably he was able to resolve his share of the pain by the very act of composing music to express an almost identical emotional triangle—for this is what his new music-drama was: the story of Tristram, torn between the love of two Iseults. In his setting of Thomas Hardy's play *The Queen of Cornwall* he achieved perhaps his finest work.

In later years he recorded the events that led him to choose Hardy's play:

The first I heard of it was in a review by a daily paper. The reviewer was of the opinion that the play was the production of a senile artist. But, fortunately, I chanced to see also a review in the *Daily Telegraph*, and that not only spoke fairly of the play but quoted some of its lines. Those lines made it clear enough that it was the other man, not Hardy, who was senile—young and modern though the reviewer was supposed to be.

Having read the *Telegraph* notice I toured the booksellers of London and at last got a copy of the first edition—a copy which the bookseller had put by for himself. And in an ABC nearby the mood was evoked which, if encouraged, meant music sooner or later. Two days later Herbert Lambert posted to me another copy, with a note asking if it were not the very thing for

Glastonbury, where we had made a persistent stand for choral drama. The only difficulty, so far as I was concerned, lay in the unrelieved grimness of the tragedy. The swift pain of the spoken word found relief enough in the exquisite songs which Hardy had given to Tristram, but when set to music the emotional expression of the bulk of the work would be nearly doubled, and those two songs would not be nearly enough to prevent the feeling from becoming intolerable. It was not a question of making the work palatable for superficial tastes, but the actual weakness which would result from continuously playing on a single series of emotions. My study of the *Alkestis* of Euripides had shown me with what amazing intuition the Greek dramatist had carried each section of his work to its separate climax, and then relieved it and prepared our emotions for the next section by a chorus which, while remaining entirely appropriate, lifted the weight of the tragedy into the less personal and nobler realm of mass emotion. (For, contrary to general belief, mass emotion is a very noble thing when the individuals of the mass are engaged in an impersonal service).

But what had been possible in the drama of Euripides was impossible in this play of Hardy's, wherein the chorus themselves were but ghosts of the common men and women he had taught us to love and to pity in his tales—ghosts, moreover, who are intent on emphasizing and not at all on relieving the tragedy. The only chance of making the play right for musical expression depended on the willingness of the author to reconsider it afresh from a point of view which he could scarcely be expected to appreciate, or on the existence of lyrics of his own which would exactly fit, verbally and emotionally, into those places where they were needed. An unlikely thing, but it happened all the same. I read again his poems from cover to cover and so discovered six which might have been written for The Queen of Cornwall itself. So I approached Hardy and asked if I might make a musical version of the play, if he would be willing for the extra poems to be interpolated, and for certain cuts to be made in the existing text. He suggested that I should call and consult him; and, of course, he proved to be the generous and friendly artist one would have expected from the simple quality of his work. He accepted all my suggestions for the music-drama, and copied into my volume of The Queen those passages which he added, subsequent to its first publication, in case they might be found suitable for the musical version also.

Later on, when the first sketch of the music was finished, I visited him again and played much of it to him. His musical tastes were folkish so he felt near enough to my own work to enjoy the lyrical parts. For the rest, he left it to me. During that visit his energy seemed unending; and in a long drive, during which he pointed out some of the places and houses mentioned in in his books, made it clear that the world he had created, beginning in realism, had to some extent passed over into the world of vision. In the evening he brought out some old music in the hope that we might find a song suitable for the drunken choruses in The Queen. We were unsuccessful, and they had to remain wordless, as I could not induce him to write even four lines of words, nor could the right sort of song be found elsewhere in his work. . . .[1]

After that memorable occasion, Mrs Hardy was able to report that her husband had said 'I think I like that man as much as anyone I have ever met.'

The Queen of Cornwall was first produced on 21 August as part of the 1924 Glastonbury Summer Festival. Wholly in keeping with the harsh, elemental nature of Hardy's play, the music is harmonically more astringent and technically more complex than any of its predecessors. At the same time there is no loss of that power to crystallize an emotional climax in a flood of exquisite melody. Of the three songs that were later published as separate items, Vaughan Williams was to write (10 January 1926):

I want to tell you that I admire your Three Hardy Songs very much and especially the middle one. I rather believe that that kind of music growing out of those kind of words is what will go down to future generations as the best type of our generation and country. Probably a Frenchman or a German would hate them, and I rather hope he would.

For the most part, critical reaction was favourable. Typical is the account printed in the *Manchester Guardian* on 22 August, but written on the previous day:

No one who has witnessed to-night the first performance of Rutland Boughton's new music-drama The Queen of Cornwall, described by Mr Thomas Hardy, its author, as 'a new version of an old story', can have failed to notice that there is much more dramatic feeling and at the same time much more tenderness in the musical version of the play than there are in the original one-act play. In the first place Mr Boughton's music, for the most part bare-boned and astringent, intensifies and prolongs the quickly-passing moments of tragedy and makes exquisitely mellow and gracious the lines that lend themselves to such treatment. Secondly, the libretto of Boughton's music-drama is, so to speak, a revised version of the original play published last November . . .

The altered version is a perfect libretto for many reasons, not the least of which is that the prolonged tragedy is broken here and there only that it may rise to greater heights. The music-drama can be staged either in two acts or given completely in one. Mr Boughton must be congratulated on his gentle daring in asking Mr Hardy to alter a work that the author must have regarded as finished and complete . . .

A composer who takes so much care over his libretto is not likely to be found wanting in his music. Tonight, as is always the case at Glastonbury Festivals, the music for the orchestra was played on the piano alone. . . . Mr Boughton played from the vocal score, but his experiences at Glastonbury have taught him to play orchestrally. There are many passages it would seem only his hands can play. His music is closely knit, saturated with that kind of psychology which the leitmotif system has made possible, and full of detail which when played on the piano must be guessed at rather than perceived. As the time covered by the actual events corresponds closely to the time of representation, the music is written with severe economy in regard to voiceless orchestral passages which, when they occur, are merely employed to further the action.

The musical characterization is so swift that on the appearance of a new character on stage he is at once stamped on the mind by the music that is associated with him. The relationship between King Mark, Sir Tristram, Iseult Queen of Cornwall, and Iseult the White-handed, though clearly defined by Mr Hardy, is made still more clear by music which seals King Mark and Tristram with a doom that cannot be evaded.

Mr Boughton's music almost glitters with psychological insight, but at first hearing one requires a quick mind to perceive even one half of the subtlety that binds the characters in past and present events, binds them and hurries them to their violent end. The lyrics, sung in a kind of deathly peacefulness, sear the heart by their tenderness. The beautiful moments of dreamy passion are woven with wonderful art into the texture of the tragedy, and the stark quality of the libretto is enormously increased by these short moments of ecstacy. It has also been hinted by the previous music that Tristram is to die, and perhaps Mark also. And the heart is wrung when Tristram sings to the Queen of Cornwall as though death were a hundred years away.

Hardy himself had a high regard for the music, speaking of it as 'a glorification of the play', and was particularly interested to find that it had made his Queen of Cornwall a more sympathetic figure than words alone had been able to do. He attended the second performance (22 August) with his wife and a shadowy figure who insisted on being elaborately concealed—'Though everyone', wrote Mrs Hardy, 'seems to know about him.' Backing steadfastly into the limelight, it was not difficult to know about T. E. Lawrence. Boughton's reminiscences of Thomas Hardy are especially poignant in that they might equally serve as a description of himself in old age:

He was as blithe and ruddy and active as a winter robin. His simple nature was one more example of Shaw's statement that 'greatness is merely one of the sensations of littleness'. There was no pose in Hardy. For him life was more important than art. His art was not a refuge from the woe of the world, but the battle-plain of a courageous spirit.

In spite of the breakdown of their marital relationship Christina continued to design for Glastonbury. With *The Queen of Cornwall* she achieved one of most remarkable effects, clothing the chorus of ghosts so that they would blend in with the walls of King Mark's castle. This, and the imaginative music, made it seem as if the story was rising from the age-old stones themselves, borne on the wind and sea down the uncounted years.

A new feature of the 1924 festival, which, preceded by the usual Summer School, ran from 9 to 30 August, was a cycle of short plays by Laurence Housman for which Boughton wrote a considerable amount of incidental music. One of the plays, *The Seraphic Vision*, he set complete, so that it stands some way between choral cantata and miniature opera. Housman, a close friend of the Clarks, had become an enthusiastic supporter of the Glastonbury scheme and rightly saw it as a suitable atmosphere for his *Little Plays of St Francis* which, after this year, became increasingly part of the festival.[2] The only other novelty was a belated first performance of *Agincourt*, the setting of Shakespeare that Boughton had made in March 1918. Lectures, with special reference to 'The Arts of the Middle Ages', were given by G. K. Chesterton ('The True Moral of the Middle Ages', 6 August), Laurence Housman ('The Pre-Raphaelite Spirit in Poetry and Painting', 9 August), H. J. Chaytor ('The Troubadours', 11 August), William Poel ('The Drama of the Middle Ages', 13 August), Sir Richard Runciman Terry ('The Music of the Middle Ages, 16 August), and Walter Raymond ('Somerset in the Middle Ages', 20 August)—altogether a remarkable series and admirable testament to the regard in which Boughton's festival was now held.

Some time before the summer meeting took place, however, it was decided to abandon any attempt to preserve the original pattern of the year's work and concentrate instead on building up a touring company, working on a co-operative basis, that would use Glastonbury as a centre from which to cover the West Country. Thus, in August 1924, the Glastonbury Festival Players were registered as a Limited Company with Rutland Boughton, Christina Walshe, Penelope Spencer, Laurence Housman, and Frederick Woodhouse as its directors. Salaries were to range between £4. 10s. and £10 a week. After these and other expenses had been deducted, each player might expect an agreed additional proportion of the profits: chorus members rating one share, principals two, and the Festival School itself one share for the use of its wardrobe, scenery, and props. Boughton was to receive one share during those weeks when he was not entitled to royalties from the performance of his music. From time to time, individuals would come together as 'The Glastonbury Festival Singers' in order to present small-scale recitals.

It was a scheme that offered positive hope for the future. All that was needed to make it work was a lively box-office, and that did not seem an impossibility. Boughton's music was beginning to pop up all over the place. *The Immortal Hour* had reached Sydney in November 1924 (Melbourne had to wait until 1932), and was about to be presented by John Tobin in Liverpool's 'David Lewis Theatre' (16 January 1925). Productions of *Bethlehem* had taken place in Streatham, Banff, Buxton, Keighley, Wakefield, and East Grinstead in 1924 and 1925, and *The Queen of Cornwall* was on John Tobin's list for January 1927. There were plans, too, for performances abroad. In 1925 the Berlin publisher Ahn und Simrock commissioned Dr Walter Jansen to make a translation of *The Immortal Hour* (*Die Unsterbliche Stunde*) and then hawked it around the German opera-houses—only to have it turned down by Dresden, Stuttgart, and Darmstadt. Early in the same year came approaches from the Province Town Theatre (February), the Greenwich Village Theatre (March), and the Cherry Lane Theatre (August) of New York, also for *The Immortal Hour*, while on 19 October Egon Wellesz wrote from Vienna to say that he had 'received the score of your splendid "Immortal Hour" and [would send] it to the Theater of Gera, where the Prince of Reuss interests himself very much for new, important works.' Even though these also failed to mature, it was clear that Rutland Boughton was the British opera composer of the day—as sought after as Balfe had been in the nineteenth century and not to be eclipsed until Britten made his debut.

At the reunion meeting on 26 February 1924, the Glastonbury Festival supporters had been told that Laurence Housman would take part in all future festivals. By the middle of the year he had set up house with his sister Clemence (in Longmeadow, Street) and was in a position to play a full part in Glastonbury activities. The formation of 'The Glastonbury Festival Players' as a Limited Liability Company changed the pattern of events. Touring now took precedence and festivals were to be held in the summer only.

The first of the tours took *The Immortal Hour* and *The Queen of Cornwall*, plus a concert of songs and dances, *The Moon Maiden*, and *Agincourt*, to Bath (16–18 April 1925), where they played to full houses, and then to Bournemouth (20–5 April) where Dan Godfrey's orchestra was again placed at their disposal and it was possible to hear the new opera in its true colours for the first time. Hardy and his wife attended the first performance on 22 April and, if anything, were even more impressed with what Boughton had achieved. The Bournemouth visit, which included three performances of both main works, was carried out at a cost of £917. 17s. 0d.—the loss a mere £85. 5s. 6½d.

The 1925 Summer Festival came as near to being 'Arthurian' as any

Figure 3   Festival programme 1925, by Christina Walshe; and incorporating the design of her 1914 'futurist' poster; black on white

Figure 3   Festival programme 1925, by Christina Walshe; and incorporating the design of her 1914 'futurist' poster; black on white

that were ever presented. Again preceded by a Summer School, it ran from 27 August to 5 September and consisted of four performances each of *The Immortal Hour* and *The Queen of Cornwall*, five of *The Round Table*, and three of *The Birth of Arthur*, together with two cycles of Housman's *Little Plays of St Francis* ('The Revellers', 'Fellow Prisoners', 'The Bride Feast', 'Our Lady of Poverty'; and 'The Builders', 'Sister Clare', 'Brother Juniper', and 'Sister Death'), each performed on four occasions, thus making a grand total of twenty-four staged performances. But it was the last time Boughton took an active part in any work at Glastonbury. The last festival of all, in August 1926, consisted entirely of Housman's plays.

The gradual decline in Boughton's contribution to the Glastonbury Festivals was due in part to the shift of emphasis to the touring aspect of the Festival Players, in part to the domestic upheavals which left him unwelcome in certain circles, and in part to the increase in his other activities: some musical, some, as we shall see, political.

Musically it was a very busy time. Besides the triumphs of *The Immortal Hour*, *Alkestis* was given at Covent Garden in January 1924 by the British National Opera Company. Illness prevented Percy Pitt from conducting the first performance on Friday, 11 January, and Boughton had to take over at the last moment. He was not surprised. One night, some forty years before, he had dreamt he would conduct his own opera in the nation's opera-house.[3] The fact that he was now doing so seemed only natural. The two Covent Garden performances were followed in March by a tour, conducted by the young Anthony Bernard, which played in Birmingham, Leeds, Bradford, Manchester, Glasgow, and Edinburgh. The London revival, at His Majesty's Theatre in July, was generally thought to be an improvement on the original production. But *Alkestis* was not a commercial success. Its austere beauties told against it in the plush of Covent Garden, and even, to some extent, in the more intimate atmosphere of His Majesty's. This was as Boughton expected, indeed hoped. For he had suggested this work rather than any other, confident that it was not likely to attract the unthinking adulation that had marred his feeling for *The Immortal Hour*.

Before letting the BNOC take the work on tour Boughton rescored it for a slightly larger orchestra, acknowledging that his original scheme had been a miscalculation. Perhaps Shaw's comments (12 January 1924) were the catalyst:

Alkestis came off with much more of the Glastonburian success than I should ever have supposed possible at Covent Garden. It was very lucky you had to conduct it. The end, being essentially orchestral, was better than at Glastonbury.

I was too close to the stage to judge whether Kathleen was hitting hard enough to make herself felt at the other end of the house. She did not betray the least misgiving herself. Alkestis failed to make the full effect of the return from death, because her ridiculous limp veil and gown made the swathed mummy rigidity of Astra [Desmond] impossible. That can be remedied.

I have no business to meddle with you technically; but the score won't do. Your style is not the pre-Mozart XVIII century style; and how much contrast could Haydn or Gluck have got without bassoons, or Mozart without clarionets? Wagner had to treble his woodwind to get tremolo from the flutes, oboes plus cor anglais, and clarionets plus bass clarinet; and the public and the grand opera conductors have become accustomed to the richness and variety of this. I have just heard, on top of *Alkestis*, Bax's new symphony in E flat. Bax has complete four part harmony in every tone colour, 3 flutes and bass flute; 2 oboes, cor anglais and heckelphone (bass oboe); 3 clarionets and bass clarinet; and 2 bassoons and 2 contrabass sarrusophones. You may call it a full band in flutes and single reeds, and a full band in double reeds.

Now in so far as these instruments are used only to make a devil of a noise I am all for the reaction to small orchestras, as Rossini and Beethoven (and Offenbach) could make as much noise as anyone can stand, and Stravinsky can do no more. But the reaction for its own sake is nonsense. If you symphonize your opera in the Wagnerian manner—and you do this—you need woodwind triads: bass clarinets and English horns and so forth, to get the necessary fullness without mixing up the colors of the single and double reeds. And the reason you mustn't mix is that you need characteristic instrumentation.

This bring me to *Alkestis*. In it you start with an exceedingly sensational contrast of a white god and a black one, of Apollo and the King of terrors. It demands an orchestral contrast compared to which the Mime Wotan contrast is Haydnesquely ancient. Death should be all double reeds and muted trumpets, and Apollo all single reeds and open pipes. But you have made no contrast to tone color at all. Both of them overwork the horns and the trumpets in just the same way. It was by the horns getting on my nerves that I first realized the stupendous and incredible fact that there were no bassoons. I said 'Either there is a strike on, or Rutland has gone mad.' Later on, the explanation appeared. You were saving up one bassoon to characterize Hercules was in the tenor register, and would have been far more appropriate to Admetus. My ears, accustomed to Strauss's Sancho Panza, itched for a cheerful roaring tuba (an ophicleide would be better if an instrument and a player could now be got) and the penny whistle with which Handel made Polyphemus enormous. You sacrificed the bassoons for less than nothing.

I admit that by saving up the drum you made it as surprising as if I had never heard a drum before, but you could probably have done as much by reserving this particular effect.

Now look at it from the sordidly professional point of view. *Alkestis* will make its way to the German opera houses and to New York. The conductors will not believe their eyes when they see the score. They will write in 'additional accompaniments' as Wagner did to Spontini. The economy will not appeal to them, because they all have to provide for Wagner, bass clarionet and English horns and third flutes are all in a night's work with them. A heckelphone and a contrabass sarrusuphone would make them feel that you are 'in the movement', and would be really useful for Death and Herakles. And what are three trombones without a tuba?

In short, you must rescore for foreign consumption on business grounds, and for Covent Garden on artistic grounds. It would not take you so long.

I believe Glastonbury, with its piano accompaniment, so horribly broken into by that military episode, has atrophied your orchestral sense, and made you a victim of the cornet in B flat. Awake, arise, and let the sarrusophone to the heckelphone speak.

In haste, ever
G.B.S.

He was, of course, mainly indulging himself with musical banter of uncertain authenticity, but there was a grain of truth in his criticism and Boughton duly took heed.

15

1924–1927

From early in 1924 Rutland Boughton was much in demand as a choral conductor, contriving at one and the same time to serve such apparently irreconcilable groups as the Civil Service Choir and the London Labour Choral Union. The Choral Union came into existence in January 1924 when Boughton and Herbert Morrison (elected to the House of Commons in 1923) combined their energies in the belief that music-making among the workers could be an invigorating element in socialist politics and culture. It followed closely on the pattern established by the Clarion Choirs Boughton had worked with in his Birmingham days. These had indeed proved a vital contribution to socialist culture, until the Labour Party decided, in its wisdom, that the slogan 'Let us Work and Sing our way to Socialism' had had its day. On 30 November 1924 Boughton had the pleasure of conducting the massed choirs at the London Labour Party's 'Great Rally for the Daily Herald' in a concert that began with the 'Marseillaise' and ended, by a steady progression through Edward Carpenter's 'England Arise!' and Parry's 'Jerusalem', with 'The Red Flag'. Quite what Elgar's 'My Love dwelt in a Northern Land' or Boughton's 'Pan' had to contribute to the meeting's revolutionary fervour is not altogether clear, but a good time seems to have been had by all.

And when he was neither conducting nor supervising performances of his own music, Boughton was busy giving lectures all over the country—lectures on music, aesthetics, politics: anything, in fact, that seemed important to his combative mind. Leisure moments he filled up by acting as music critic for the *Daily Herald*. Inevitably he came to blows with his editor—this time because he attacked Covent Garden's policy in bringing over foreign singers at the expense of native artists. He pointed out that the orchestral players had retained their right to take precedence over foreign invaders only because they had been organized strongly enough to make resistance effective. He argued his case at length and on every occasion, drawing at last, on 24 June, an exasperated plea from the editor, Hamilton Fyfe:

Will you *please* write about music? That is what the readers want. We get quite enough Trades Unionism in other parts of the paper! I cannot have long screeds about it on page four. That is a general reading page.

Your articles on music are really valued. The enclosed would merely annoy those who are looking for musical instruction. I'm sorry, but I know I'm right.

Boughton, of course, refused, and refused again:

No my dear Fyfe, I do *not* mean that I'd rather not write for you. I mean that if I'm not allowed to write on the economics as well as the aesthetics of art I should, as things are today, soon become what most of the other poor devils of critics become—gasbags.

I'll wager that you love literature more than journalism; nevertheless you are using yourself up on political journalism instead of writing books because you know at this juncture of the class war no true literature is possible unless related to the Socialist fight.

His uncompromising response to Fyfe's perfectly reasonable appeal is typical of his determination to pursue his convictions no matter what the cost, either in material or personal terms. To Boughton, Fyfe's restraining hand appeared only as another example of the official Labour Party's willingness to compromise on all-important issues. And when, at the end of 1925, twelve of the Communist Party leaders were arrested on a doubtful charge of seditious libel he felt obliged to declare unequivocally in their favour. He explained his position in a letter which the *Worker's Weekly* published on 12 February 1926. It set out his reasons for joining the Communist Party in the following terms:

Because I believe that the only hope for a decent world lies through its organized control by those who do the work.

Because, while I believe in the delegation of governmental work (foremen in factory or minister in Whitehall) to suitable individuals, I do not believe in individual political leadership.

Because the Communist Party seems the clearest as to its objective and the most sensible, straightforward and courageous in its methods.

He then proceeded to outline his position as an artist:

I am a musician who is continually robbed of the real benefits of my work, and at the same time offered charity. So let me develop my standpoint a little more fully:

There is no room for the fair appreciation of beauty until all human beings are properly fed, clothed, housed and educated. That will never happen unless the workers themselves govern.

The arts can only grow under conditions of freedom for all. The human mind cannot be free while the tiresome old body has to worry about its needs. Art needs spare time and lots of it. . . .

Wage slavery will never allow the spare time and the personal consideration necessary if the arts are to be cultivated properly. It offers no certainty of money to spend upon them, and robs men of the freshness of body and mind which are necessary if the world of beautiful things and thoughts is ever to be properly explored.

True, the Capitalist system of today allows a few artists to make things for the enjoyment of the masters and their associates—for bored wasters with private incomes, for the hypocritical mysticism of the churches, or for the speculative stupidity of musical commerce. And it is cute enough to provide a certain amount of artistic dope for those of the middle-classes who live near the edge of the proletariat.

This narrowing down of the creative spirit of art reduces the artists themselves to the position of parsons and footmen. And so I was pushed into Socialism as the constructive theory opposed to the steady and increasing destruction caused by the irresponsible control of capital. But because State-Socialism seemed in many ways as careless of individuals as a Limited Liability Company of its workers, I called myself a 'Communist'.

Pointing out that he was obliged to reconsider his position when the Communist Party of Great Britain was formed shortly after the war, he refrained from joining it because of the discipline involved. Now, he says, it is obvious that it is this very discipline that has enabled the Party to stand firm, while the Labour Party merely dithers helplessly.

The core of his letter comes in a paragraph which reveals the passionate humanist, trusting that other men have the same noble aspirations: 'The real leaders of Communism are those men who have led us to it by seeing parts of it from afar—men like Plato, Jesus, More, Ruskin, Marx, Shelley, Morris, Hardie, Lenin, Upton Sinclair, and Bernard Shaw.' And finally, after excoriating the leaders of the Labour Party for their indecision and occasional tendency to 'betray the workers by sabotaging proletarian action', he concludes:

One of my strongest reasons for trusting the Communist Party is because it is not silly enough to pay its political agents a wage which removes them from the proletariat. The Communist Party officials prove their honour because they are pledged to poverty and danger so long as the rest of the workers are in poverty and danger.

I do not think that it will necessarily give clever service to the workers. But it will give honest service. And, as Bernard Shaw has shown, it is possible for the cleverest leader to rise into the clouds on his own wit. The workers of the world, as they grow conscious of their equal humanity, will insist on running the world for their own benefit. Their leaders need be no cleverer than the rank and file, so long as they are trained for their jobs and are of proved honesty.

One other reason for joining the Communist Party: I have so much to learn.

The C.P. is apparently the only party that will acknowledge its mistakes; it is therefore the only party that is fit to teach.

The editor of *Worker's Weekly* felt obliged to preface Boughton's declaration of faith by disclaiming complete agreement with such 'characteristically unorthodox opinions', but nevertheless extended a cautious hand of welcome. The rest of Fleet Street greeted the news with delight, revelling in the thought that 'Comrade Boughton' had now come into the open and would, henceforth, be a sitting target for snipers. Shaw was not impressed:

6 March 1926

My Dear Rutland,

No: I am not the anonymous donor the £100. I don't hold with the Communist Party. I happen to be a Communist; but as far as I can make out the party's speciality is not Communism, but Coup d'Etat-ism, the coup d'etat to have no ulterior aim, but to be struck for its own sake as a sort of Marxian ceremony, and also to be announced beforehand in the manner of Mr Winkle, who took off his coat slowly declaring that he was just going to begin. By sensibly omitting this preliminary, Mussolini, General Primo di Rivera and Lenin got away with their coups d'etat, the circumstances being favourable. Our coup d'etatists, having piously insisted on it in wildly unfavourable circumstances, got twelve months, which served them right for their hopeless failure to measure the situation.

The Labour Party made a serious mistake in expelling them as Communists. It should have simply imposed a test for membership, like the signing of the Fabian basis, which would have excluded all the coup d'etatists without repudiating Robert Owen, William Morris and myself.

There is something pathetic in hearing these Clydeside men, all quite genuine in their social indignation and revolt against the capitalist world, spouting absurd, muddled translations of Marx's 1848 German in the firm persuasion that they are uttering important economic and historical truths of the most advanced school. One man whom I heard doing this presently obliged the company with a song, and turned out to be a vocal genius with a rare basso-profundo. I asked him why he insisted in talking obsolete foreign stuff that was neither English, nor German, nor good broad Scotch, and had absolutely no meaning for his audience, when he could sing so well. All this Marxian resurrection pie is pitiable, almost tragic. They must be laughed out of it. Even Trotsky, a writer of splendid ability, and a powerful realist critic, makes the most appalling gaffes occasionally because he really believes what Marx was struggling to formulate in his academic way 75 years ago, and that the phase of capitalism (employer's capitalism) that confronted him, are still not only fresh, but startling and ultra-advanced and that Webb is a belated Thiers, Wells a petit bourgeois Volney, and I, who passed through this

Marxian phase forty years ago, and had it so acutely that I know the symptoms thoroughly, a benighted farceur to be spared the epithets he applies to
Macdonald only because I am so sound on the subject of Trotsky's genius,
and stood up for the Soviet right through its worst ill fame, until I let
Zinovieff have it with both hands as a pretentious and mischievous noodle
(one of Z's feats having been the rustication of Trotsky).

When you intimated that instead of composing you intended to devote
yourself to a polemic against me, I was out of all patience; so much so that I
had no time to tell you off in black and white. Fortunately you appeased me
by sending me that new Whitman setting, which seemed quite healthy as far
as I could judge by reading it away from the piano. I was away in Cornwall
for weeks. When I play it over and sing it, I shall let you know more about it.

One Kathleen Davis writes to me from Harrow to speak there. How is *your*
Kathleen? Chris also wants to see me about something—not you this time,
some journalistic stunt. She implies that I owe her reparation for your mis-
deeds, which I consider rather hard on me.

I note that as you know a hundred times as much about music as I, and I
know a hundred times as much about Communism than you, you listen with
the greatest docility to my views on orchestration, and are eager to undertake
my reproof and instruction on Communism. Human, human, all too human!

Ever,

G.B.S.

Politics dominated the whole of 1926, and the General Strike, which
Boughton in common with many others at first believed to be the dawn
of revolution, provided the climax. Eager for the barricades he placed
himself at the service of the Party and was promptly put in charge of
'amusements'—though exactly what form they were to take was not
made clear. In the end he was forced to amuse himself by making
brass-band arrangements of 'The Red Flag' and the 'Internationale'. He
did, however, take part with Kathleen in a number of fund-raising concerts, and had the pleasure of seeing his 'Song of the Labourer' as the
centre page of Lansbury's *Labour Weekly* on Saturday, 13 February. 'A
New Song by Rutland Boughton', exclaimed the headlines, but it was
not. It was at least twenty years old and scarcely calculated to stir the
revolutionary heart. The Comrades would probably have done better
to put their trust in 'Irving's Yeast-Vite Tablets', whose fulsome advertisement occupied pride of place on the paper's front page.

Much of the summer was given over to the composition of a workers' ballet which he called *May Day* and for which he devised his own
scenario. It was to make clear the exact nature of the capitalist conspiracy and thus be a welcome addition to future Hyde Park rallies.
Unfortunately no one seemed willing to come forward with money

enough for a stage production. The work received only one performance—on 2 March 1929, as an orchestral suite under the less than
revolutionary skies of Bournemouth and the aegis of the imperturbable
Dan Godfrey. It is not a work of any value: the music means little
without the scenario, and that is a bundle of painful naïveties.
However, the feeling that prompted Boughton to write it cannot be dismissed quite so lightly. Of all English composers he was the only one
who thought it important to declare any sort of understanding and
sympathy with the events of those tragic days.

Even so, it is instructive to observe the difference in the response to
the challenge of creating a socialist art between Boughton, and those
few of his compatriots who thought it important to commit themselves
(Alan Bush, for example), and their German contemporaries.
Admittedly such composers as Kurt Weill, Hans Eisler, and Paul
Dessau were that much younger than Boughton and had had to
weather fiercer political upheavals than had ever been visited upon
Great Britain. They were, moreover, fortunate in being galvanized by
such stimulating authors as Bertold Brecht. But the British response,
almost pastoral in its mildness, lacked any kind of edge. A glance at
the *Twelve Labour Choruses*, edited by Leonard Pearce and Alan Bush
and published by the Independent Labour Party in 1930, to which
Boughton contributed two original songs and two arrangements, is
sufficient to explain why their cause was doomed.[1] Instead of the burning venom of Brecht, the British socialist is urged to action with such
soporific exultations as 'From out thy drugged and ancient sleep,
Democracy, arise!'[2] Nor does the music—a turgid four-part setting in
the manner of *Hymns Ancient and Modern*—do anything to offset the
gentlemanly pallor of this futile call to arms. Small wonder that the
revolution never came.

One expression of Boughton's socialist sympathies that, for a while
at least, had teeth was the Workers' Theatre Movement. As early as 11
June 1925 he had spoken in favour of a co-operative drama movement,
organized along the lines of the London Labour Choral Union. It seems
that an early version of what eventually became the Workers' Theatre
Movement took root in 1925 with Christina Walshe and Rutland
Boughton on its committee. It mounted plays by Upton Sinclair
(*Singing Jailbirds*, October 1925), and the Fifeshire miner Joe Corrie (*In
Time of Strife*, March 1926), but by 1928 had become more or less
moribund and ripe for transformation by Tom Thomas into 'The
Workers' Theatre', and eventually into the Unity Theatre. Boughton's
involvement appears to have been fairly superficial, and probably
undertaken more out of deference to Christina's enthusiasm than anything else.

In any case, he was much in demand for all kinds of public occasions. Not all of them were to his liking—as when in October 1926 the *Daily Express* asked if he would join the organizing committee of a 'National Community Singing Movement' to be launched at the Albert Hall on 20 November. His response was prompt and stinging: 'How can the Workers of Britain sing on empty stomachs? And how can the Slackers of Britain sing with the knowledge that there is so much suffering among those who do the work for them?' The editor did not reply.

Altogether the events of 1926 were a disappointment. The General Strike was effectively undermined; and revolution, if it was ever to occur, was postponed indefinitely. Boughton was mortified:

Let us face the fact. We were not equal to the event. We failed to take the power which paused before us. The evening of May 4th 1926 found a government which had lost its powers, which dared not use its military forces and was so unsure of its police that it grabbed schoolboys to act in their stead. That government was faced with the 'leaders' of the people, the officials of the trades unions who ran away from the victory their men had won. And though they tried to infect the masses of their people with their own fear by means of their contemptible news-rag, even then the men stood firm and solid, waiting but for the words which we Communists should have spoken. We did not speak. Some of us were in prison, for the government knew its real enemies. But surely some of us were free to say the word our men were waiting for— those brave, straight, decent men who, with their families, faced starvation and worse in the hope of a better world to follow. Of course they will never trust their ridiculous trades union 'leaders' again; but if they are to learn to trust us we shall have to learn to act on the British event, and not go on stammering the Russian theory. The Russians did not win by the literal theories of Marx, but by revising those theories in the light of the Russian reality they call Leninism. We too must revise Leninism, do our own thinking and not hope for a Bolshevik angel to help us.

There is more than a touch of arrogance in Boughton's pronouncements at this period. He was a famous man and believed absolutely in the validity of his theories about life and art. These he expressed forcefully and with something of the manner of an Old Testament prophet. What he did not know was that he had reached the zenith of his public career and had already begun the slow descent into comparative obscurity.

It would not have appeared so at the beginning of 1926. *The Immortal Hour* commenced its second successful run in London on 30 January, this time at the Kingsway Theatre. The critics were still kind—though one (in the *Sunday Times*, 7 February) went so far as to suggest that Miss Ffrangcon-Davies's other-world gestures had become 'a little too mechanical' and that he feared that 'if she does this or that only once more with her hands and arms [he would] get up and scream'. But his reservations were felt only by a minority. It was still convincing for the author of *Elizabeth and her German Garden* to have the hero and heroine of her 1925 bodice-ripper, *Love* (or *I Never Should Have Done It*), meet as a result of their obsession with the work.[3] And letters of appreciation continued to pour in—sometimes from surprising quarters, as when Armas Järnefeld wrote to Boughton's friend Miss Agnes Thomas on 23 February 1926:

I have just spent an immortal hour at the Kingsway theater and found it beyond all my expectations. I have seldom heard such a charming musik, and the change or substitution of the old opera manner of moving hands at random for a conventionalized style is very happily found and has in my opinion a great future. For this I liked particularly Miss Gwen Ffrangcon-Davies. I love her. Much was lost for me for the sake of my bad understanding of English poem language, but the music made clear to me that here was a strong touch at the motive, which always has on my mind the relation between sensuality and divineness in sexual love. The music was *wonderful*, and I shall write to Sibelius about Rutland Boughton in order to transmit my enthusiasm. I wonder why this great composer is not more widely known abroad?[4]

Boughton, by now fairly convinced that the world had ways of ganging up on him, would doubtless have been happy to provide an explanation.

His perennial suspicions received a considerable boost with the events leading up to the New York production of *The Immortal Hour*. This opened on 6 April at the Grove Street Theatre in Greenwich Village, a house specially designed for intimate opera and built for The Opera Players, Inc. It was their first production, and almost their last. The critics attacked the work with such glee that it seemed as if their malice had been orchestrated—Boughton supposed by foreign opera interests intolerant of a British effort. After the end of the run, Enrica Clay Dillon, who was responsible for the production, wrote:

Of one thing let me assure you: that the Opera Players, despite everything that may have been said by the critics, are your devoted slaves. They have a deep reverence for you as the writer of this work that would touch you deeply.

The Hour, as it went on from evening to evening through the whole four weeks of the run, gained so many friends and ended in such a blaze of glory that your heart would have been satisfied had you been with us on that last evening. . . . We would have kept it on had not the financial losses been so immense, and I am convinced that had the Board of Directors seen fit to lose a little more we would have repeated the history of its success in London.

There is no doubt that the critics killed the box office. The essential thing is that people here are beginning to wake up to the fact that The Immortal Hour was an exquisite, beautiful thing and that something has happened in New York. . . . They took it for granted that a new Theatre and a new Company meant youngsters, amateurs, inexperienced people—anything you like. To add to this situation a new opera was more than the venerable fathers could endure. This little village of New York cannot conceive that anyone should dare to give opera outside of the Metropolitan Opera House.[5]

She went on to explain that the orchestral players were so incensed that they wanted to bring an action against the critics and wished to testify that 'Rutland Boughton is one of the few great writers in the world today' and they consider *The Immortal Hour* to be 'a great work'.

Her partisan views were understandably and eagerly swallowed by Boughton, who does not seem to have set the same store by a retrospective review that his press agency sent him from the Baltimore *Sun* (18 April 1926). The writer, one Henrietta Straus, had seen the opera in London and understood the importance of a thoroughly professional production if so delicate a work was to make anything like a true impact:

Such an enterprise, therefore, as the Opera Players had everything in its favour and yet, by the time the curtain fell, one merely thought of it, if one could think of it at all, as an opportunity wasted. For the curtain fell on a performance so hopelessly amateurish that it was impossible to take it seriously. The soloists had neither voices nor ability to speak of. The chorus—the best feature of the score—left one merely grateful that it had hung together; while the score, itself a tenuous imitation of Wagner thinly coated with Celtic 'atmosphere', lost whatever purpose it had under the baton of Alberto Bimboni, who gave it a frankly Italian reading that drowned its subtleties along with the voices of the singers.

This should have been enough to alert Boughton to the possibility that his music had been treated as if it were Puccini, and suffered accordingly. But unfortunately the production had been attended by all manner of difficulties, and some of them looked remarkably like wilful obstruction. The business of supplying vocal scores and orchestral material had only been completed after endless delays. Neither Stainer & Bell, who handled the Carnegie Collection material, nor Ricordi, their agents in New York, seemed able to cope with what should have been a routine matter. Rehearsals were therefore delayed by several weeks and the Opera Players forced into spending money they could ill afford. Eventually the material was delivered, but only after a series of infuriated letters and telegrams had winged their way over the Atlantic, and by that time Boughton had lost all confidence in his publishers and was ready to believe the worst.

It now seemed obvious that if capitalist society was as corrupt as communist dogma would have it, it was also capable of conspiring to deprive him of the means of making his living and voicing his opinions. Once the idea had been planted in Boughton's mind it struck deep roots and he became clamorous in making his suspicions known. It was almost as if he had decided on nothing less than martyrdom. His disgust at the treatment of the miners after the General Strike was to give him his chance.

He had not been involved in the 1926 Glastonbury Summer Festival, which had been devoted to two cycles of *The Little Plays of St Francis* and five performances of the Housman/Granville Barker play *Prunella*. Housman then went on to tour the works from Bristol to Newcastle upon Tyne, and even managed to make a profit of some £300. The winter tour of *Bethlehem*, however, was Boughton's. It was scheduled to begin on 6 December in Bristol and end with a two-week run (20 December–1 January) in London at Church House, Westminster. Boughton decided to drop the traditional costumes and decor for the London performances and substitute a contemporary setting. Christ was to be born in a miner's cottage, and Herod, supported by uniformed police and soldiers, was to appear as the cigar-smoking, evening-dressed embodiment of capitalism.

Unfortunately his plans were well advanced before his fellow directors realized their significance. Naturally they were disturbed and annoyed—the more especially because Boughton insisted on his absolute right to do as he pleased with his own work, regardless of whether the Glastonbury Players were supposed to be a corporate, co-operative body or not. Politics, they argued, were a personal matter. But Boughton refused to accept that he had done anything that could possibly deserve censure. The Board met on 6 December, and after a considerable amount of discussion he got his vote of confidence. But, as Housman recorded, his victory was won only at the price of virtual blackmail:

I came away from the Board feeling that, as at present constituted, it had not the will to stand against you when differences of opinion so grave as to involve your possible resignation came up for debate. And that brought me face to face with the weak point in my own position—that, while feeling bound to oppose you because of the risk involved, I did not think that I or anyone else could replace you if you resigned.[6]

A letter from Kenneth Curwen, written immediately after the meeting, suggests that Housman was right in attributing to Boughton almost hypnotic powers over his fellow directors:

I have never felt so much admiration for you as today when I looked through the door afterwards and saw you at the piano. I felt in a certain degree as if I had let you down in not finding a formula that they would agree on; but I believe it was better that each should have to realize exactly where he stood.

His proposed solution was as devious as any communist or clerical campaigner could have wished in the circumstances:

I believe it will be wiser that the Church House people should hear from you rather than through the press what is proposed. The formula is a little important: 'Returning to the original method of the mystery, it will be presented in contemporary costume' lets it sink in gradually perhaps. I don't think it really matters.[7]

The news, of course, did leak out and was so elaborated by the press that the Church House authorities took fright and were only calmed down by G. K. Chesterton giving an official assurance that: 'If Our Lady were to appear today it would probably be in a miner's cottage.'

Financially the production was a disaster. Despite a number of sympathetic reviews (the *Manchester Guardian*, 24 December, 'greatly preferred this production to the theatrical one of a few years ago'), the expensive seats remained obstinately empty, and it was small comfort that the inexpensive seats became increasingly filled. The week ended with a handsome deficit. Rather than place the burden on the touring company, Boughton obtained an advance from Barry Jackson against future productions of *The Immortal Hour* and met the deficit himself. But it was the end of the Glastonbury Festival Players. Housman and Woodhouse resigned, and on 7 July 1927 the company went into voluntary liquidation.

The modern dress *Bethlehem* also effectively brought an end to any further association with Glastonbury. When the idea of a fresh beginning was raised, one of Boughton's few remaining local supporters warned: 'Several people have spoken to me about the production of *Bethlehem* in London, as no one would like anything of that character to enter into a Festival here.' Having nailed his colours so firmly to the mast, it was out of the question for Boughton to consider trimming his sails to the wind of other people's sensibilities, though perhaps, in his less confident moments, he recalled the advice Shaw had given him on 9 January 1925:

Indeed you had better not fash yourself at all with politics. The human animal, under competent direction, is capable of performing an opera passably,
and even, in very special cases, of composing it. Of solving the political problems caused by his own multiplication and aggregation he is utterly incapable.
When a race arises with that, it will go after him with a gun. Even Frederick
the Great (sogennanten) was awestruck by Bach, a monumental success among
a chaos of political failures. So do not introduce Glastonbury to Moscow
which will only let you down out of sheer clumsiness. Bach, Hadyn, Mozart,
Beethoven, Wagner, Strauss, etc., wont.

But it was not advice likely to appeal to a man determined to see
everything in terms of black and white. Boughton had dug his own
grave and would devote the next few years to pulling in the earth on
top of himself.

16

1927–1929

Once the links with Glastonbury had been severed Boughton became more acutely aware of the need to find a place in which to settle down. During the fifteen years since leaving Birmingham he had known far too many homes, none of which he could in any real sense call his own.[1] Now he was no longer young, no longer willing to make such frequent changes. Moreover, his relationship with Kathleen had brought about a tranquillity that would undoubtedly flourish even more satisfactorily in a permanent setting. It thus seemed as good a moment as any to attempt that part of his dream that Glastonbury had never realized. Accordingly, in November 1927, he and his family moved to a smallholding at Kilcot, a tiny hamlet near Newent on the Gloucester side of the Herefordshire border. Here he remained for the rest of his life, and in its unspectacular way the house and land and the whole manner of living provided as deep a satisfaction as anything he had known.

The move was made after a brief visit to Moscow at the invitation of the USSR Society for Cultural Relations as part of the tenth-anniversary celebrations of the Revolution. Boughton's understandable determination to admire the new way of life had a hard time. Instead of revolutionary operas he was treated to performances of Tsarist masterpieces and a rather unproletarian production of *The Love for Three Oranges* (Prokofiev, he decided, was rather like Ernest Newman: 'always interesting, and never convincing').[2] Even more exasperating was the state of Soviet plumbing, but he was impressed to find that Soviet artists were assured of a reasonable livelihood—something he knew he could not look forward to when he returned home. A speech written and delivered in Moscow on 11 October made very clear his jaundiced view of British musical conditions:

The musical world of Great Britain is in an increasingly state of artistic decay and economic chaos. There is nothing in its mental life great enough to need expression; so antiquarian revivals and performances of acknowledged masterpieces are all there is worthy to be called music.

Musicians themselves occupy an entirely parasitic position in the state, where luxury thrives and need increases. Orchestral players have a strong

Trades Union which is able to enforce a fairly good minimum wage. A player in a London orchestra, for example, will not receive less than seven guineas a week, while the leaders of various sections will receive ten and twelve guineas. It has even been known, when a player is in special demand, for his wage to reach forty pounds for a week's work. But the instrumentalists are strongly organized. For the singers, who are not organized, there is no minimum. High salaries are often paid to men and women with leading parts: but a starvation wage is given to the chorus of some companies, more especially those which tour the provinces. A few years ago rank and file singers were fairly well organized; and their Union succeeded in winning a minimum wage which, at all events, saved girls from earning part of their living by the sale of their own bodies. But, owing to a false tactical step, the Union was broken by a Masters' organization called The Stage Guild, and the last named is now the only Union available.

The difference in the power exercised by these two sections of musicians was exemplified about four years ago when it was proposed to bring to England the Vienna State Opera in the height of the London season. Macdonald was in office, and was reported to be in favour of the proposal for political reasons. The Musicians Union prevented the orchestral players from being thrown out of work, and the Vienna Opera left its orchestra behind. But the chorus and even some of the principal singers were unable to hold their own. At the time I was acting as music critic for the *Daily Herald*, and pointed to the necessity of establishing a British Union for all kinds of artists, on the Russian model. But the editor was Macdonald's man and refused publication. Since then things have been getting worse all round, until today it is fair to say that the only musical *life* in Great Britain is that of the amateurs.

Even there is degeneration. The industrial districts have always had fine choirs; but the deepening cleavage between the middle-class and the workers has taken the guts out of the pre-war musical organizations, which were generally officered by bank officials, lawyers, and parsons. Today even the more popular organizations are often turned into gambling events because of their association with the competitive commercial spirit. The large competition festivals for choirs, brass bands, and other combinations, are weakened as to their musical value because they do not originate in a pure musical instinct.

In spite of this depressing tale, there are signs of hope beneath the surface of things. The workers have begun to form musical organizations of their own. The most successful so far has, I think, been the London Labour Choral Union: a group of twenty choirs which sing for Labour functions in various parts of London, and combine for important occasions in large concert halls. It is by no means an ideal organization, but its spirit was shown last May when the reformist Labour Leadership tried to prevent a big demonstration in Hyde Park. The great majority of the choirs rejected the influence brought to bear on them, and joined the demonstration. A similar group of choirs has also been formed in Newcastle. In Glasgow and South Wales districts there are many choirs, but they have not yet joined together for mutual support. And the Welsh miners are given to singing bad hymn tunes. But all choirs suffer from a lack of suitable music. The sort of words they want to sing is impossible of publication in capitalist quarters. The Independent Labour Party has a number of choirs and theatre groups with a left-wing tendency. Some of them incline to lose their fighting political value, serving rather as dope than stimulus; but in art as in politics the earnest members sooner or later move in the direction of Communism, where only a real Worker's Art is likely to develop in Great Britain. The C.P.G.B. is now taking steps to organize its worker-artists; and it is to learn from the example of the Russian Workers, more especially the musicians, that I am now here.

Quite what the Welsh miners would have had to say in defence of their hymn tunes is best left to the imagination.

In order to move to Kilcot, capital had to be found—a matter of £1,000. Although he had earned very acceptable sums during the London runs of *The Immortal Hour* Boughton had been unable to save anything. This was partly because he had a large family to support, including the ever-watchful Flo; partly because he had ploughed back some of his earnings into the work at Glastonbury; and partly because sudden affluence had tempted him to take lodgings in some of the more expensive parts of London (there had been addresses in Parliament Hill and Emperor's Gate) the better, perhaps, to provide Kathleen with a setting worthy of their love. The necessary capital sum was provided by one of his most loyal Glastonbury supporters Miss Agnes Thomas ('Tommy')—on condition that the Boughton family looked after her for the remainder of her days. Since she already counted as one of the family, having frequently looked after the children, chaperoned Kathleen, and witnessed the painful details of their domestic upheaval, it seemed a sensible arrangement. It meant, however, that the choice of farm was hers.[3]

She might have been better advised. The land produced cider apples from trees well past their prime. Replanting was out of the question. Even if they had been able to afford it, it would be years before the new trees grew profitable. They decided, therefore, to leave things as they were and to make room for small livestock—pigs, goats, hens, rabbits—and turn over at least some of the land to vegetables. Thus, while there was little hope of making a living, at least a large family would be supplied with some of its more important needs. And for Boughton there were the compensations of the gentle Gloucestershire countryside. The house stood high on a hill side. To the north were the Malverns, blue in the distance, rising like mountains. From his desk at the window of a long, low, book-lined room he could see fields spread thick with apple orchards, dotted casually with neat cottages, soft and untroubled. It was a landscape of warmth and comfort, and above all peace.

Gradually a daily pattern was established. Boughton rose early, often just after dawn, busied himself with household chores and prepared food for the animals. After breakfast he would retire to his study and begin work, first ordering his mind by playing Bach fugues on the piano. When particularly engrossed in the problems of a new composition he would go straight from his bed to the study, leaving the family to arrange the household duties among themselves. And then he would emerge only for lunch, sitting absorbed, speaking to no one, living entirely in his own creative world. He would work on, far into the night, indifferent to sleep. For his family it cannot have been an easy existence. Shatter the precious crystal of his creative mood and his anger was quick and wounding. The children learnt to play quietly and at a distance. Not an easy life, but not one of them would trade its memories—the mornings when they woke to the sound of his piano, as softly in the growing light he played through the music he had written while they slept.

His method of work was that shared by a great many composers. He would sketch out the complete composition, rapidly, in skeletal form, on two staves if purely orchestral, three if voices were involved. Although only a few indications of particular instruments are included in the sketches, it is evident that he had a very clear idea of the orchestral colours involved even at this stage. The manuscripts give an impression of speed and fluency: the notation sloping to the right and often legible only to the composer, as if his pen could scarcely catch up with the flow of his thoughts. The appearance of the final score is quite different: the notation bold, upright, and unequivocal.

As the years went by, the Boughton family had good reason to be thankful there was the farm to fall back on. Gradually his sources of income were running dry. Some were too far away to be exploited, others he chose to abandon. He resigned his conductorship of the Civil Service Choir, and then handed over the London Labour Choral Union to Alan Bush. Both were difficult to serve at such a distance, though in the case of the Choral Union his disgust at the ineffective policies of the Labour Party was a deciding factor. A letter to Herbert Morrison, which the *Sunday Worker* published on 7 July 1929 under the heading 'Boughton Breaks with Morrison', made his position quite plain:

As you know, the pleasure of working with the London Labour Choral Union has been mixed with many heart burnings.

Figure 4 Sketch for Symphony No. 3 in B minor, fourth movement

It is not merely that the officials of Eccleston Square have sabotaged the proposals which you and I made to them for musical development on a national scale; but, more vitally, that the Labour Party itself has steadily turned its back on that policy of Socialism which was originally the cause of my association with it . . .

I have much admired your genius for organization, and at one time hoped that the day might not be far off when that genius might be placed at the service of a socialist state. Had a militant socialist policy been pursued by the Labour Party during the last ten years, I believe you would have been found today on the side of the workers.

Now you also are pledged to the conditions of a Capitalist constitution and the sheer joy you have in handling your machine will, I fear, place you in open opposition to the welfare of the workers of this country and of the world.

For Boughton pragmatism was as much an unknown word as opportunism.

How deeply his own work had been appreciated by the members of his choirs can be judged by a letter, signed by some forty of them, which expresses exactly the sort of loyalty he aroused among those he served:

Toynby Hall,
20 September 1929

Dear Comrade,

Having heard with deep regret the news of your resignation as Musical Adviser of the London Labour Choral Union, the undersigned members of the rank and file of the union take the liberty of sending you this message of appreciation and gratitude for the great work you have done for the Choral Union since its inception. Those of us who have attended the rehearsals regularly and have been privileged to sing at various functions under your magic baton realize to the full the great debt of gratitude we owe you for the stimulation and inspiration received under your conductorship. Our taste in music has been raised, and you have opened up to us a new world of beauty and enriched our hearts and minds with imperishable treasures.

We are proud to have been associated with you in your endeavours to serve the cause of Music among the Workers, and in particular we shall always remember your self-sacrificing labours on behalf of the Miners during the tragic lock-out of 1926.

May we never prove unworthy of your inspiring leadership and of the generous spirit of comradeship in which you have always met every one of us.

It has to be said, though, that some Comrades got very short shrift when they managed to do something silly—as when, in November 1926, a group of them organized a concert without taking the trouble to engage an artist and then expected Boughton to bail them out. 'Dear Comrade', he wrote, 'You are incorrect in saying that I promised to supply an organist. If you will make the mistake of announcing an organ recital before you are sure of a player it will of course lead you into difficulties, but that is not my fault.'—an answer worthy of G.B.S. for its succinctness.

If for the moment Boughton was taking a less active part in musical life he had by no means withdrawn himself altogether. He kept his colleagues nervously aware of his existence by means of the articles he contributed regularly to such magazines as the *Musical Times* and the *Musical Standard*, and, most important of all, by the series of pieces he wrote between 1927 and 1931 for the *Sackbut*, then under the editorship of his friends Ursula Greville and Kenneth Curwen.[4] Forcefully expressed, and in some respects the most interesting contributions to a magazine whose importance had declined sadly since the days of Peter Warlock, his articles for the *Sackbut* inevitably raised hackles—though, a Ursula Greville reported on 29 June 1929, there were nods of approval from unexpected quarters: 'I thought it would amuse you to hear that Arthur Bliss was in and said to Kenneth that he thought your writings simply magnificent. As Arthur is one of the most conservative people I have ever met with . . . !'

Three targets claimed his special attention: the whole problem of British opera; the composer's right to adequate payment and the need for effective organization; and the problems raised for musicians by powerful monopolies like the BBC. Since some of these questions have yet to find satisfactory answers, Boughton's point of view is of interest even if the form his attacks took was often exaggerated and not always well argued.

Throughout his career Boughton wrote on the subject of British opera. But the major explosion of articles occurred in 1931 as the result of the general outcry that accompanied Philip Snowden's bungled attempts, as Chancellor of the Exchequer, to provide an opera subsidy. It had all started with Beecham's grandly named 'Imperial League of Opera', which had been founded in November 1927 to provide an outlet for British artists through opera seasons in London and six major provincial cities.[5] Beecham had somehow calculated that there were at least 150,000 opera lovers in Great Britain, and that if they each subscribed 10s. a year for a period of five years he would have at least £60,000 to play with. Eventually he hoped to build and endow suitable opera-houses in London and other important centres. By July 1929 some 40,000 subscribers had pledged their support. He was then invited by Ramsay Macdonald to discuss the whole question of 'national' opera, and did so to such effect that he was promised £30,000 a year in the event of the Labour Party being returned to office. For his part Beecham promised that this would be matched by the League with a similar sum.

At this point Beecham entered into negotiations with the Covent Garden Opera Syndicate, telling them that if the subsidy came to pass it would be shared between the Syndicate and the League. But on being returned to office, Macdonald went back on his word and in June 1930 announced that the subsidy would not be granted. In November the Chancellor indicated that a subsidy would be granted to Covent Garden alone. In the mean time, Beecham had been negotiating with the BBC for a contract to broadcast the Imperial League's performances, as and when they took place—unaware that the corporation had already entered into a similar agreement with Covent Garden.

Towards the end of 1930, however, the original Covent Garden Syndicate disbanded and reformed itself as the Covent Garden Opera Syndicate, Limited (1930). It was revealed that five-sixths of its resources were to come from the Government and the BBC, and that the BBC would be the 'conduit pipe' through which the Government subsidy of £17,500 a year would be paid. It also became clear that the BBC would have a majority shareholding in the Syndicate and that, as Boughton put it, the new body was 'in reality, if not name, the British Broadcasting Corporation Opera Company'. On learning this Beecham withdrew, but after Covent Garden's disastrous 1931 season he was persuaded to reconsider his position. In July 1932 he joined the Board (Bruno Walter had been unceremoniously ditched) and, after a few skirmishes, remained triumphant until the outbreak of war.

In all this, Boughton's concern was for British artists and British opera. It seemed obvious to him that with foreigners on the Covent Garden Board, and with pledges of foreign money, British interests would not be well served. And if Covent Garden's activities were to be linked with those of the BBC, British opera composers would indeed be in for a thin time:

It will be argued that British composers have produced no operas fit to rank with those of Mozart, Verdi and Wagner. Very true. But other works of less value are performed at Covent Garden. Nor shall we ever produce a great opera composer of our own while fashion and finance combine to his exclusion. Further, until regular encouragement is given to a national school of opera, with the English tongue set to its own natural music, our English singers will labour under serious disadvantages.[6]

Enraged by the Byzantine intrigues, real and imaginary, of the London operatic scene, Boughton became active in the formation of an Opera

Group within the British Music Society. Its express purpose was to ensure that British opera and British artists might get a fair deal from whatever machinations were afoot. He found his fellow composers somewhat less than wholehearted in their response. Neither Holst, Vaughan Williams, nor Dame Ethel Smyth, the three composers after himself with the greatest operatic experience and ability, felt able to support his proposals. On 14 January 1931 he took Dame Ethel to task: 'For the Lord's sake don't suggest again that my activities are those of an idealist. Apart from you and Dunhill, I'm the only *practical* British composer alive. Its talkers like Holst and Gatty who idealize things'. And by the same post he made a last appeal to Holst and Vaughan Williams for support. But all three remained lukewarm in their attitudes, and in the end all he succeeded in doing was to convince them that he was what they had always supposed him to be: a major thorn in the flesh of a complacent musical establishment. Not even the equally paranoid Holbrooke could be persuaded to take an interest. Writing from a hotel in Monte Carlo on 25 February 1931, he made his position very clear:

Dear Boughton,

Ta for yours. Please leave me out of *any* 'B.S.M.' activity—or any other. Let the treatment of my work continue as it has done for 15 years. After a lifetime helping every composer at my 6 years of London Chamber Concerts (including yourself) there has not been *one* musician to raise his voice on the BBC's treatment of J.H. That is clear enough to me that no one wants any of it. As for 'guineas'—I belong to no Society, and have no intention, now, of doing so.

Regards from Holbrooke.
I'm also too deaf to be interested.

And there the matter had to rest—Boughton firmly convinced that the scales were weighted against British opera and its composers and were likely to remain so.

In support of the rights of composers to a decent return for their labours his attacks were no less trenchant. Here he drew freely upon his own relations with publishers, which had not always been peaceful. An article written in 1928, and probably intended for, though not actually published by, the *Musical News and Herald*, is a fair sample of his attitude:

As compared with executive instrumental musicians, composers seem a hopelessly foolish lot. The Musicians Union has organized the orchestral players into a body of self-respecting artists, whose material interests have been so well protected that they have been able to live without serious concern for the future; but many composers have fallen victim to the humbug which has placed them in a class apart, to their own undoing. First, they have been ridiculously flattered as creatures upon whom occasionally descends a divine spirit. They are advised that the divine spirit acts most freely in conditions of material poverty. So, when they are asked to guard their interests in a world organized to exploit fools and unselfish persons, the well-tutored composers answer, with their noses in the air: 'We are not concerned with such vulgar things', or 'We leave all matters of business to our publishers.' They do, including a greater and unfair share of the profits.

With great accuracy he then exposes the besetting sin of British art-amateurism:

A publisher with whom I was discussing the point (a knowing old bird) met my argument with the complacent remark that 'Composers are not expected to live by their compositions, but in other ways—teaching, and so on.' That is largely true as a statement of the position of British composers; and that is why we have had so few of the best kind during the past two hundred years.

After inveighing against the stupidity of expecting a man to do a full-time job and then be able to carry out the full-time labour of composition, he concludes:

Give a composer only reasonable holidays and daily spare time, and he will need practically all the rest for his work. He might with some advantage to himself be required to do an hour or so daily as a navvy or a farm worker; but if, apart from the details of composition, he must be occupied with hack work of a musical kind, he will do that badly, and at the same time use up the nervous energy which ought to be devoted exclusively to composition. Even then he will not earn a decent living, unless he falls into line with the schemes of the commercial world and strikes lucky in one of its gambles.

Nor is patronage any solution to the problem. Why should a man doing honest work be placed in the position of a parasite? Even in so great a nature as Beethoven's the position of protégé developed undesirable traits of character. In smaller men the same relation develops appalling results.

The only solution is to admit that when a man has made clear his right to serve the public by musical composition (whether by light or deep-probing music) he shall be able to maintain himself in a position of independence by the ordinary channels of work.

It was the same argument he had advanced in 1909 in *The Self-Advertisement of Rutland Boughton*—less light in touch, perhaps, but equally difficult to gainsay. Nor has society even yet been able to find an answer to the challenge he posed.

While in these instances Boughton's attacks, in general terms at least, are unexceptional, his attitude to the BBC is less happy. During the late 1920s he became convinced that a degree of boycott was being exercised against the broadcasting of his music, mainly, he believed, on political grounds. After an inquiry, Sir John Reith emphatically denied his assertions. But in the years that followed it did seem as if his music was receiving more attention, and this, in his eyes, was proof enough. He returned to the attack whenever it seemed necessary, and for various reasons the occasions became disconcertingly frequent.

At the root of this uneasy situation lay not politics or boycott, but the plain fact that not all of his music was suitable for frequent broadcasting, and that much of the stuff that might have been used regularly is of relatively inferior quality. It was a delicate situation, and one all too easily aggravated by his complete inability to do business with the Corporation's officials. He misunderstood their oriental urbanities; saw concrete promises in what was only polite waffling; and on pressing home the apparent acceptance of his work drew the refusal that had been implicit even in the opening skirmishes, had he been able to read the signals. Had the BBC officials been a little less ready with empty courtesy, a little more like Boughton himself, misunderstandings might have been avoided. As it was, they were speaking different languages.

To Boughton one of the most irritating things about broadcasting (and here he had a perfectly valid point) was that it had destroyed a number of the composer's traditional sources of income and in their place substituted a monopoly in the hands of a very few men. 'I do not think the BBC staff is a collection of corrupt officials', he wrote, 'but just a number of second-rate men who are, under existing conditions, dominating their superiors.' Moreover, it seemed to him that the BBC was managed 'as if it was a branch of the Secret Service'. Particularly unconvincing was its method of selecting music for broadcasting. Writing to Reith on 20 November 1930 he put his objections forcefully:

Dr Boult is a man of great ability and wide sympathy, but it is clear that he cannot be expected to read all the music that is offered to the BBC for performance; nor will a *committee* of readers to winnow the suitable from the unsuitable command any general confidence. To satisfy the mind of the musical public on that point—and I recognize that on that point nothing will satisfy it completely—would it not be possible for it to be publicly known who are the winnowers? You asked me to send for consideration—of whom?—the pieces already approved by Sir Henry Wood; but, frankly, I should not have cared to let them go unless I had known who it was who would decide whether they were fit for performance. I would not dispute the right of certain musicians to turn down my stuff if they think it bad, or even of a conductor in a public position to refuse it if he had a personal distaste for it; but I am sure it is unnecessary for me to suggest that the BBC as a national organization should discover some method whereby every shade of acknowledged music might find a place in its programmes. If that is done I think it will be found that my works will suffer less than under the methods which have prevailed during the last four or five years.

Could we not be told: who is the man who weeds out the big orchestral works before they go to Dr Boult; who makes first choice of light orchestral works; who of big choral works, and who of lesser choral music; who of serious songs: who of light songs: and soon? A musician in such a responsible position, and therefore more open to public attack, would surely serve the BBC more effectually than one not so open.[7]

Sir John's reply, that the selection was made by a committee chosen for their 'catholic taste and their experienced musicianship', but that their names could not be disclosed, was exactly the kind of oblique excuse calculated to increase Boughton's suspicions. The fact was that he had hit out at a difficulty that has not been solved to this day, and is probably insoluble. An anonymous selection committee is bound to look like a cabal, while a named one is open to lobbying.

Boughton's belief that critical opinion hardened towards him as soon as he declared his communism is impossible to prove. When projected performances of his works failed to mature it was not because of some capitalist conspiracy, but simply the luck of a very tricky game. There seem to have been very few occasions when critics dismissed his finest music out of hand, and what can now be seen as weaker pieces seem always to have been criticized on purely musical grounds. Individuals may well have disliked his politics, felt queasy about his free-wheeling sex life, resented his outspoken attacks on the establishment, but it seems unlikely that their reservations evolved into a deliberate conspiracy. The plain fact was their neither Boughton nor his music was to everybody's taste. But it was not to be expected that a man who had scaled such difficulties, known such enormous success, and then witnessed the collapse of all his schemes, could accept the turn of fortune's wheel in utter silence. He was not, and never had been, the silent type. Moreover, he was alarmingly capable of giving offence even to those who were sympathetic to his cause.

A magnificent instance was to occur in 1934 when he managed to provoke Edward Clark, who was then employed by the BBC. Though an ardent and perceptive advocate of the latest continental developments in music, and on familiar terms with such masters as Bartók, Kodály, Hindemith, and Dallapiccola, Clark seems to have retained a soft spot for Boughton. He had worked with him as a conductor of the London Labour Choral Union choirs, and even conducted *The*

Immortal Hour from the BBC's Newcastle studios (15 October 1925). On 6 June 1934 Boughton wrote to Clark inviting him to attend a private recital of a new music-drama, *The Lily Maid*, at the Wigmore Studios. Unfortunately he did not notice that he had muddled the dates, inviting Clark for the 6 June instead of 15 June. Clark, who received the invitation probably on 7 June, naturally assumed that Boughton had simply mistaken the month and on 14 June replied that he would 'certainly be at Wigmore Studios on July 6th'.

Boughton, unable to imagine that he might be the author of a genuine misunderstanding, leapt to the conclusion that Clark's reply was a gratuitously well-timed insult and promptly made his feelings clear:

16 June 1934

Dear Clark,

I do not understand your note of June 14th. If you will refer to my previous letter you will see that you were invited to a recital (of my new opera) given yesterday, Friday. Of course, I am not surprised that you didn't come. The unfair treatment of my work by the BBC is apparently becoming obvious to the general public. As you have long been in a position where you could have corrected that treatment, I should prefer you no longer to make private and personal professions of interest in my work. It will survive the rubbish you chiefly have been instrumental in foisting on the London and wireless public.

And now a *good* bye to you. I can't waste my time with a grievance because you are incapable of knowing music from muddle and muck.

Yours faithfully,
Rutland Boughton.

Clark, wisely, seems not to have replied. But Boughton evidently took his grievance to Shaw who, on 24 June in an attempt to undo the mischief, wrote out a draft letter to be sent to Clark:

My dear Clark,

I was so convinced of the justice of my last letter to you and the infamy of your treatment of me that I showed a copy of it to Bernard Shaw for his approval—after I had sent it.

He assured me:—

1. That I am a bloody fool.

2. That my works will never be performed until I am dead and unable to insult everybody who is interested in them

3. That I have no right whatever to expect you to force my music on the BBC to the exclusion of all other composers.

4. That my renunciation of your friendship must be a most blessed relief to you.

On reflection I find I am unable to controvert these propositions. Will you therefore consider my letter unwritten, as well as all other letters of the same kind that I may have written or may hereafter write to you. They are only symptoms of that painful and most unamiable disease called the artistic temperament.

Kindly intimate that I am forgiven.

R.B.

Boughton, however, could not believe he had anything to apologize about. The nearest he came to proffering an olive branch came on 28 March 1936:

My dear Clark,

Today I read in the *Telegraph* that you have resigned your broadcasting job. If that is so I trust it means you have found other work more to your liking. Your influence at the BBC has, in my view, been all to the bad, as I have already hinted to you; but I have very happy memories of your association with the London Labour Choirs, and to you personally I wish all good things.

Yours always,

Rutland Boughton.

But it was arrogantly and ungraciously done and merely served to confirm the general opinion that Rutland Boughton was his own worst enemy.[8]

17

1929–1933

By the middle of 1929 Boughton was obliged to admit that his relations with the Communist Party were as unsatisfactory as his relations with almost everyone other than his own family. On 14 June he made his views known to Party officials:

Since joining the party I have been gradually squeezed out of the musical world, though at the time of joining almost any kind of recognition that was could have been mine for the asking. I do not complain that my old capitalist friends have given me the cold shoulder; but I do complain that the Party has made practically no use of me unless I pushed myself forward.

I was asked by Comrade Winteringham in 1926 to make choral arrangements of the 'Internationale' and the 'Red Flag' for choirs to sing unaccompanied or with brass accompaniment. I did so, but they have not been published, though there certainly would have been a large sale for them.[1]

When I have published music of propaganda value it has, of course, been ignored by the capitalist press; but it has also been ignored by the Party press, though I have had review copies sent. William Paul informs me that he wrote a review of my *May Day* ballet for *Workers' Life*, but it was returned to him and he was refused reasonable space in the *Sunday Worker*, though that paper can find room for an anonymous personal attack on me.

Now either I must be used more effectually, or I must take steps to make myself less obnoxious to the capitalists who rule musical conditions—though to be quite frank I should scarcely know how to start the latter line of procedure.

His complaint brought about no apparent change and on 21 November he resigned his Party membership, though at the same time making it clear that he still believed that communism offered 'the only honest leadership to the workers of this country'.

As if in echo of his unsettled and unsettling political adventures, the compositions of this period are of startlingly mixed quality. A fine suite of choral songs, *Child of Earth*, written on arrival in Kilcot as if to celebrate the event, is followed almost in the same breath by a dismal pot-boiler, 'Burglar Bill'. A number of delicate songs and partsongs go hand in hand with plans for a musical comedy to a libretto by Eimar O'Duffy, to be called *Butterflies and Wasps* and presumably designed to

capture the Noël Coward market. After a few uneasy experiments with
blues chords and fox-trot rhythms the matter was quietly dropped,
while a proposed three-act romantic opera entitled *The Hunchback*, to
words by Gladys Morton (a fearful mixture of sentimentality, Grand
Guignol, and religiosity) mercifully got no further than nineteen very
sketchy pages. Apart from *Child of Earth*, these years produced only
two works of consequence: a three-act ballet, *Deirdre*, and the music-
drama *The Ever Young*.

The ballet came about when Terence Gray, the founder and director
of the Cambridge Festival Theatre, sent Boughton a volume of 'dance
dramas' and asked him to write music for one of them:[2] 'The one on
Deirdre made the strongest appeal and I set about it. When I had made
the first sketch he brought Ninette de Valois down to hear it. She
spoke enthusiastically at the time, but I thought he responded to it
somewhat chillingly.'[3] At all events, nothing more was heard and
Boughton decided to cut his losses and turn the whole thing into a
purely orchestral work that, with a little stretch of the imagination,
might reasonably pass as a symphony. To do this he simply removed
some 67 bars, most of which were concerned with a scene-setting commentary by a 'Bard'. The result was a thirty-six minute work that
depended heavily on a 'programme'—a work that was not quite a symphonic poem, and certainly not a symphony in the usual sense of the
word. Boughton suggested it might best be thought of as 'A music-
drama without action', but pointed out that the underlying concept
was symphonic in the wider sense. The original ballet was completed
on 17 March 1926, and he inscribed the score with the words: 'St
Patrick's Day, and an end of all my romantic music.' There he was
wrong.

In some respects *The Ever Young* is a pendant to *The Immortal
Hour*, for it reverses the basic situation and presents a mortal woman
wooed by an eternally youthful god. But it is in no sense a sequel to
the earlier work, even though it contains a few musical cross-refer-
ences. Boughton began work on it in the summer of 1928 and completed the score in July 1929. He wrote his own libretto, evolving the
tenuous plot from a paragraph he had come across in a book of Irish
legends which tells the story of Aengus, the Celtic god of love, and his
wooing of the mortal maid Caer. It occurred to him to wonder what
would happen when the mortal woman grew older, and wiser, while
the immortal god remained forever young. Something that on one level
was merely an attractive legend could, on a deeper level, become a
parable of human development—how men and women outgrow each
other, as he had good reason to know. Dedicating it 'To Kathleen', he
came to regard it as one of his most important works: 'probably the

most consistent example of the kind of stage work I had hoped to make—a work which places this life and this earth as the focus of human happiness, but accepts the inexplicable influence of the mystical world as real and in fact not without vital import for our present ill or well-being'.

With a new music-drama awaiting performance, Boughton's thoughts naturally turned to the possibility of recreating the conditions that had existed in Glastonbury. The earliest indication of new plans came in February 1930 when Lord and Lady Londonderry sponsored a performance of the third act of *The Round Table* in the long gallery of their splendid London home. The Londonderrys, and more particularly Edith, Lady Londonderry (1878–1959), had become keen supporters of his music during the successes of *The Immortal Hour* and Boughton, after a preliminary qualm or two, had responded to their friendship. Indeed, it would have been difficult for a man as susceptible as Boughton to the charms of beautiful women not to respond to the blandishments of 'Circe'. And for her part Lady Londonderry had all the pleasure of taming an unusual 'lion', who became even more intriguing when it turned out that he was 'red' on top of everything else. Their friendship blossomed, and, by talking about his music among her own circle and attending performances of his music, the Marchioness of Londonderry did much to bring his name to the attention of people who mattered. As a consequence, even as late as 1933, it was possible to find a recital of Rutland Boughton's chamber music at the Aeolian Hall attended by a prime minister and a member of the Royal Family. Such an occasion was the performance of the *Round Table* extract.

It took place on 5 February 1930 and provided the gossip columnists with irresistible material. 'Now it seems that Comrade Boughton is becoming a favourite in Society', sniffed the *Evening World* for 30 January, in eager anticipation of the event. 'When I last heard of him he was active in the Communist movement, and it looked as though he had given up opera for uproar!' Nor can the guest list have done much to clarify the enigma, including as it did the Princess Helena Victoria ('who followed every note and word intently'), the Prime Minister and Mrs Baldwin, Sir Austen and Lady Chamberlain, Lady Curzon of Kedleston, Lord Berners ('who is an accomplished musician'), Mr and Mrs Winston Churchill, Sir Barry Jackson, and Mr C. B. Cochran—not to mention assorted American and French diplomats who were attending the Naval Congress that happened to be in progress. Whether they appreciated King Arthur's plea for peace is to be doubted, but we are told that 'Mr Boughton himself accompanied the singing at the piano, and, as was to be expected, played with exquisite sympathy and sensibility', so it can be assumed that some pleasure was had from the occasion. Lord Berners, however, found himself obliged to write to Boughton after the newspapers had had their say, pointing out that he was actually 'a great admirer of your work' and that 'the whole account of the party at Londonderry House was extremely offensive and obviously designed to make mischief, and no doubt inspired by the fact that Lady Londonderry has already had trouble with the newspaper in question'.[4]

For Boughton the relationship also had its awkward moments, as when Lady Londonderry invited him to a supper party in honour of the 1932 revival of *The Immortal Hour*.

During supper Lady Cunard called out 'You are a communist, aren't you Mr Boughton?' I answered 'Yes'. There was a silence, broken by Mr Sacheverell Sitwell who drawled 'Why are you a communist?' A proper answer would have meant a sermon on Christianity, so I said the first thing that came into my head: 'I believe that Russia will be the salvation of Europe.' And then, turning to the unhappy hostess who was sitting by my side, I said 'If they ask me, I must tell the truth, musn't I?' 'Say what you like', she answered. 'Its a free country. The only one there is.' It was the last effort of the many she made to establish me in the goodwill of her friends who held her political faith, but none of them her generosity of spirit, and none of them her beauty.

But she had done her best. Working indefatigably from the moment in 1923 when, on 4 March, after expressing her delight in *The Immortal Hour* and begging that he would 'excuse her presumption' in writing to him out of the blue, she was also able to report that she had been able to get Mr Heseltine to sing certain extracts 'the night the King and Queen came to dine with us', adding 'And I told the Queen she ought to hear it and support an entirely British opera.'[5]

Her enthusiasm for *The Immortal Hour* must, however, have caused Boughton some embarrassment—as when she invited Mr Heseltine, Miss Ffrangcon-Davies, and Mr Arthur Cranmer to sing excerpts in the grounds of Mount Stewart, her County Down mansion. How he must have squirmed as he read the report in *The Northern Whig* (5 July 1926) which, after referring to the opera as the 'bonne-bouche' of the event, went on to describe it in terms that were almost beyond mere prose: 'Free from the realities, the delicate melodies of this music floated away with an unstressed, yet poignant, melancholy which wounded the soul with a touch as light as the passing of a song.' And it may have been with some relief, if indeed he read on, that he found the reporter on firmer ground when it came to enumerating the charms of Miss Ffrangcon-Davies who 'looked most attractive with her Duchess of York fringe, and her pretty frock of bois de rose georgette and caramel-toned lace, adorned with bands of bois de rose velvet'.

As to his own first visit to Mount Stewart, he suspected that Lady Londonderry had imagined he would

seal the holiday with a work of genius made hot on the spot. But of course I just loafed, sat silent, or made resentful and awkward conversation when clever people like Lord Hugh Cecil and John Buchan talked Toryism and agreed with Lord Londonderry as to whether the miners were human beings or not, and was only tolerably comfortable with Circe herself.

Boughton's relations with the Marchioness of Londonderry must have puzzled and even dismayed his more austere socialist friends. But he would not have been human if he had not felt in some degree flattered by her attentions—she was, after all, a beautiful and intelligent woman. He may even have cherished the the belief that she would absorb some of his political convictions if only he argued his case with sufficient force. He therefore made his views crystal clear, while she listened politely and was amused. Men of genius had to be indulged, and she was secure in the knowledge that the miners, whose cause he insisted upon advancing, were 'devoted' to her mine-owning husband and his family, 'and had been for generations'. 'I feel only grieved', she wrote, 'that the brain which wrote *The Immortal Hour* is mixed up and occupied with politics, which without wide knowledge and training can easily be misled and make things infinitely worse, in this imperfect world, than they are already.'[6]

Something of his fascinated exasperation can be felt in a letter he wrote in November 1933:

My dear Circe,
    Was Odysseus a lunatic? I think not. The lunatics were really those who left undeveloped all the loveliest parts of Circe's nature by allowing themselves to be led up the garden and changed into pigs, hinds and other pathetic creatures.
    Odysseus refused to become a boar, though he might remain in some measure a boor and even a bore.
    The eternal beauty of Circe will be remembered, not because of her pigstyes, but because Odysseus was devoted to her and yet hated some of her ways. A certain work is in process of germination which records what is eternally lovely in Circe's namesake. It will help to balance the rather one-sided view maintained by Byron and Shelley.
    It is the irrevocable nature of art and not the wobbling opinions of history that carries fact into the future; and you know it.

But Circe, sunning herself in Cairo, only replied haughtily and with aristocratic disdain for the colloquial: 'Dear Mr Boughton, I have not the slightest idea to what you refer.'[7]

The work 'in progress of germination', a music-drama on the legend of 'Circe', never, in fact, took root. Nor did an idea to develop the story of 'Fand' and her magic garden, though this undoubtedly shared the same source of inspiration. Despite her title, her husband's estates and hateful coal mines, Edith, Lady Londonderry, made a deep impression on Mr Boughton.

Shortly after the performance at Londonderry House Boughton began to consider the possibility of an Irish Festival. On 25 February 1930 he wrote to Lady Londonderry:

Barry Jackson called here on Saturday and proposed that he should put on *The Immortal Hour* at Malvern next July for four performances. I agreed only on condition that a London revival followed and urged him to follow that with a production of The Round Table. But I have little hope of that maturing as he is a timid man. What is most important is for me to gather my crowd again as at Glastonbury.

Is there a suitable hall at Newtownards where such shows could be given? If so, would you resent a holiday school such as I used to have at Glastonbury? Then I could convince you of my right to control my own destinies and not be left to the mercy of the commercial theatre. If you are willing to think of such a plan, please do not talk of it until it is decided.

Lady Londonderry took the proposal seriously, but her idea of an effective time-sequence for such a complex undertaking was not Boughton's, and on 10 May he wrote to explain the difficulties:

You think that I should have far more time 'to organize now for the end of July' than 'to organize in May for the middle of June'. But there was never any suggestion of a June festival. I always organized the Glastonbury shows in August, and that was the month I originally suggested to you. Then, when you told me that July was the Irish holiday month, I agreed that we ought to try and arrange it then. Nevertheless you will recall that I said that if there was not enough time to arrange for this year, the work done would serve for next year. You see, with several years' experience of the game I have learned that the printed notices ought to be issued to the public at the end of March if they are to be effective; and the preliminary work in this case would be extra-involved as I have dropped threads to collect—artists to regather, and modified conditions to adapt myself to.

Of course I am very sorry that you do not think The Round Table important enough for a London production; but I would like you to know that it is the second of a series of music-dramas which were regarded by most of the leading men in British art, music, and letters as important enough for them to issue a special appeal for a theatre to be built for their adequate production. I trust therefore that you may still be able to use your influence on its behalf.

You know I have no desire to bother you; and if you find you are unable to do anything I'll just thank you for having tried, and lay aside my creative work to write a record of the whole business from 1913 (when the appeal was issued) to this day. That will be easier because of my business training which has caused me to keep all letters and documents. It will be a nice way of paying some tribute to the many fine friends who have helped my work, and I shall have some pleasure in exposing the dirty tricks of those who have tried to hinder it.

He was not the most tactful of letter-writers. Typically, he had misinterpreted her reservations about *The Round Table*. What she had said was 'I do not think that *The Round Table* would be a success in London, at least I strongly advise you to try it somewhere else first. I know and appreciate how fine it was, but I think it is not sufficiently important to run a season entirely on its own.'[8] A less self-absorbed artist might have conceded that 'entirely on its own' was a perfectly reasonable debating point. But paranoia is hard to gainsay. There was a certain satisfaction to be had in suspecting that yet another friend was not really as wholehearted in her admiration as she had claimed to be. In the face of such wilful touchiness, their plan came to nothing.

Nor had an earlier idea for a Welsh Festival at Mountain Ash fared much better. This had been hatched in 1929 in partnership with Holst and Vaughan Williams, as an undated memorandum from Vaughan Williams shows:

After you left, Holst and I had a long conversation which is summarized in the enclosed suggestions which I send to you in case they are of any use to you—

Memorandum and suggestions for opera scheme by G.H. and R.V.W.

A. *No splash*. e.g. Mountain Ash Festival to start with—the opening performance to be on the same small scale as what we propose to continue with. Also, *under the same circumstances*. A show by a small company intended for co-operative hall should not be judged by the critics under the circumstances of a London theatre; therefore the invited 'Press' show should take place in the provinces at one of the co-operative halls.

B. Performances, though small in scale should be as nearly perfect as possible in *quality*. Therefore we do not advise operas which audiences may have seen on a large stage with all the glitter and tinsel of a large crowd and showy costumes etc. There must be no feeling of makeshift e.g. 2 men and a boy to represent a whole regiment of soldiers. The operas must be chosen with regard to this.

C. We suggest that under a good Director, local *amateur* effort might be enlisted in each town visited, to walk on and represent crowds etc., or even, where possible, to help in the chorus.

D. We suggest everything on a small scale to start with—no orchestra but only a piano (N.B. a good piano should be bought and carried round, not to depend on the local piano). A company of about 20—everyone willing to take a hand at everything. e.g. the principal soprano of Monday would walk on on Tuesday, help mend the costumes on Wednesday and be noises off on Thursday—but every one of these artists must be first rate and hard working.

E. We suggest as two quite separate propositions that *The Immortal Hour* is a good opera to start with and R.B. would be the best Director—and to avoid any suspicion that R.B. was made Director and then chose his own work, the work should be selected first and R.B. then invited by a Committee to direct it and incidentally any other operas in the repertoire.

F. We suggest a repertoire of not more than 3 operas—besides *The Immortal Hour* we suggest Gluck's *Orpheus*, Mozart's *Cosi Fan Tutte*, Verdi's *Rigoletto* (Also Stanford's *Shamus O'Brien* and Holst's Golden Goose Ballet—these last two added by R.V.W. since conversation with Holst).

G. We think it a mistake to pose as 'British' opera. British opera is at present very naturally suspect; it will we are sure have to be introduced in small doses.[9]

This idea also came to nothing—one can imagine Boughton bristling at the idea of Verdian melodrama (mere 'entertainment'), and being unwilling not to go wholeheartedly for British products. But it was an idea with possibilities, as later twentieth-century practices have shown.

Thwarted in his attempts to mount a new festival, Boughton turned aside from composition to write a new version of his 1907 *Bach* book which had proved its worth by running into three editions. The invitation came from Sir Landon Ronald, editor of a 'Masters of Music' series for Kegan Paul, Trench, Trubner & Company. The 154 small pages of the original volume now became 302 pages of larger format, and the content was very different. Whereas the small *Bach* had concerned itself mainly with the music, the larger volume saw Bach as a victim of his times: a man forced by circumstances to support Lutheran orthodoxy, but whose art continually reveals him to have been in sympathy with the Pietists who were opposed to that orthodoxy. Bach, in effect, was the champion of the downtrodden masses.

It was reviewed widely by the critics and at a length commensurate with their astonishment. Bach as a crypto-revolutionary—his inner dissent hidden, like a cultural time bomb, in his use of the Chorale (the 'people's' music)—was not for them. But they reviewed it fairly and gave credit to its good points. And though they could not agree with his interpretation of history (his dismissal of the aristocracy as 'vampires of the worst description' caused particular alarm), they allowed that his point of view was stimulating and could not quite be rejected out of hand:

Mr Boughton's main thesis being what it is, the reader will not be surprised to find him taking what may be called a Clydeside view of the social conditions of the period. He has the proletarian sheep and the capitalist goats neatly divided—the virtuous 'folk' of course, under the heel of the vicious rulers. With less exaggeration he would have made out a better case, for he has some telling historical data that need no underlining.

Mr Boughton has only himself to blame if his reviewers spend their valuable space attacking the main point of his book, and so are unable to do more than glance at its merits. They are considerable—more so than might have been expected, for you cannot turn a biography into a socialist treatise and make a complete success of both jobs. As often as the socialist drops into the background and the musician comes forward we get first-rate critical writing. (*Musical Times*, December 1930.)

Ernest Walker's assessment, in the November issue of the *Monthly Musical Record*, sums up the general reaction and is as fair as could be expected from any political adversary:

As a study of Bach, this book needs plenty of supplementing. In the main, it is narrow and exaggerated, the hasty work of an obsessed man; and its high-pitched querulousness wearies. But it is sincere; and it is alive. We turn the last page, indeed, feeling that, although we have been told something about Bach, we have been told a great deal more about Mr Boughton.

The enigmatic dedication: 'To Circe, in grateful revenge', he left unremarked.

Despite the first performance of his 'Celtic Symphony at Bournemouth on 25 January 1933 (Boughton himself conducting Dan Godfrey's orchestra) it had begun to seem as if the only works which anyone would take seriously were *Bethlehem*, a firm favourite with amateur societies (who had mounted five productions in 1932 alone), and *The Immortal Hour*, which Sir Barry Jackson had revived in 1932. Irritated, Boughton refused to allow Jackson to include it in his 1933 Malvern Festival, thereby sabotaging what might have proved an attractive and viable solution to his search for a new Glastonbury. Malvern would have offered a fully equipped theatre of a size (900 seats) admirably suited to Boughton's operas, and the lure of Shaw's plays, as well as Sir Barry's expertise.[10] But by this time he felt he could afford to take a high hand. There were other proposals in the air, and in token of the promise the future seemed to hold he began work on the third part of the Arthurian cycle.

1934–1937

In March 1934 a neatly printed 'Note from Rutland Boughton regarding the Production of his Music Dramas' warned his friends that a new festival was in the offing. After a brief reminder of his work at Glastonbury and the subsequent London productions, he explained the events that had made the new situation possible:

Last summer I had a visit from two young orchestral musicians who had played in the last revival of *The Immortal Hour*, and had taken part in some rehearsals of excerpts from *The Ever Young*, my most recently composed music-drama. They said they represented a number of such players in London who were of the opinion that my work was not getting a fair show in the musical world. They said that those players wished to organize themselves on a co-operative basis and give *The Ever Young* a London production. They thought they would easily find the necessary backing. I suggested that instead of *The Ever Young*, which offers fresh problems to the producer, they should start with *The Immortal Hour*, and then follow it up with *The Round Table*. This was the music-drama which (next to *The Immortal Hour*) had been most attractive to the Glastonbury Festival public. They agreed, but found themselves unable to get the necessary backing. Even so, the proposals of these young musicians has heartened me. As in New York, so in London, the demand for my work comes most definitely from those musicians who are highly skilled in their work and, at the same time, hold a comparatively obscure position in their relations with the public. Singers have their meed of direct public approval, but members of the orchestra and chorus only through the personality of the conductor.

So I spent the first six weeks of the present year in London looking into the possibilities of a co-operative production with their help. I found that a West-end theatre would work with us on a sharing basis. Some of the foremost English singers were willing to be associated with us. The financial returns of the last revival of *The Immortal Hour* were such that the project would have succeeded with considerably lower takings. It would also have employed a number of musicians who otherwise remain unemployed. One of the leading musicians in London said that in his opinion 'it would be a fine thing for British music'; and he later introduced me to a well known and wealthy supporter of music. But the latter, though friendly to the plan, thought that this was the wrong time to launch it. I could not persuade him that entertainments will regularly continue through the most difficult times, though he readily enough admitted the importance of finding employment for British musicians in a sphere which generally employs foreigners.

Therefore I must make up my mind to do what I can in a small way in the immediate future; for you will realize that, having been practically excluded from the musical world for several years, my own livelihood is at stake. My present hope is to hold festival performances of at least two of my works in some pleasant place next August. To that end I need financial guarantees. A guarantee of £250 will enable me to give a week of small works—chamber opera, and ballets. A guarantee of £1,000 would make possible a fortnight of music-dramas with orchestral accompaniment. For the last two festivals at Glastonbury it was not found necessary to call upon the guarantors; but starting in a new place in more difficult times, such a favourable result could scarcely be anticipated at first. I need hardly say that if anything is to be done this year the urgent attention of my friends is required.

His friends gave their attention. Stroud, a small town some 15 miles east of his home (as the crow flies) was fixed upon as the 'pleasant place' and a festival arranged for the second week of September.

The performances (five of *The Immortal Hour* and four of a work he had just completed, *The Lily Maid*) took place in the Church Institute—a hall no bigger than the Glastonbury Assembly Rooms, made even smaller, from the audience's point of view, by the presence of a chamber orchestra: twenty-two players, led by Tom Jones and including such eminent artists as Frederick Grinke, Winifred Gaskell, Joy Boughton, Enid Simon, Adolph and Emil Borsdorf, Cecil James, and John Cozens. Anthony Collins conducted *The Immortal Hour*, and Boughton *The Lily Maid*. Costumes were designed by Christina Walshe, and the scenery by Walter Spradbury. The stage director, introduced by Adolph Borsdorf, was a bright young man, Christopher Ede, who rounded off the festival by becoming engaged to Boughton's favourite and most talented daughter, Joy.[1]

Despite the cramped conditions in a hall that quickly grew stuffy and airless, the week was a success. Even Boughton's suspicions may have been allayed by the press comments on *The Lily Maid*:

At the end one comes away feeling that one has experienced genuine tragedy. Not many composers can achieve that, using a musical language that claims little from novelty, eschews sensation, and is quite frank in its appeal to the human emotions. (*The Times*, 11 September.)

The last ten minutes of the second act is worthy of all praise; it is not too much to say that Mr Boughton here surpasses all his previous work, and to add that no English operatic writer has reached that level . . . What distinguishes him from all other writers of today is that he can think in the old-fashioned diatonicism without cliche or convention of any sort, and is not afraid of beauty—real, simple, corporeal beauty. (*The Observer*, 16 September.)

Exactly how far *The Lily Maid* may be considered a 'new' work is difficult to decide, for Boughton had written the libretto as long ago as 1917, designing it as the third part of the Arthurian cycle. There exist a few pages of musical sketches, probably dating from the same period. But it seems that the bulk of the composition was accomplished in 1933, the full score being completed in July 1934. *The Lily Maid* was of particular interest to Shaw, who had been associated with it since 1919 when Boughton had asked for his advice over the libretto. Then he had written:

23 January 1919

Dear Rutland Boughton,

I have sent *The Lily Maid* to you from Ayot: no doubt it has reached you. This letter had to wait to be typed from my shorthand.

From the literary point of view the poem is out of fashion. It is written in that curious language faked by Sir Walter Scott out of Chaucerian English, Euphuism, and old ballads, and never spoken by mortal man at any period of the earth's history. Wardour Street English it came to be called at last. But it had its vogue; and lots of quite good verses were written in it. You could not publish *The Lily Maid* as a poem without producing the Rip Van Winkle effect on a generation which scoffs at Scott and Southey; but as everything connected with music is privileged to be a century behind the times, no one will be surprised at a libretto being coeval with the battle of Waterloo.

This being understood, I see no reason why you should not set *The Lily Maid* to music. You will have to touch it up a little. Some bits are laughably bad—bits cobbled in for the sake of a rhyme or to complete a quatrain; and you have calmly assumed that old English had no grammar, and written in sober earnest the things that Artemus Ward wrote as comic blunders, such as methinketh and so forth. I have signposted these spots with derision and red ink. But in the main the lines are presentable and emotional enough for music in spite of your frequent use of blank verse, which is always rhetorical rather than lyrical.

Of course you are challenging comparisons recklessly all through, not only in the verse with Tennyson and Morris and the Rossettians, but with Wagner who has not left much for you to say in music about the Holy Grail. However, you knew what you were doing; and on your head be it.

When I first read the thing, I thought it was unfinished. I now find the end very good and sufficient.

You had better make Hod a young man with a red head and a bright robust tenor voice, if you can rewrite his part in possible terms. Lancelot, if a tenor at all, must be a melancholy one, the brother, who is rather a bore, should be an alto, so as to have two trebles in the score instead of one. I presume Sir Bernard will be a basso as well as a nuisance, and Gawain the usual villain-baritone.

For an entirely unscrupulous amateur you do very well with your pen. And anyhow, perfect word music would be not only thrown away on a libretto, but spoiled by a superadded score.

I am sorry to have kept it so long; but it took a little time to go through it with care enough to be of any use. I think it is quite successful in what is after all the main thing: the production of a convincing story and picture.

Ever,
G. Bernard Shaw.

On examining the libretto Boughton found that, true to his word, Shaw had adorned it with cheerful advice:[2]

The dews of morn **are on** ⟨*are over*⟩ the
  mead,
The mist **half veils** ⟨*lies over*⟩ the golden   (The tidditytum spoils the quietness)
  brake
Shadowy deer in the quiet feed
Hardly the swan moves on the lake   (Catalogue incomplete without swan)
⟨*On the lawn and drink at the lake*⟩   (Lord!)
The noonday sun is hot in the sky   (Damn it all! What about the dews
  of morn three lines back?)

His horn the hunter bravely winds,
Over the heather the arrows fly
And many a mark he finds   (No, no no! How could the shadowy deer in quiet feed with the huntsman blowing his horn and shooting arrows all over the place simultaneously. Just try to blow a horn and shoot an arrow at the same time.)

But after a play-through on 15 June 1934 he wrote:

24 June 1934

Dear Rutland,

I have been too hardly pressed by rehearsals to say anything about *The Lily Maid* until this evening.

I was greatly struck by the complete consolidation of your harmonic style. Now that Elgar has gone you have the only original English style on the market, free at last of all survivals of the Wagnerism and Debussy-Ravelism that give an exotic flavour to your rivals. I find that I have acquired a strong taste for it. So much so that I even suggested to Kathleen that you should compose a Symphony so as to let your harmony breed undisturbed by words or stories. I thought the unisons of the little chorus quite right, like the unison of the two armed men in *Die Zauberflöte*, and equally unexpected. This work will give you a considerable lift: its quality is more intense and solid than anything you have yet done as a whole.

In haste, ever
G.B.S.

Of the Stroud Festival, which he attended with Sir William and Lady Rothenstein, Shaw made equally pertinent remarks—some of them remarkably prophetic of later twentieth-century developments:

5 October 1934

Dear Rutland,

It is my business to tell you what is wrong and possibly remediable. You may infer that what I dont complain of is right. Next time you must try for a really young Elaine, without an obvious wig. And you must try to overcome your Elgar-like hatred and contempt for singers. A singer is really a musical instrument and not a dustpan. Why not learn to sing yourself? You have a useful voice. It makes an audible noise; why not make the noise agreeable?

I was sorry not to hear the *Hour* again; but I damaged myself slightly (I dont know how) and had to take extra care of myself that week.

I never enjoyed the London performances of the *Hour*; they were unendurable after Glastonbury. There was the same quality at Stroud. It is possible that the success of the Malvern Festival in bringing money to the town may start a fashion among the little municipalities, and induce the Stroud people to give you a decent hall and supplement you with jazz and fireworks and a festival club with rosettes in the buttonhole and all the rest of it. Only, you will have to receive the guests at the club receptions and read the lessons in church at least once.

There is nothing between a week's rehearsal and six months'. A rough and tumble performance with genius on the bridge has a quality that a conventional preparation misses. These desperate adventures are somehow the only memorable ones.

I return the cheque completed. If a few shillings will save you from having to ask the performers to pay up I do not see why you should not send the hat round a selected circle again.

Still one of the faithful,
G.B.S.

It did not, in fact, prove necessary to pass the hat round again. The nine performances cost a little over £618 and the loss came to a few

1. William and Grace Boughton, with their children (left to right) Rutland, Muriel, and Edward, c.1896

2. Rutland Boughton during his Birmingham days, c.1906

3. Florence Hobley, the first Mrs Boughton, on the eve of her departure for Canada, 1911

5. Boughton, about to enter the Glastonbury Assembly Rooms, c.1916

4. Christina Walshe, the second 'Mrs Boughton', 1911

6. Kathleen Davis, the third 'Mrs Boughton', as Mary in *Bethlehem*, 1922

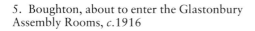

7. A dancing class in the Victoria Rooms, Glastonbury, *c.*1916

8. Margaret Morris as the Wicked Queen in *Snow White and the Seven Little Dwarfs*, 1916

9. *The Birth of Arthur*, Glastonbury Assembly
Rooms, 1921

10. *The Immortal Hour*, Act 2, Glastonbury;
Gwen Ffrangcon-Davies as Etain, and Sheerman
Hand as Eochaidh

11. The Regent Theatre, King's Cross, November 1923

12. Gwen Ffrangcon-Davies as Etain, 1923

13. *Alkestis*, Glastonbury 1922: Kathleen Davis, Handmaiden (left), Astra Desmond, Alkestis (centre), Steuart Wilson, Admetus (right)

14. *The Queen of Cornwall*, Glastonbury 1924: King Mark (Frederick Woodhouse) stabs Tristram (Frank Phillips) before the horrified gaze of Queen Iseult (Gladys Fisher)

15. *Bethlehem*, Bristol Folk Festival School
1922: the Angel Gabriel (Kathleen Beer) greets
Mary (Kathleen Davis)

16. The modern-dress *Bethlehem*, Church
House, Westminster, 1926: Herod, Edward
Nichol (centre), Herodias, Dorothy D'Orsay
(right), Dancer, Penelope Spencer (left)

17. Kilcot, *c*.1930; the lady is the
Boughtons' friend Miss Agnes Thomas

18. Kathleen Boughton, *c*.1950

19. Rutland Boughton, passport photograph,
c.1950

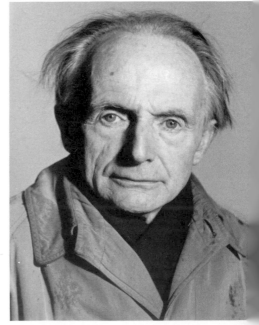

shillings less than £200, which existing guarantees more than covered. But the loss would have been greater had the artists who came to work for Boughton wished to charge anything more than bare expenses. They felt as Steuart Wilson when he wrote, on offering his services: 'You don't know how much we owe to you, and you must realize that we *want* to show it.'

One unexpected thing happened while Boughton was preparing the Stroud Festival. He returned home one evening to find a telegram from Ernest Irving asking him to make a simple setting of a poem by R. D. Blackmore for Basil Dean's film version of *Lorna Doone*. The song, 'Love, and if there be one', was to be sung by Victoria Hopper (Basil Dean's wife) in the role of Lorna Doone to John Loder's John Ridd. Background music for the film was being written by Armstrong Gibbs, who presumably was quite capable of setting the poem itself. It looks, therefore, as if Irving's invitation was by way of being a disguised act of charity—though, in the event, Boughton's song turned out to have such melodic charm that Irving also used it as background music in an orchestral version that completely outclasses Dr Gibbs's contribution. Boughton wrote the song overnight, posted it off next morning, and had the satisfaction of earning £30 for what seemed like very little work. That £30 represented a quarter of his income for the whole year!

The success of the Stroud Festival encouraged Boughton to hope that a touring company might be formed to cover the West of England—a project about which he felt especially deeply now that it was clear that Covent Garden under Sir Thomas Beecham had little intention of opening its doors to British composers, even if it did advertise itself as an 'international' opera-house. Carrying on the Glastonbury practice, Christmas performances of *Bethlehem* were given at Bath, Stroud, and Ross-on-Wye (27 December–2 January), using professional soloists and local choruses. But plans for a permanent touring company did not mature. Neither Steuart Wilson nor the 'two young orchestral players' of the previous year (Adolph and Emil Borsdorf) were able to come to a working agreement with Boughton, for Boughton had very firm ideas on the matter and refused to budge an inch. A letter from Steuart Wilson, written on 24 November 1934, gives a clear idea of the dilemma the would-be collaborators faced:

I am not trying to lessen any of your monetary advantages in the draft contract, only to spread them out a little. I wonder very much whether any manager has ever signed a contract with you in which you reserve a 'Dictator's' position in respect of every 'artistic' condition. We couldn't do business on those terms; all we are left to do is find the money and stand the racket while you impose upon us any production and any artist you like. It is really hardly a serious attempt at business as I understand it, and I'm reluctantly forced to the conclusion that you don't want our plans to go through. If you were really anxious to see the *Hour* toured by us, and that play and *The Lily Maid* put on in London by us you would have grasped at some of the many compromises we have put forward. So I must respectfully draw the conclusion that you want something else. The essence of doing business is the willingness to give way here and there in order to effect a compromise—to put forward a scheme which cannot be varied is impossible—a sort of Versailles treaty to the Germans.

But 'compromise' was not a word that figured in Boughton's vocabulary. And if he was awkward to deal with it was more often than not because he stuck out for things which do credit to his sense of loyalty, if not his business sense. On this occasion one of the difficulties lay in his insistence that Christina should design all the sets and costumes. So far as financial agreements were concerned, he based his requirements on the bitter knowledge that 'composers have to live longer than performers without earning anything; and it is partly due to the willingness of many English composers to compose for glory and teach for a living that they have been recognized as teachers rather than composers'. Such attitudes made reasonable compromise impossible, and plans for Easter performances of *The Immortal Hour* and *The Ever Young*, followed by a Summer Festival featuring *The Round Table*, *The Lily Maid*, and the *Queen of Cornwall* had to be abandoned. In their place Boughton substituted a six-day event organized entirely on his own terms.

This time he chose Bath as his festival town and the 'Pavilion' as his theatre. No sooner had he laid his plans than a major problem arose. The Musicians' Union had negotiated an increase in orchestral players' fees, and Boughton, a long-term admirer of militant solidarity, found himself having to pay more for his orchestra than he had bargained for. The irony of the situation was not lost upon him, but he went ahead and the festival began on 9 September 1935. Six performances of *The Immortal Hour* were given, and four of the work he had written in 1928–9, *The Ever Young*. The orchestra was the largest he ever used: thirty players, led by Louis Willoughby and including such rising names as John Francis and Millicent Silver. This time Boughton shared the conducting with Boyd Neel, while Christina Walshe designed the costumes and Walter Spradbury the scenery. Christopher Ede, having proved his worth at Stroud, was again entrusted with the production. Among the soloists were to be found Astra Desmond, Elsie Suddaby, Steuart Wilson, and Augustus Milner. It was, in short, an ambitious and highly professional undertaking. William McNaught's account, which the *Musical Times* published in October 1935, gives a lively picture of the conditions under which the performances were given and the effect the new work made upon a rather bewildered audience:

The Pavilion is a large, bare, single-floored building, square and rather like a drill-hall fitted with seating. One end was curtained off, and behind the long curtain was everything the building lacked as a theatre—stage and background, wings, storage and dressing rooms. The distinguished company was little better than a troupe of barn-stormers; but, as one of them remarked: 'We will do anything for Boughton, he is so unpractical, always making something out of nothing.'

The orchestra, slightly to the right, was on the audience's level. On the other flank of the stage was a large bleak emptiness which helped the rest of the building to resist illusion, intimacy and magnetism. This was the kind of thing that Boughton and his art set out to defeat. Not to be entirely frustrated by it was a victory in itself. Boughton's idealism is, apart from what it performs, of pure and sterling quality. His world remained intact; the frustration occurred within it.

It is difficult to argue with a man who has his head in the clouds. otherwise one would take Mr Boughton to task for his prodigal waste of good music upon a nebulous and inanimate subject in which nobody can take any interest. . . . The most attractive parts of the work (*The Ever Young*) were those in which the drama stood still and the production became a Masque or a Pageant. . . . The Second Act is almost entirely Masque-like. The Moon-goddess sits on her throne and dispenses her influence upon virgins, children and old people who have left their earthly bodies to visit her. This vision from some other art than opera is a unified and lovely scene and, in its very distance from opera and its freedom from incongruity, is the best part of the work.

The music of *The Ever Young* is full of colour and fire and imaginative energy. In the orchestra it is elaborate and restless and often richly expressive. The score as a whole, while containing some of Mr Boughton's best music, is lacking in big, significant passages that clinch matters and provide holding-places for the mind; and it occasionally drops into bathos. In this it matches the libretto.

Financially the week was a disaster. The increased orchestral fees nearly doubled the anticipated expenses, which eventually totalled £1,090. 19s. 10½d. Of this, £473. 6s. 11d. went on the orchestra. The festival ended with a loss of £557. 19s. 10½d. Shaw was alarmed and immediately sent another £50:

21 September 1935

My dear Rutland,

What an appalling result! Evidently it does not run to an orchestra. And though the *Hour* does well enough with a piano (it was never really itself in London as at Glastonbury) *The Ever Young* would lose heavily without all the delightful descants which continually play around your cantus firmus and which constitute the special art of Rutland Boughton orchestration. And to think that you never knew there was an art of orchestration until I mentioned it to you!

You still have some discoveries to make. Your rough stuff is naive rough stuff and nothing else; it has never occurred to you to seek an artistic form for it. The same with your comic stuff. If one of your characters has to fall downstairs and make an ugly noise you say, very reasonably but not quite artistically, 'Falling downstairs makes an ugly noise and a scramble, therefore all I have to do is make an ugly noise and a scramble with the band.' Mozart would have made a musical staircase like the leaning tower of Pisa out of it. When your male characters are not making love or describing the magic of fairyland, and are being rumbustious, you write prosaically appropriate music for them: that is, horrible music, and make them sing it horribly. I try to imagine how you would have composed the part of Polyphemus in *Acis and Galatea*, or of the ghost in *Don Giovanni*, and thank God that Handel and Mozart averted that horror. You are still only half awake to your art.

Before you print and publish the vocal score, get some literary friend to go over the words with you and straighten up the bits of reckless doggerel in which it abounds. And remember that tryst rhymes, not to kissed, but to Christ.

However, in spite of your ridiculous bits of string the chain holds and the work is a masterpiece.

I have suggested amendments in the circular, which, as it stands, is like a tradesman's letter demanding payment for an overdue account. You must always write to your guarantors as if they were each your most special and intimate friend. 'At your earliest convenience' is an appalling gaffe. You've no manners.

You might at least have offered to raffle a goat. You are practically asking them to make up the £164 deficit; and this is not the moment to tell them that you 'do not choose' to be dependent on their charity. Damn it all, £557 19s 10½d is worth touching the hat for.

Composers cannot be choosers.

Ever,
G.B.S.

In the end, after guarantees had been paid and donations collected, Boughton was left with a personal bill far in excess of his total income for the year. Not an enormous bill, it is true; but then, his income had dwindled to less than £100.

The situation was hopeless. He made one more attempt to organize a series of subscription performances of *The Immortal Hour* and *The Ever Young* in London, 'if you are not quite tired of me and my musical adventures', but the response was inadequate. The Bath Festival was the last Boughton ever held. His personal debt, however, was cleared up in a manner that must have seemed little short of a miracle. Boyd Neel got him the job of rescoring the music of *Pagliacci* for a Richard Tauber film version that was about to be shot. Boughton must have found the task uncongenial, so great was his dislike of Italian verismo, but Leoncavallo's banalities were worth a precious £400 and he set to with as much enthusiasm as he could muster. Irony piled on irony. Tauber decided to give the job to a personal friend and Boughton's score was never used. He got his money, though; and so, presumably, did Tauber's friend.

With the Bath Festival safely paid for, Boughton resigned himself to the impossibility of holding another and, rather than give himself the temptation, turned his attentions away from music for the stage. During the next few years he wrote a number of large-scale orchestral works, including two Oboe Concertos, a Flute Concerto, a Trumpet Concerto, a Concerto for Strings, and, possibly recalling the suggestion Shaw had made to Kathleen in 1934, a Symphony in B minor.

Of these, the Oboe Concerto No. 1 in C, written expressly for his daughter Joy and first performed by her in Oxford on 6 May 1937, achieved instant recognition and soon became a standard for every soloist. Later that year Boyd Neel, who had conducted the first performance, took it to Salzburg as part of a programme that included the first performance of Britten's *Variations on a theme of Frank Bridge*. Leon Goossens was the soloist, and Boyd Neel was able to report: 'A real triumph for the Oboe Concerto. Spontaneous applause after the slow movement . . . the whole concert went wonderfully'—terms which echoed the reception it had received in Oxford. Boughton had been unable to hear either performance: even Oxford, a mere 50 miles from his doorstep, being too far away for his purse.

He was able to hear a one and only performance of the Symphony No. 3 in B minor which he had completed in the autumn of 1937. This was at a private concert arranged by Adolph and Emil Borsdorf at the Kingsway Theatre on 1 January 1939 as a belated sixtieth-birthday present. Critics were not invited, so the occasion passed more or less unnoticed and Boughton seems to have made no attempt thereafter to bring it to a wider audience. It disappeared from view until Edward Downes and the BBC Philharmonic Orchestra broadcast it from Manchester on 10 September 1983. That performance proved a revelation. Though cast in a traditional mould and employing a musical vocabulary that was distinctly Elgarian (and thus scarcely in tune with the 'modern' music even of England in 1937), it emerged from its hibernation as a full-blooded romantic piece—virile and positive in its thematic content, satisfying on every level. Performances by Edward Downes and the BBC Philharmonic revealed that the Second Symphony, the 'Celtic', was also a work of power and beauty, if less immediately satisfactory in its formal construction. It became clear that an injustice had been done in considering Boughton's gifts only in terms of his music-dramas. Though in no way of Vaughan Williams's stature, it is probable that, in the right circumstances, he could have made a significant contribution to British symphonic literature.

None of Boughton's instrumental works, nor the occasional amateur production of *Bethlehem* and *The Immortal Hour*, nor yet his regular musical column for *The Railway Review*, brought in sufficient money for a decent livelihood. His income, an unprecedented £1,946. 0s. 3½d. in 1923 during the first London run of *The Immortal Hour*, had sunk in 1935 to a mere £93. 4s. 5d., and now, in 1937, seemed likely to fall below even that pitiful sum. Out of whatever he earned, however little, he was pledged to pay Flo her share. Fortunately Shaw was always willing to come to the rescue at moments of crisis:

15 April 1937

Dear Rutland,

I must bother you for a moment with your domestic trouble. Florence, as you know, occasionally borrows a pound from me when she is very hard up, and pays it back with the most desolating honesty when you are able to cash up, in spite of the broadest hints from Miss Patch that she had better keep it for future emergencies.

She has now found peace with The Society of the Sacred Heart, an Anglo-Catholic order which take in young and old women who want to lead a religious life and can contribute £80 a year (in round figures) for their board and lodging. She begs me to induce you to enable her to join by promising to let her have your contribution with clocklike regularity. I have pointed out that as you have no clockwork income you could not keep such a promise if you were rash enough to make it. I have had a talk with the Lady Superior (Sister Faith) who is quite willing to take Florence on as a postulant if she can pay her way. She has to be a postulant for six months and a novice for two years before she dons the long veil and becomes a full fledged societaire.

Not knowing what else to do I forked out a cheque for £80 to dispose of Florence for a year. After that, the deluge. Meanwhile she will not starve, and, as the enclosed letter shows, she is for the moment in the seventh heaven. If your fortunes improve you can take her on at that rate at the end of the year. If not, you can easily acquire sufficient virtuosity on the piano accordian to make a living on the highway with a placard on your breast 'I am the composer of *The Immortal Hour.* . . .'.

always yours,
G.B.S.

Needless to say, Florence did not remain in her 'seventh heaven' ('postulant objected to hour of prayer and insisted on tea instead') and soon found ways of troubling Shaw with further applications for help (for which she compensated him with pots of home-made jam), and her husband with legal threats for her share of money that he had not earned. Shaw's verdict (the final paragraph of a letter written on 12 March 1938), 'As usual is nothing whatever against her except that she is impossible', summed up everybody's feelings. She remained importunate for the rest of her life, perversely outliving everyone, even Kathleen, and never apparently able to grasp the realities of her husband's financial situation. A letter Boughton wrote on 26 October 1947, when she had been pestering her son, Arthur Boughton, sets out the whole sorry tale:

To Florence Boughton:

Arthur has now told me why his kind addition to your income was delayed at the beginning of the month; and so I learn that you are treating him to the same kind of consideration and unwise pressure that you have accorded me from time to time.

From the day you failed to take advantage of Lady Isabel's help to enable you to start a new life, to the day you exposed our affairs in a magistrate's court, you have acted in a manner not merely unfair to the family wherein you might have lived a decent and happy life, but also in a way that continually made your own conditions of life less agreeable; for the public actions you have taken from time to time have of course had a bad effect on some people's opinion of me, and so affected my actual income. And when Shaw came to the rescue when I was completely stranded, you and your religious friends positively tried to get your income twice over! I had hoped that the result of your action would have taught you a lesson; and inasmuch as since then your behaviour has been less inconsiderate it almost seemed that you had learned that lesson.

I have always tried to be fair to you, and when my pension was slightly raised in the Spring of last year I increased your allowance without any request from you; though you had not tried to relieve me when you were earning good money during the war. That last fact I learned from Owen. He said you were secretive about it, but we hoped that might be because you were saving money against more difficult times. Now those times have come you have again begun to act in a way that will make them MORE difficult for yourself as well as for us. I shall continue to do my best for you and see that your allowance is equal to the amount upon which I myself live, namely £2 a week. Arthur says he will go on sending you £2 a month, and I will send £6 13s 4d a month; but you must realize that if my income goes down yours must also suffer, while if mine should ever go up (of which there were slight signs recently) you will also benefit.

Yours faithfully,
Rutland Boughton.

By 1937 it was clear to Boughton's friends that he was in financial difficulties, and, in July, Steuart Wilson began to lobby for a Civil List pension. The petition was supported by Sir Granville Bantock, H. C. Colles, Sir Walford Davies, Edward J. Dent, Ralph Vaughan Williams, Herbert Morrison, Bernard Shaw, Lady Londonderry, and Dame Elizabeth Cadbury, and was successful in obtaining for him, for the first time in his life, a small, regular, assured income.[3] On 22 May 1938 Walford Davies, Master of the King's Musick, wrote from Cookham Dean:

My dear Boughton,
   I'm so very glad! It is a modest figure, but outstandingly huge among Civil List allowances to the men and women who give to this country; and I feel it is a more significant encouragement to you, dear man, than appears on the surface. I feel so grateful to Steuart Wilson for his pilot-wisdom in it all.
   Heaven bless your pen and get it busy,

Yours ever,
Walford D.

When his engineer brother, who had spent the years since their childhood amassing a considerable fortune, heard that a Civil List pension was being canvassed he wrote (on 10 March 1938) in great alarm:

Why, Rutland, didn't you let me know just how impossibly small your income had become reduced to in the last few years? I would have liked to have helped you, and conditions were improving for me then—but of course you may not have known that. . . . If you would accept it, and as long as it is possible, we should like to augment the pension to bring it to, say £300 a year, as we feel that to give your best, and there must be still lots of it, you need more than the pension named.

He rounded off his letter with a significant postscript: 'Kick those damned bailiffs out quick.'
   In later years Boughton was inclined to dismiss his Civil List pension on the grounds that he was merely being given with one hand what was being denied him by the other because of the lurking political boycott. Though the suspicion was unworthy of him, it is easy to see how

it could have arisen in the frustration of not finding an outlet for the work of a lifetime. But it was an attitude that perplexed and irritated many of his friends, who felt as Shaw did when on 28 May 1948 he sent a characteristic postcard:

I am only warning you not to pose as a man with a grievance: the worst of public nuisances.

The pension was a miracle of public recognition and leaves you nothing to complain of. Though Elgar was Master of the King's Music I had to give him £1,000 to get him out of money troubles. Of course it was nominally a loan; but I never got it back. You are on velvet comparatively, and should hold yourself up as a precedent for state recognition.

So whatever you do, dont grouse.

G.B.S.[4]

And eventually Boughton's attitude did mellow and he was able to acknowledge the good fortune he had known. 'People have always been kind to me', he wrote. 'I have had more luck than I deserve.'

19

1937–1945

The year 1937 began with one of those acts of faith that Boughton seems always to have drawn from his friends and which are in themselves confirmation of the quality of the man who inspired them. This time it was Steuart Wilson who took the initiative. He arranged to present *The Lily Maid* for a short season beginning on 12 January at London's Winter Gardens Theatre.

Apart from the simple pleasure of seeing his work on a London stage again, Boughton had the peculiar satisfaction of knowing that the money behind the venture came entirely from the BBC, against whom Wilson had waged a successful libel action. Unfortunately the offending remarks (a matter of whether or not his singing was marred by an instrusive 'h') had not gone deep enough for their cash value to sustain a loss-making production. On 19 January Wilson wrote in despair:

My dear Rutland,
the financial position is pretty bad. I cannot go on after the end of this week, because all the money I can put up will have gone and the returns are not going up. They have been—roughly:

	£	s.	d.
Tue:	44	10	0
Wed:	11	14	0
	18	19	0
Thur:	18	8	0
Fri:	15	15	0
Sat:	14	18	0
	24	0	0
TOTAL	£148	4s.	0d.

(On yesterday evening they fell to £10)
There are few signs of choral societies coming in: not more than 10 parties have applied, making a total of about 150 seats which is about £20 in value. *If I can raise more capital and we could say that we could hold it for 3 more weeks we might survive*, but it is useless to hold out for one more week. The running losses are about £600 (the accountant is on the figures as I write), the production was about £700; the fact that we only took £150 in the first and

are likely to take *less* in the second week has made the position untenable except for a man whose capital to be risked is not less than £5,000. The performances are improving. I took the bull by the horns and put in your cuts *at once* and we now finish at 11, which is all to the good.

I have written to G.B.S. to tell him the position—no amount of publicity *alone* can save us. It must be capital plus publicity.

The orchestra and everybody else would help I'm sure, *if* we could promise 3 more weeks, but I can't do that even on a reduced salary scale. I have not lost my faith in the music, not have any of us who are in it—BUT we cannot go on without more money.

Where is it to come from?

yrs,
S.W.

At the end of the two weeks, ironically on Boughton's fifty-ninth birthday, *The Lily Maid* had to be withdrawn. But it was clear that the work had life, for there were signs that, like *The Immortal Hour*, it had the power to make people return a second and third time. Wilson's own opinion never wavered, and after the run he wrote to reassure Boughton:

I think you have produced a much better work than any of the others (I still cherish the memory of *Alkestis*). The musical power of it seems to me to be infinitely greater and you haven't lost the power of knocking us down with a fact stated with such simplicity that at first it seems too simple.[1]

There were to be other disappointments in 1937. In June, Frederick Woodhouse reported a definite American interest in broadcasting *Bethlehem* coast to coast on the Ford Sunday Night Radio Hour. He wrote to Shaw about it, proposing a helpful puff. But Shaw was not impressed and on 7 July scribbled a conclusive reply: 'The idea is not original. In America it is endemic. Tell him you don't need blurbs from Bernard. They wouldn't make twopence difference. Hang it all, I can't get my own works performed; but they think I can get other People's. You will have enemies enough of your own without having mine on your back as well.' Nothing came of the idea. Nor did anything come of revivals of *The Immortal Hour* planned for Covent Garden and Dublin in 1939—though the success of the Royal Academy of Music's 1938 production suggests that they might well have proved popular. But war had intervened and such ventures were out of the question.

Accepting the inescapable, Boughton turned his attention to farm work, adding cows to the varied livestock and even turning a couple of fields with a hand plough in order to plant potatoes and, like everyone else in the country, 'Dig for Victory'. Members of his large family removed themselves to his quiet corner of Gloucestershire, bringing with them a generous supply of grandchildren. There had been eight children in all: three by Flo (Ruby, Estelle, and Arthur, born in 1904, 1907, and 1910); two by Christina (Joy and Maire, born in 1913 and 1916); and three by Kathleen (Sheerman (usually called 'Peter'), Brian, and Jennifer, born in 1924, 1927, and 1928). Nearly all the girls sang (Ruby and Jennifer professionally), Brian studied the trumpet, and Joy had won a reputation as one of the country's leading oboists. On the whole they got on well together. Each 'wife' had taken over her predecessor's children, and all three families were loyal to each other. They had, inevitably, their differences—but these were largely fuelled by the challenge of living with a strong-willed father.

Once the family had assembled, music became part of their daily life and Boughton busied himself with arrangements to suit their varied talents. Kathleen began to learn the cello, and at certain periods they could muster a violin, a viola, a clarinet, a trumpet, Joy's oboe, and, of course, Boughton's own idiosyncratic pianistic skills. These, and the fact that they were accustomed to sing madrigals and partsongs, helped fill the wartime hours. The days flew by: farming, firewatching, Kath helping in Gloucester's 'Club for American and Dominion Service Men', and, for some of the children, service life itself.

Sometimes acknowledgements of Boughton's former musical standing came from unexpected corners of the world:

The Union of Composers of the USSR

Moscow
47, 3rd Miusskaya ul.d.4/6
24 September 1942

Dear Mr Boughton,
    We thank you for the letter in which you and other Musicians of Britain send us greetings and express their readiness to work together with us for the overthrow of Hitlerite Germany.
    We believe that the heroic past of the Soviet and of the British peoples is a guarantee of our victory.
    We send you greetings as the creator of a remarkable cycle of operas about King Arthur the conqueror of the Germans, who waged war against the usurpers on the Islands of Britain.
    May we express the wish, dear Mr Boughton, that you will write, in the near future, an opera about the overthrow of the Germans on the Continent, and we believe that it will be just there where the British Army will deliver

mortal blows to the enemy, similar to the blows which the Red Army is inflicting upon him in the East.

The Presidium of The Union of Soviet Composers
Vano Muradeli, Dimitri Shostakovich, Dimitri Kabalevsky

As it happened, the beleaguered Russians had anticipated the completion of the Arthurian cycle, for it was not until 1943 that Boughton felt able to take up the two remaining parts. Suddenly he understood how they should be. The librettos were written during that October: 'The personalities involved became so clear and intense that I had to work hard to be relieved of them.' He plunged straight into the music of *Galahad*, working sometimes from 4 in the morning until 11 at night, scarcely bothering to eat or sleep. By the end of December the short score was complete and he began the orchestration and elaboration. But the obsessional quality of his efforts proved too much for his health and he suffered the first major illness of his life. Oddly enough he made no great effort to recover and was only nursed back to health by Kathleen's devoted persistence. By August 1944 he was able to resume work and make a start on the last drama of the cycle, *Avalon*. For a while his ideas limped, but during the following year they sprang to life again and long before the war had come to an end he wrote the last pages of the complete cycle.

Looking back on the five music-dramas, Boughton was astonished at the way in which the cycle had developed. Writing to Roger Clark on 25 September 1946 he said:

The Arthurian Cycle has been my life for forty years and if I see its production nothing will be added. The legend is the big peg to which I have clung, but Malory's funny people are not those I have tried to express.

The whole work is indeed an epitome of Christian Civilization during its rise and fall. At the end of *The Round Table* the ball seems to be caught by the Roman Catholics; but the drama of *Galahad* shows how the naïve boy who takes Christianity literally is at once opposed by the official and ecclesiastical world. Then comes *The Lily Maid* with the fascist Lancelot.

The last drama—the most important—shows the collapse of that life because of its inherent contradictions. I hope it also shows that what was true in Christianity remains true.

What he did not say, although he was well aware of the fact, was that the cycle reflected his own mental and emotional development; that in the main characters he had seen echoes of himself and of his friends; that trivial incidents in his life had suddenly found an appropriate place in the dramatic structure; that even poems written fifty years before had found a new home in this work of his old age, his maturity, and his youth. The Arthurian cycle is in every sense a summing up of Rutland Boughton's life.

The last two parts of the cycle were written with little hope of production, and it is, perhaps, significant that in them he reverted to the large, quasi-Wagnerian orchestra of *The Birth of Arthur* and made stage demands of the most elaborate kind. They were written, too, without much encouragement from his friends. Shaw, to whom he had always turned for advice, was as sceptical as ever:

2 September 1944

My dear Rutland,

I thought I had sent you back the enclosures and written about them months ago. I am too old to distinguish between my intentions and my performances.

And now I have really nothing to say. Like Wagner in the *Ring* you are combining differing arts: verbal music, like mine and Shakespeare's, with vocal and orchestral music, and epic with dialogue and drama. They don't always mix very well. Sometimes the result is an immortal masterpiece, like *Messiah*, or *Acis and Galatea*, or *Alexander's Feast*, or *The Immortal Hour*. Sometimes the composer gets the better or worse of a poet, so that the book becomes a libretto that doesn't matter. Rarely, except in simple ballads, does the contrary happen. Anyhow the work cannot be judged except as a whole: nobody can tell what effect the music will make without the words or what the words will sound like with the music. Your epics in rhymed metrical dialogue are full of liberties taken with them because the music will cover them. What, then, can I say except that the epic will be good enough if the music is good enough; so go ahead and see what you can make of it. If you cannot help composing it so much the better. My plays really write themselves: if your scores dont compose themselves, leave them uncomposed.

I am a bit dashed as a critic by the fact that the subject does not attract me as it does you. I can dip into Malory and read ten pages or so, but I have never read him straight through and never shall. I could stop and look at a picture of Arthur, or Lancelot, or Guenevere, of the Lady of Shalott, but I could never write a play about them; and as to the Grail, I would buy it for five shillings if I found it in a shop and it was ornamental enough to put on the mantlepiece, but it would not inspire me to any artistic activity.

Then there is the obligation of having to add something to what Wagner has done, to get Arthur in on top of King Mark, and Galahad on top of Parsifal and Tristan. It would be easy to vulgarize them; but that is not your object; you can keep that for Dagonet, who is not in the least funny and cannot be made so by a Will o' the Wisp leading him into a bog and spoiling his clothes.

The story of these Round Table people is to me hardly more interesting than round tables are as such, whereas *Alkestis* and the *Hour* are dramatic not epic, and therefore are music-dramas and not this unmanageable hybrid, music-epic. The *Ring*, though an epic, is in four fairly dramatic sections, each with its set of strikingly dramatized characters; and you will have to find some new message to compose if your work is not to be wasted as a mere aftermath of Wagner.

You see, I cannot honestly encourage you to go on trying to revive the corpse of King Arthur, who to me is as dead as Titurel. But heaven forbid that I should discourage you: this is to me the worst of crimes. You must do what comes to your soul, advisedly or not. I am really too old now to be of any use; and romantic epic was never my line. I can only say my senile say, of which I advise you not to take the slightest notice.

In art it is no use asking other people's opinions.

G. Bernard Shaw.

In view of the nature of Boughton's transformation of the Arthurian legend it seems possible that Shaw did not read the libretto he had been sent very closely—understandably, for he was nearly 90 years old—and had simply taken it for granted that it was a rehash of the traditional story. But upon the main events of the tale, Boughton had grafted the history of his own emotional and intellectual development. Beginning, in *The Birth of Arthur*, with a vision of salvation through the coming of a national hero—part Wagner, part Carlyle, and perhaps a subconscious reflection of what Boughton himself hoped to achieve through his music—it progresses, in *The Round Table*, through disenchantment with the established Church to an impassioned, humanistic plea for peace. *The Lily Maid* is largely a digression: a story of innocent love betrayed, of intense personal significance to Boughton. His claim that it shows Lancelot acting as a fascist is little more than a desperate attempt to disguise the fact that the story has nothing to do with the main thrust of the cycle. The eponymous hero of *Galahad* is Boughton himself, dismayed that only lip-service is paid to Christ's teachings, but hopeful of salvation when men learn to think for themselves and act as brothers: the true prophet of these conditions being the artist-musician. The various strands of his thinking come together in *Avalon*. Corrupt religion is overthrown, along with the rest of a disordered, discredited society, and communism (a new vision of the Christian ethic) is accepted in its place—just as it had been in Boughton's own life.

All this is implied in his reply to Shaw's understandable but rather cavalier dismissal:

No, you don't discourage me, though I had hoped to get the sort of criticism you gave to *The Lily Maid* in its first form—that helped with Lancelot, who grew as a result.

I must go ahead in spite of you, because you look at my work through Wagner's spectacles and that I won't have.

There is a drama of emotional and mystical life that can only be expressed with the help of music. The Greeks had a hint of it, their dramas roll forward with musical form as a result. Wagner's musical passion lifted his work out of the dramatic into the symphonic form.

I don't care much for Malory, and care nothing for Tennyson; but I do care for the fight Christianity has made against the world, even wherein Jesus has proved ineffective; and so I must carry through my experience of that fight— with less of Wagner's musical passion, with more than Wagner's awareness of the religious nature of what I am doing.[2]

Now that it was complete there remained only the possibility of performing the cycle. Yet even this seemed of no great importance compared with the fact that the work had at last been completed. 'Should I fail of an opportunity of supervising the first production of these works', he wrote, 'there will be no reason for complaint, inasmuch as I have been enabled to enjoy in peace and quiet these last years of their making; and experience has taught me that any production of this sort attains only in a lesser degree that top of the world feeling associated with its conception and growth.'

20

1946–1960

Now that his life's work was complete the days moved quietly for Rutland Boughton. But if they had less of the old urgency, they still had much to offer. He continued to compose—mainly now for his family and friends. For Kathleen he wrote a Cello Sonata (1948) and a set of short pieces (1955) whimsically entitled *Kilcoteri*; for his friend Charles Pope, conductor of Aylesbury's orchestral society, he wrote the three-movement *Aylesbury Games* (1952). In 1945 he completed a second Oboe Quartet for his daughter Joy; and in the same year, as a very personal message to his whole family, he wrote his last full-scale orchestral work.

He called it the *Reunion Variations*, and in a letter to a friend, written on 29 March 1947, he explained its origins:

At the end of 1943, when I thought I was dying, a tune came to me complete with words. I wrote it down, and after my recovery (due solely to the insistence of Kathleen) I made a set of variations for orchestra in which the different members of my family are characterized, though there is no separate variation for Kathleen herself, because she is the central tune of us all, and so the work comes to an end in the actual singing of the words. Here they are:

My love to all my dear ones, wherever they may be,
When I have passed to Nowhere, of all the earth made free
I go whence came my melody, where love to music burns.
In music I come back again, and so my love returns.

For music comes from Nowhere and into Nowhere goes,
But love still burns as sunshine in the secret heart of the rose.
See! Ariel is laughing in the mirror of the sun,
And Ariel with Raphael and Gabriel is one.

If Raphael bring heaven's brotherhood and Gabriel bring heaven's word,
'Tis Ariel brings heaven's architones to every child and singing bird.

'Tis Ariel who has whispered some undertones to me;
'Tis Ariel who is calling through that star-enchanted sea.

He weaves the waves to mingle in paths of rainbow sound,
'Till Nowhere leads to Everywhere, and earth to heaven is bound.

> So sing your songs love-hearted throughout the stelline blue,
> When you on earth make music I shall be there with you.
> My love to all my dear ones, wherever they may be,
> Or on the earth yet singing, or in the air with me.

Now I have an object in introducing this matter to you. When I really go, my dear, I want you to let Kathleen know the nature of the score which is called 'Reunion Variations', and perhaps arrange for its performance. I have sometimes thought of trying to get a performance as by an anonymous composer, but so far haven't found a means of doing it without giving the show away.[1]

It was his last attempt to come to terms with the 'metaphysical element that [had] rejoiced and confounded [him] throughout life'. Without it he would 'never have been a musician at all', but he still found it hard to explain what it was that constantly 'beat into music', and why the numinous should be so powerful.

Whatever Boughton may have believed about the 'neglect' of his music, the truth was that it was a comparative neglect of the kind only to be expected of a largely operatic output. There were, in fact, a surprising number of performances, mostly amateur, of *Bethlehem* and *The Immortal Hour*, and sometimes even of *Alkestis* and *The Queen of Cornwall*. There was even the occasional broadcast. And after each occasion came letters of gratitude from unknown admirers, from friends and old associates.

I went on Saturday to hear *The Immortal Hour* at Walthamstow. . . . It brought back memories of the exciting times at Glastonbury, and I still find it as beautiful and moving as ever—indeed I think I find it more beautiful and original than I did before. People in the old days used to day it was Wagnerian, but that has all faded away now and it really is very much all of a piece: a complete musical unity and completely and uniquely your own. I really began to think it is one of the best operas ever written, and when I say that you must bear in mind that I judge all operas as a complete whole on the stage, not by details of music or libretto . . .

wrote Edward Dent on 16 December 1946.

And sometimes the response revealed wounds that went as deep as his own:

It is a great pleasure for me to see that you are HEARD again, unlike many of us. Your fine work, I'm told, has been ignord by the idiot BBC? I hope not. In these days I do not read much on music. Some of us have done our best in a moribund land?, and OF COURSE the 'new' ones get the reward, or DO they? Not one of the BBC coterie seem able to write a melody or interest any of my friends! So much for wealth in music? Go to it, lad, and do your bit. Britain never lets her artists starve? they only ignore them?

A FINE OPERA, yours,

Joseph Holbrooke.[2]

But more often it was a matter of simple celebration, as when, on 21 June 1953, Vaughan Williams wrote of the Sadler's Wells production of *The Immortal Hour*:

It was a great pleasure to hear *The Immortal Hour* again; there are some lovely things in it and you are not afraid of writing a tune. On the whole I liked the first act better than the second, except for the druid's song; and as to the luring song, it is like the 'Marseillaise' or the National Anthem, 'hors de combat'.

I first heard it with you playing magnificently on the pianoforte years and years ago at Glastonbury, when you gathered together a noble company of enthusiasts. I so well remember that we did not take ourselves too seriously to be frightened of relaxing in the evening when Clive Carey and Johnstone Douglas used to improvise on pianofortes that were slightly out of tune with each other, and Fellowes was there with his new-found Madrigals, and Bernard Shaw propounded truisms from Cherubini, imagining that he had discovered them himself; and Katie and Bea Larpent set up a restaurant which they called the 'Cramalot', which rather shocked the more pure minded of your audience. Well we were young then, but I believe you remain young still.

And if to have a mind actively concerned with all the problems of a post-war world is any sign of youth, Boughton was young indeed. In 1945 he rejoined the Communist Party, believing it to offer the only sure path to a new and vital civilization. In this spirit he associated himself with the 1948 'Peace Conference of Intellectuals' at Wroclaw, even though the journey there, and on to Czechoslovakia, was taxing for a man of his age. It is less pleasant to record that, once there, his enthusiasm for peace led him to see symbolic depths in quite trivial incidents. His description of a presentation made to the Wroclaw delegation is a case in point:

The present was a cast iron statuette of a foundryman with his ladle. In cast iron! What nasty forms those words convey. But the statuette was as finished and delicate in form as a Rodin marble. And think of the moral significance of that production: one of the great munition works of the world (the works we helped to hand to Hitler in 1938) was turning its attention to the works of peace and art.

It is difficult to credit Boughton with the belief that the Skoda Works had actually decided to flood the market with cast iron foundrymen and the like, yet this is what his words imply. Did he really believe in the symbol? Was he totally unable to recognize sentimental propaganda when he saw it?

It is possible to deplore naïvety and still admire the honesty that can allow itself to be so misled, but it would be more reassuring had Boughton been able to find it within himself to assess the political facts with the clarity that moved Vaughan Williams to reply to his suggestion that he should support the 'Congress of Peace, Trade and Friendship with the USSR' at the St Pancras Town Hall on 12 June 1949: 'In reply to your letter of April 5th, as soon as the people of Russia are allowed by their tyrants to know anything about European conditions, and are allowed to express an opinion of their own, I will gladly support any movement for peace and friendship with the *people* of Russia.' But for Boughton communism could do no wrong. In 1948 he had seen in Zhdanov's attack on formalism and decadence among Soviet composers (spelt out on 10 February in the Central Committee's resolution: 'On the opera *Velikaya Druzhba* by Muradeli') no more than a confirmation of his own misgivings about the confusions of twentieth-century music. Writing to Alan Bush on 18 February he declared:

The Russian C.P. criticism of composers is, as you can imagine, a great satisfaction to me. Having struggled with a lone hand and under the implications of being a musical reactionary, it is good to see daybreak ahead. It will also be good for you who, having been born into a musical decadence, yet struggled against it when faced with a living job—e.g. the Co-operative Pageant at the Crystal Palace. And it seems to be of vital importance for our younger men (even those younger than you) that we analyse the matter more fully for ourselves. . . . The younger men mustn't feel that they have been misled or deceived, but understand that our mistakes are due to our time and conditions.

In 1950 his views were unequivocal. Writing to Kingsley Martin on 30 November he said:

That Russia is naturally sympathetic to Korean Communism is as inevitable as Capitalist Governments' sympathies for Rhee; and such reactions apply no less to Europe. But what right have you to suggest from time to time that there is actual behind-the-scenes Russian aggression? If that could be proved, the West would not be divided.

The fact surely is that all aggressive action, military or economic, originates from the political Right, and the people instinctively know it, even when bewildered or forced to bow to it.

But six years later the events of the Hungarian revolt could not be explained away quite to glibly. Stunned and shocked, Boughton at first wondered if there had been some mistake. But the proof was beyond doubt. He acted immediately and with all his old courage: he resigned from the Party he had trusted for more than thirty years.

The letters he wrote at the time (November and December 1956) reveal how terrible a decision is was to reach:

It hurts horribly to be detached from the Party which contains three of my closest friends. But as I see it today the adhering members are putting Party before Communism itself. When I seceded in the twenties it was on a purely business and personal point and was with the friendly understanding of both Harry Pollitt and John Campbell, and I can truly say that in all years that followed I remained true and openly true to communist principles as I understood them. The second half of my work is bound up with those principles; but the Party never had any real use for me. They reveal the reason when they refer to me as a 'shining decoration'—which only a fool would want to be. We believed the Party because we had so much evidence that the capitalist press reports were slandering the Russians—we should have been told the truth.

In answer to a similar cry of pain, Vaughan Williams replied (9 December 1956) with words of common sense and comfort:

I feel much touched at your taking the trouble to explain your new position to me. It seems to me that all right-minded people are communists, as far as that word means everything should be done for the common good. But communism has come to mean so much other than that, which I, for one, cannot subscribe to. Myself, I see no reason why the Russian atrocities should prevent your remaining a communist in the ideal sense of the word.

The Russians are a strange mixture of artistic ideals and barbarism, and the barbarians seem to come out top, whether you call them Czarists, Karenskyists, or Stalinists.

To the end of his life Boughton retained the belief that some form of communism was the natural goal of Christian civilization, and that peace and goodwill would only be attained in a world where people could live in conditions of approximate equality. He did not become embittered by the betrayal of these ideals, only more tolerant of the possibility that there were other and less dogmatic ways of reaching the same ends. Not least of them, though he may never have recognized it, was the example of his own integrity—an integrity that had become so intimate a part of the fabric of his life that, to his own undoing, he expected to find it wherever he looked.

Beside this one event all other incidents in Boughton's last years seem unimportant. Indeed, it is of comparatively little significance that *The*

Immortal Hour was successfully revived at the People's Palace in 1949 and 1950, and by Sadler's Wells in May 1953, that the BBC broadcast *Bethlehem* in 1949 and *The Queen of Cornwall* in 1950, and that there were two dozen amateur performances of *Bethlehem* between 1950 and 1958 in places as far apart as Australia, Ceylon, Rhodesia, and the length and breadth of England. Nor does it matter that each successive birthday became a reason for producing his work, or broadcasting, or writing about him. What matters is that he enjoyed life as vigorously and actively as ever.

Little escaped his attention, either in world affairs or in the achievements of his fellow musicians. Just as he had been quick to welcome George Lloyd's opera *Iernin* when it reached London in 1935 ('*Iernin* rings like a well-cast bell'),[3] he was generous in praise of Inglis Gundry's *The Partisans* (1946) and Sir Arthur Bliss's *The Olympians* (1949)—'proof enough that there are younger men intent on the goal towards which I have aspired, and with a better chance of reaching it'.[4] More importantly, he was deeply impressed by *Peter Grimes*:

3 July 1945

Dear Benjamin Britten,

Hearing the interlude from 'Peter Grimes' at your Cheltenham rehearsal, I was compelled to run up to town for the complete work, and I rejoice in it even though my old ears cannot always accept your dissonances, though my 3 days of Cheltenham give me hope that they (my ears) are still capable of education. I am ordering the vocal score of *P.G.* from B & H, in the hope of a more real understanding of the work. The relation of voices to orchestra I thought was completely satisfactory, & the scene with the fog-horn very fine.

Don't you think the lighting might be increased with advantage in some of the other scenes? It is only on a very dark night that one cannot see much better than your producer seemed to think.

With very real congratulations

I am

Sincerely yrs,

Rutland Boughton.

P.S. Is there likely to be a London performance soon after Sept 1? On that day I am lecturing on opera at the Regent St. Polytechnic & would like to refer to *P.G.* if those references could be followed up by a performance.[5]

It is pleasant to note that Britten, though utterly different from Boughton in operatic achievement (even if he too had felt the need to establish a 'Glastonbury' at Aldeburgh) was willing to acknowledge the older man's importance in the history of British opera. Writing to Brian Boughton in 1963 he said: 'I had the highest admiration for your father's principles & courage, & I think he is often misjudged. I hope his music will "come back" when fashion changes (as it <u>always</u> does!)[6]

Britten's operatic method, however, was far removed from Boughton's ideal. Much more to his taste was Vaughan Williams's operatic morality *The Pilgrim's Progress*. On 12 May 1951, after hearing the BBC's transmission from Covent Garden, Boughton wrote to express his satisfaction at his friend's conversion to something akin to his own operatic convictions:

Dear Ralph,

The OUP tells me that there is at present no published vocal score of your *Pilgrim's Progress*. That is a pity, as I'm sure I'm not the only musician who wants to get at it more intimately than is possible by radio. However, what came over was enough to assure me that it is the complete YOU, disembarrassed of your age. Apollyon was less terrible than your Satan in *Job*, and I was sorry that you had left out Mr Valiant-for-Truth; but what gave me special satisfaction was the fact that in casting off the absurdities of opera you had discovered the realities of music as dramatic expression, as indeed you did in *Job*.

I wrote to Willcocks of Worcester, and to Steuart, asking if they could not arrange to have the Covent Garden crowd in the Cathedral about the time of the next Three Choirs Festival, for the dramatic values of the work would come through more completely in such a building, and what was lost of lighting, grease-paint, and the usual operatic milieu, would be a positive advantage. But Willcocks says it couldn't be arranged in time; and Steuart says that so much of the production (and even some of the actual music!) would have to be altered, and that says little of the C.G. production staff; for a producer who is not prepared to put on any work in the courtyard of a pub, a Greek theatre, a cathedral, or even a modern theatre, simply doesn't know his job. But such excuses are just part of the topsy-turvy art world which prepares *Hugh the Drover* for concert performance and does Chester Plays in a hall. The anti-Christians have got such a hold on our life that they are afraid of Christian vitality in Christian buildings (and by the way, your Pilgrim seems to be afraid of his Christian name), still more of admitting Christian thought into our theatres. That is where you have scored in the superpersonal sense, for you have carried the thought to Covent Garden in spite of them.

However, until C.G. is controlled in the service of the people wholly and without reserve, and not chiefly by Ricordi-interests, they being controlled in their turn by American finance—until that time *Pilgrim's Progress* will continue to be a magnificent anomaly there, and will not yield up its true spirit until the whole set-up of our national opera is honestly national. Fortunately there are signs that the better thing is coming, the most important being the wholesale resignations at the Old Vic school. Another sign of the better appreciation of reality was your splendid radio talk on the birthday of Sibelius. For that, and for the proof in *Pilgrim's Progress* that you meant what you said, British music will owe you a deep debt.

Yours ever,

Rutland.[7]

Though he took a special interest in the achievements of those composers who shared his political beliefs, receiving visits from and corresponding with, among others, Christian Darnton, Bernard Stevens, and Alan Bush, he often found it hard to tolerate their allegiance to a musical style that he could only regard as mere cacophony. Occasionally he detected signs of repentence, as Alan Bush recorded in 1960:

Boughton did not care particularly for my compositions, though he recommended those that I wrote for the London Labour Choral Union to them. So I was thrilled to receive an enthusiastic letter about a choral piece I wrote to my wife's text called *Our Song*. In 1948, when I started serious work on *Wat Tyler*, I went to live in Newent, the neighbouring town to Boughton's remote home. I used to bicycle over to see him once a week and play through what I had written. This proved rather depressing for me. I never received a word of praise from him and, on one occasion, when I had spent the week working six or seven hours a day on Wat Tyler's meditation, one of the most important passages in the whole opera and one I think to be among the best things in it, after hearing it sung through he remarked in a surprised tone: 'Is that all you've done this week?' In the light of this it is not difficult to imagine how I rejoiced at his real enthusiasm for *Men of Blackmoor*.[8]

But if Boughton found it impossible to assimilate the way music had developed in the twentieth century, he had no reservations about the art of Paul Robeson who, after visiting him in 1959, wrote on 24 November from Stratford-upon-Avon where he was playing Othello: 'I did so enjoy the day with you, your family, and friends. For many years you and your work have given me great courage and faith. So thank you for receiving me as a friend.'[9]

In his personal relations it is pleasant to record that as they both grew older both he and Flo were able to deal more gently with each other. For her part, Flo seems to have abandoned that immediate recourse to solicitors whenever she fancied herself short-changed in her claim on his supposed income. And Boughton, in consequence, was able to treat her with the same friendly concern that had always informed his relationship with the equally 'wronged' Christina. Letters in the last two years of his life prove the point:

24 January 1958

Dear Flo,

it was very good of you to send me 10/- for a birthday present. I fear it means you will deprive yourself of things you ought to have. As things are just now it will go on stamps to acknowledge all the kind messages that arrived yesterday—21 telegrams, and letters and cards so many that I shan't be able to reply to all of them before my eye op[eration] which has been postponed till Feb 3rd. One very nice letter came from Christina who now lives in France, teaching in a school.

Our gratitude and love to you. May you be rewarded as the blessed one who gives. I shall hope to spend something on myself later on, as you suggest.

R.

31 July 1958

Dear Flo,

Herewith August cheque.

We have just been to a school where the girls performed *The Moon Maiden*—the piece in which dear Ruby used to be the chief figure. It was beautifully performed, and brought tears to my eyes We can't manage a holi day yet, but I should like you to have one. If you do, let me know and I'll send a little extra pocket money, as I shall get a fee for *The Moon Maiden*. With love to you from

Kath and Rutland.

As for Kathleen, his relations with her were as loving as they had ever been. As Shaw had guessed, she had been able to 'hold' him despite the difference in their ages. Something of the quality of their love can be seen in a letter, typical of many, that he wrote when she was in hospital in 1952 recovering from a serious operation. In it he excuses himself from the challenge of choosing something for her to wear:

Thirty years ago you were only just in love and everything I did you saw with angel-eyes. Even then I chose one thing (a snaky piece) which you wore only a few times though you never complained of it. But now, when you're every bit as lovely but much more particular, I don't want to choose something which is in any way unsuitable. In those days you had a seraphic little nose that went with everything; but now you have stronger features which need stronger stuff to decorate them.

I do hope you slept better last night. I did! You know, that effect you had of slipping away from yourself was really your mind trying to slip away and come to me; but it couldn't unless your body came as well, so the sooner you decide that your husband is the one best suited to have charge of that precious being the better for us both. Being in charge of the house without you in it is like being in charge of a casket without any jewel inside, nevertheless, I'll do my best by the casket till the jewel is put back.

Your

R.

Kathleen recovered and returned to him and the sanctuary of their tranquil home. And in these last years he seemed to take a keener pleasure in talking of the past and meeting old friends. But now there were fewer of them. Scarcely a month went by without some link broken— Shaw, Christina.[10] Soon it would be his turn.

The idea did not alarm him: 'At my age death does not seem such a bad thing.' He had done what he set out to do, and in all things had given his whole heart. Now he knew the tranquillity that comes only of fulfilment. If there were any fears, they were natural to a man of his age. In 1952 (22 April) he had written to his friend and medical adviser, Dr Tomlinson, in typically forthright terms:

Dear Pip,

Legs inclined to give way at times. Physical work followed more frequently by giddyness and eye-haze.

It occurs to me that such may be premonitory symptoms of a stroke. I don't want any dosing or doctoring; but if my suspicion should (ever) prove correct, kindly speed the lingering guest.

Kathleen does not know of this letter . . .

Fortunately he was spared the indignity of physical and mental decay. He continued to carry out the daily duties about the house— chopping wood and bringing in the coal, tending the chugging petrol engine that drove the household's electric dynamo. His eighty-second birthday came. On the following day he left Kilcot, travelling with his beloved Kathleen to London and his daughter Joy's home, to celebrate in the company of his children and their children. The party was a happy one.

Sometime during the night that followed he died—so quietly that he cannot have known where sleep ended and death began.[11]

Rutland Boughton
His Work

21

Prose Writings

Throughout his adult life Rutland Boughton read and wrote. He read because he needed to make sense of the world and its affairs, of himself and his feelings, and in particular of the religion he had been brought up in. Although he often wrote in order to supplement his income, he also wrote in order to clarify his understanding and share his ideas.

Boughton's journey of self-discovery was made almost entirely in isolation. As widely and deeply read as any university man, he conducted his education without the leavening discourse of other seekers, eager with argument and counter-argument, ready to mock pretension and expose false assumption. Moreover, he was by nature too passionate and involved ever to be truly objective. Truths that might more reasonably have been accepted as provisional were all too often claimed as absolute—at least until dethroned by new 'truths' or, less palatably, by bitter experience. But though he could be rash in his judgements, Boughton was never dull. His errors sprang from enthusiasm and total commitment. Where others might have been content merely to acquiesce, he fought for his beliefs.

In various attempts at autobiography Boughton charted those books that had the most profound influence on the process of his self-education. This is how he saw it:

Having read [Darwin] I felt as if I had traced through a long catalogue of facts that were incompatible with many things I had been taught at school and Sunday school, but I did not understand to what desirable end this mass of facts was being co-ordinated. I was glad enough to get rid of the God of Adam and Abel, but he seemed less important than the God who revealed the Kingdom of Heaven in the hearts of men. In some troublesome way those two Gods have centaured in our minds. So long as music existed, the Kingdom of Heaven existed. On that half of the old centaur Darwin threw no fresh light . . . and at that point I was struck for a year or two.

One day I picked up a copy of Ruskin's *The Two Paths* and thought I had found the right gospel at last—an idea which received support in the approval of my friends. We agree that even more important than the composition of music was the need to spread Ruskin's gospel of art. In spite of Darwin, and putting Ruskin's anti-Darwinism on one side, it seemed irrefutable that art was concerned with something more important than mere beauty. We could now recover that moral basis for our work which had seemed to disappear with the religion we had discarded. . . . But Ruskin could not face up to Darwin's evolutionary theory, and to that extent I felt he was missing something important.

In 1906 I came across Edward Carpenter's book of free-verse, *Towards Democracy*. This gave me just the material I wanted for the musical expression of my faith. His poems accepted Evolution as a basis of life, or at least they were not incompatible with it. I then read his *Art of Creation*, and this gave me the mental stability I needed by first paying tribute to scientific fact and then showing that the artistic impulse was not a merely mechanical thing. It helped me to resolve the contradiction between my acceptance of Evolution as the impulse of life and my feeling that music arose in a greater power than that contained by the individual composer. But it did not resolve the contradiction between my emotional adherence to Christianity and my feeling that churches were forces of reaction.

In 1909, however, Chesterton's *Orthodoxy* appeared and its message helped to reconcile the contradictions I felt. The fact is that although I am a Communist and Materialist, I belong to a civilization in which all that is good has been developed under the banner of Christianity. The symbolic figure of Jesus as the nearly perfect man, and the English of the Bible must always arouse my admiration and satisfaction, in spite of the humbug that has been associated with them. Chesterton's book expressed those feelings with a cleverness that made it a crucial part of my early manhood.

A year or two later, at one of the chief crises of my life, I came across Allen Upward's *The New Word*, and that was to be my bible for the next thirty or forty years. By appealing to various branches of science and then dismissing them at the point where he believed they revealed their limitations, Upward seemed to give surer ground than Carpenter for the resolution of the contradiction between Materialism and Creativity that I was always seeking. Moreover, whereas *The Art of Creation* remained chiefly in the mental realm, *The New Word* pressed forward into the world of reality and action.

Then came the personal influence of George Bernard Shaw, whose books on Ibsen and Wagner, and essay on *The Sanity of Art* I already knew. His advice was never given in dogmatic form, but by suggestion. It was never until I acted upon it that I realized how great his influence had been. He was one of those men who, like Jesus, lead us by suggestion and parable, always refraining from imposing his own personality beyond what he saw might be acceptable or endurable.

Allen Upward wrote no subsequent book of equal value to *The New Word*. In *The Divine Mystery* he moved towards a mystico-theological point of view, though he insisted on the mortal manhood of Jesus. For me creative art, and especially music, was still predominantly a 'mystery', but it was the 'divine' element in Upward's book that made it useless to me. Shaw's idea of a creative evolution, as expressed in *Back to Methuselah* and *Man and Superman*, seemed more satisfactory.

Then came Havelock Ellis's *The Dance of Life* and its reference to Sir Flinders Petrie's *The Revolutions of Civilization*. It was Petrie's analysis of the progress of the arts in succeeding civilizations that governed my thoughts for years to come, though I rejected his conservative outlook which imputed the downfall of civilizations to the envy of the 'have-nots' rather than to the greed of the 'haves'. On thinking about his ideas there seemed to be not merely a determined order in the succession of human achievements—Sculpture, Literature, Music, Mechanics, Science, Wealth, and then the downfall of Civilization—but also within each individual Art a similar rise and fall. First a steady progress in the development of the art: then a decline due to a love of beauty itself, separated from the need to express reality; then a revolt from mere prettiness in an effort to re-establish the original function of living artistic expression; then another and more serious decline; and finally a decline in which the arts disintegrate and fall into absurdity and ineptitude. This analysis of the rise and fall of the arts was henceforth the basis of my understanding.

It was not until I was nearly seventy years old that I met with a book that cleared my mind of all that was vague and mysterious—Christopher Caudwell's *Illusion and Reality* (1937), a book that shows the true relationship between the man in the scientific laboratory, the man in the factory or field, and the artist in the concert hall or studio. In this book I was able to get the essence of Karl Marx without trying anymore to plough through *Das Kapital*. It enabled me to go ahead as an artist with self-respect for an activity which indeed has often seemed as useless when I thought of the abyss between the artists and the men and women they most need to serve.

For me Caudwell's book is so grateful an achievement because it clarifies the problem of vitality in art as it issues in the course of material evolution. He accepts what so many Marxists seem to deny—that there is a point where the scientific explanation of life necessarily comes to an end. He then shows how that point is beyond our purview just because as life pressed forwards in terms of science and art, the expanding universe need present no anxiety to those who live in the reality of the present and help that very expansion by being honest in their ideas of truth and beauty.

All that lies beyond our understanding seems 'mysterious'. In its origins the word seems to have a meaning that made it fitting enough to express the musical impulse that closes its eyes on the outer world and sinks into the self. But just as that original sense of the word was later qualified by men who used it to refer to the secrets of their work as craftsmen, so 'mystery' became 'mastery' also. Caudwell showed how the mystery of closed eyes in the production of music is of no avail unless related to that life which exists only outside the individual—the social life of mankind, which can only be realized by ears and eyes that are very open indeed.

Though Caudwell is chiefly concerned with the sources of poetry, his appreciation of the function of music is true and deep. He leaves only one problem unclarified—that of Music Drama. The relation of music to the dance he deals with; but the more complicated problem of a drama that can only be expressed through music he does not touch upon. That, however, can be deduced from the principles he enunciates. Finally, he places in clear relief the weakness and humbug of the art we call Futurism and Surrealism. He shows just how and why art takes such forms in a period of human decline, explaining it in terms that are kind and generous to the artists who are unlucky enough to be caught in its toils.

Thus Boughton's Progress in This World to That Which He Believed Might Come.

Upward's ideas would have confirmed Boughton's disgust with the Church as amply as his admiration for the man in whose name it had been founded. They would have given new heart to the idealist in him, even while confirming his expectation of travail and misunderstanding. They may also, as Nietzsche and Shaw were to do in the 1920s, have contributed to a 'superman' sense of being set apart from ordinary mortals and answerable only to his own vision of what was good, beautiful, and appropriate. They were, in short, admirable fuel for a man intent upon following his own star.

Where Petrie's Spenglerian revolutions of civilization confirmed Boughton's belief that he was living in a period of decadence and that it was his duty to fight against the fads and fashions of musical 'modernism', Caudwell's analysis offered both additional confirmation and a way out. 'Social relations', declared Caudwell, 'must be changed so that love returns to the earth and man is not only wiser but more full of emotion.' He saw history as a series of revolutionary movements that broke down the barriers to the self-realization of the individual, yet at the same time raised barriers of a different kind. Thus the feudal system gave men more freedom than they had enjoyed as a mere tribe, but less freedom than they were to enjoy under the bourgeois system that replaced it. So also would communism break down the barriers that thwart the development of the individual in bourgeois society. Logic enabled him to see what Boughton, at least for a time, could not see: that communism in its turn would create barriers that would have to be scaled in subsequent revolutions. Fortunately, when it came to the test (in 1956), Boughton recognized that the need for genuine freedom far outweighed the apparent truths of Marxist dogma.

Apart from their clear intelligence and insatiable desire to find a moral basis for a world that seemed to be growing less and less rational or obliged to any morality whatsoever, Boughton's mentors had one thing in common: they were all possessed of a fearful fluency. With the acceptance in 1898 of his first (unpaid) published articles, Boughton discovered in himself a similar capacity.

Even allowing for the fact that from the turn of the century his writings became a useful way of supplementing his income, the flow is remarkable. Besides his contribution as a working journalist, as music critic, briefly, on the *Daily Mail* (1903), and, at greater length, on the *Daily Herald* (1923-4), and as a regular contributor to the *Railway Review* (1933-4), he wrote something in the region of 200 substantial articles for the leading musical journals of the day, besides any number of one-off essays for such papers as the *Clarion* and *T P'S Weekly*, and such magazines as the *Scallop Shell*, the *Musical Quarterly*, *Theatre Craft*, the *Philharmonic Post*, and the *Aria*.

He found his chief outlet in the *Musical Standard* (1898–1911), *Musical Opinion* (1898–1925), the *Music Student* (1908–16), the *Musical Times* (1910–31), the *Musical News and Herald* (1923–8), and the *Sackbut* (1924–31)—the last named being of particular importance at a time when he was moved to examine the relationship between art and politics. In all these magazines he expressed himself vigorously, and often idiosyncratically, on all manner of musical topics—from Beethoven to Gottschalk, Loewe to Richard Strauss, Early English Music to the music of Algernon Ashton and William Baines—constantly probing the state of music in Great Britain, and the position of the artist in a world grown fat and philistine.

Then there are his larger published writings: ten in all. Three are full-scale books: the small *Bach* of 1907, a brief, factual account of his life and an enthusiastic commentary on certain aspects of his music; the larger *Bach* of 1930, in which the composer is strait-jacketed by Boughton's political beliefs; and *The Reality of Music* (1934), which sets out Boughton's credo in as much detail as he ever committed to paper. Next come two short books: *Music Drama of the Future*, written in 1910 in collaboration with Reginald Ramsden Buckley, and *Parsifal*, a study of Wagner's masterpiece, published in 1920; and finally four pamphlets: *The Self-Advertisement of Rutland Boughton* (1909), a wonderfully entertaining annotated catalogue of his works to that date; *The Death and Resurrection of the Musical Festival* (1912); *A National Music Drama: The Glastonbury Festivals* (1918), a reprint of an address delivered to the Royal Musical Association on 17 December 1917; and the brief *Music and the Co-operative Movement*, written in 1929 in collaboration with Fred Hall.

Add to these publications the innumerable subjects on which he was willing to lecture—everything from 'Edward Carpenter' to 'Edward Elgar', explanations of his own music to accounts of 'Music in Soviet Russia'—and it will be seen that Rutland Boughton's mind was in ferment and that he was eager to share his ideas with the widest possible audience. To read every word he wrote is to risk becoming alienated by the things he got wrong. Some, admittedly, originated in the aesthetic perceptions of the period that have since been amended by musicological research, but a great many were the outcome of his own fondly held prejudices which are perhaps less easy to accommodate. This, however, would be to miss the greater number of stimulating ideas he had on offer. We may not always agree with them, nor are they necessarily original, but their liveliness and interest cannot be denied. Of all British composers of his generation he is the one most deeply concerned with the philosophy of aesthetics and its social application.

Though they reflect the opinion of the times all too clearly, some of the ideas put forward in his earliest writings are hard to swallow. In *Britannia Singing!*, a series of studies of British music from folksong to the present day (1905), he allowed his dislike of Roman Catholicism to blossom into galloping hyperbole.[1]

All Churches are the enemies of the people, and Rome is the mother of them all, and will be the last survivor of them. Rome is a monstrous phantom of the night, born of cruelty and fear. Rome hates the dawn with all the blind cruelty of the owl. Elgar is a bleeding lark in her claws, and she is carrying him, or rather his work, with her into the shades. He may yet be loosed from her clutches, but just as his best work is least pervaded by Roman doctrine, so he will never write good music—to say nothing of noble art—under the shadow of her undemocratic wing.

So much for *The Dream of Gerontius*! But before we dismiss Boughton's prejudice as unique, we should remember that Stanford objected to the same work on the grounds that it 'stank of incense', and that it would be 1910 before Gloucester's clergy would admit an unexpurgated version to their Three Choirs Festival meetings.

Rather closer to our acceptance, though probably disconcerting to the average reader of the time, were some of the ideas expressed in Boughton's 'Studies in Modern British Music', a series of articles that appeared in the *Musical Standard* in 1907 but were written in about 1903. They were revised in 1905, and again in 1910 when they were assembled in potential book form and dedicated to Edward Carpenter 'Chiefest of British prophets, poets and musicians'. Having identified 'beauty' as 'that which is pleasing to the senses', Boughton assumes the mantle of Carpenter:

Now the sense-pleasure which we call Beauty also seems made up of complementary sex-elements, masculine and feminine qualities, strength and delicacy; and it is only reasonable to suppose that in the mingling of these is born an approximate manifestation of the ideal of Beauty we seek. Not to become too deeply involved in this large question, let us only note two measurable facts: the bisexuality of great artists, and the approximate balance of sexual characteristics to be found in the greatest art-works.

Whether the actual sexual abnormality of the artist-nature is as pronounced as it seems is difficult to decide in view of the secrecy generally adopted in regard to this matter. It is, of course, possible that the artist, as man, is of the same sexual nature as other men. There is, however, some evidence which seems to reveal him as of a specially bisexual nature.

A challenging proposition to offer the average reader when the memory of Oscar Wilde's shameful incarceration was still green.

Further into these articles we find statements that are central to Boughton's beliefs:

To attain the ideal of beauty the composer must not think of it in terms of tone; he must live it—balance his life by vigorous action and tender affection. Without reality in life there can be no reflection of Beauty in art. And, anyhow, the art will be certain to figure forth the real nature of the man. True music reproduces a man's nature, and the components of beauty there existing, as surely as a photograph reproduces his features. Beauty, or sense-pleasure, can only result from the facts of life—from intimacy with the earth, the sun, and the wind, from the tilling of the soil and the jostle of human intercourse, from a friendship with animals and a mastery over tools, from sex-love and the wonder of parentage—these will move a man to strong and tender actions: move him, may be, to such a passion of strength and tenderness that he must find relief in the joy of musical expression. Of such passion Beauty is the flower.

Then, working on the extraordinary premiss that 'a masculine expression preponderates in the Teutonic, and a feminine expression in the Celtic nature; while those whose works proclaim a mixed racial element are the same who attain most nearly to artistic bisexuality', he classifies British composers: Parry as masculine, Elgar as feminine, Vaughan Williams as bisexual, and so on with varying degrees of improbability.

In 1920 Boughton published a fifty-page booklet which he dedicated 'To G.B.S.' It was a study of *Parsifal*, a work which seemed to Boughton to exemplify some of his most deeply held beliefs:

What distinguishes *Parsifal* from the rest of [Wagner's] music-dramas is not its religious quality, but the idea set forth in the very libretto and worked out in the scheme at Bayreuth that art is the only trustworthy form of religion and the artist the only trustworthy priest. Of course, this idea did not originate with Wagner: it has been implicit in the life work of all the great artists who have ever lived. It was this feeling that caused Blake to say that Art was Christianity and Christianity was Art.

Whether or not we can agree with it, Boughton's analysis is, to say the very least, stimulating and thought-provoking. Thus:

The drama of *Parsifal* is the fight of a natural healthy being with those matrimonial conventions which deprive him of mental life; of an artist who insists that instinct and creative power are the only forces at all likely to reveal to men the wonder and mystery and beauty symbolized by the Grail. The victory of the creative genius is also the salvation of religion and womankind. . . . [Wagner] was a man, and necessarily and rightly wrote from a man's point of view, demanding to be freed from the domestic woman, with her tyranny, her blackmail, and her meanness. Had he been a woman he would have made the demand from the other side, and required to be set free from the vanity, brutality, and grossness of domestic man.

Boughton sees Parsifal as the simple, instinctive artist who grows to an understanding of himself and the religious nature of art through the events of the drama. Gurnemanz he characterizes as 'a type of good-natured mediocrity', Titurel as a man of piety but little insight, Amfortas as a sceptic incapable of religious revelation, Klingsor as the upholder of materialistic sciences, and Kundry as 'essential womanhood' with all her intuition and age-old wisdom.

From this framework of symbolic personalities he discerns a drama that mirrors his own preoccupations and experience of life. They may or may not have been Wagner's also:

[Kundry] does not expect to win a man of Parsifal's nature by her mere femininity . . . If she cannot capture [him] by her wifeliness, she will capture him by her motherliness. If she cannot be his mate, she will seduce him into the prison joys of domestic life.

Kundry's problem vitally concerns the life of modern domesticity; for although Wagner's Parsifal is not himself so original a creation, we yet find him in a new relationship. Here not only the man is concerned with remaining a sane, free individual, but the woman also is brought to consent to (or to demand) an independent physical and mental life. This is why *Parsifal* is such an important contribution to the thought of our time . . .

The average man's objection to feminism is foolish in his own interest, and likely to result eventually in a period of female supremacy every bit as bad for the Kundry's of the future as the present pretence of male superiority is bad for the Amfortases of our time.

Add to Boughton's understanding of women (both enlightened and defensive and more than a little in advance of his time) his reflections on the religious nature of art, and the picture is complete:

Klingsor pins his faith to materialistic science . . . He has captured the spear, the sceptre of mental sovereignty, from Amfortas, as Huxley, Haekel and Co. have captured it from the Christian Church . . . [and now] fully believes that Parsifal, the simpleton, will fall as Amfortas fell and be bereft of his supernatural pretentions in the hard facts of human nature . . . Thus we reach the climax when the scientist must launch the spear of his authority at the artist or lose his influence. And lo, the spear floats into the artist's hands and we know that beyond all the hard facts of materialistic science there are mysteries which only the artist in a flash of intuition or an ecstasy of dream can reveal!

There is, of course, much more to Boughton's analysis than is encompassed by these quotations—the work's relationship to Wagner's other music-dramas and so on—but they may be a sufficient indication of the general drift of his ideas and the degree to which they relate to his own life.

Similar personal preoccupations gave Boughton's second *Bach* book its individuality. But here, as we have seen, it was politics that ruled the analysis. According to Boughton, Bach, heir to a long line of craftsmen-musicians, must have been instinctively drawn to the Pietists' concern for the downtrodden masses. 'By means of the chorale, the Christian song of Christian people in an unChristian age, Bach concentrated and voiced what was noblest in that age. It was a thing from which he could not escape if he were to have any real life as an artist, or any self-respect as a church worker.' It was through his use of the chorale, and the quasi-pictorial musical symbolism that so often accompanies it, that Bach was able to find ways of 'suggesting the realities which most men feared or denied'.

Stimulating as *Bach* was, Boughton's sweeping statements, coupled with his ability to see only one side of any question at a time, merely convinced his contemporaries that his judgements were fundamentally unsound and that there was nothing whatsoever to be learnt from his ideas—a conviction that his final book, *The Reality of Music*, did little to alter. Offered as an examination of 'the influence of real life upon music, from its primitive manifestations, through folk music, to its development during Christian civilization', Boughton's book rehearses once more, but in greater detail, his theory that music's vitality depends upon its being in close touch with the realities of ordinary life—by which he meant the reality of ordinary people in ordinary circumstances. The ivory tower, whether self-selected or imposed from without, is thus seen as nothing more than a prison and a sure recipe for decadence.

By attributing all that was great and good to the masses, Boughton was bound to come up with a number of startling conclusions. The critics duly, and very properly, begged to disagree—as may we, even though, once again, the theories he put forward are never less than stimulating. Typical is the question he raises about the Elizabethan madrigalists and the cloistered unreality of their art. It is not a reservation touched upon by most scholars even now:

The year 1597 was one of fearful famine. In 1598 Weelkes published his 'Ballets and Madrigals to Five Voices'. In it we find the following:

> Sing shepherds after me,
> Our hearts do never disagree
> No war can spoil us of our store,
> Our wealth is ease, we wish no more . . .

and other shepherd-fantasies of a similar kind.

Had those composers been in any sort of natural contact with real life their art would prove them to have been monsters. They were not in such contact. They were shut away from the world like church mice or pet monkeys, and as little able to express the realities of their time. That is probably one reason why the galaxy of Elizabethan musical talent produced no genius equivalent to Shakespeare's: they lacked the inspiration he had in his closer contact with the common life of his time.

True enough, up to a point. And it raises the crucial question of how art should relate to morality and social responsibility. For Boughton there was only one answer. Thus to those who were able to accept his point of view, *The Reality of Music* came as one of the most penetrating musical studies of its time—indeed, the only worthwhile study ever undertaken by a British composer. To those who could not agree it remained a fascinating, wrong-headed, and totally exasperating exercise in personal dogma.

Probably the most controversial of Boughton's views was his sweeping condemnation of the avant-garde of the day—Stravinsky in particular being categorized as the helpless pawn of degenerate times. *The Rite of Spring* aroused his special scorn:

Stravinsky would have us believe that this chaotic noise expresses the kiss given to Mother Earth by the worshipping Russian peasant. What it really expresses is Stravinsky's own ignorance of, perhaps contempt for, the earth—the sort of primitive embrace imaginable by a man-about-town, with concrete slabs under his feel instead of turf or ploughed land . . .

Stravinsky was cunning enough to know what kind of ballet would astonish a faithless people; cunning enough to realize that, if he were extravagant enough, his barren musical mentality would be overlooked in the notoriety he would provoke. So he proceeded to make noises which combine the ineptitudes of the savage with the extravagance which has always characterized decadent civilizations.

But it was hardly to be expected that the man who had struggled with the fearful odds of Glastonbury would be appreciative of anyone who had contributed to Diaghilev's cynical bandwagon, even though he was a giant among composers.

And yet *The Reality of Music* is not a book to be dismissed out of hand. It is full of worthwhile perceptions: ideas that cast fresh light upon individual composers and their works, on the periods and movements in music's history, and, above all, on the role that music, and art generally, has to play in the drama of evolution. For Rutland Boughton there was only one way, and he made it clear in the book's final paragraph:

Neither life nor music can be fine without the acceptance of the communal principle. Great masters have known that instinctively, as we have seen especially in the cases of Bach, Beethoven, and Wagner. So, also, no fresh and adequate expression of music can be made until musicians generally realize that their art depends upon real life for its inspiration, and turn to the communal principle to relieve them from the meagre life of dividuality.

22

Orchestral, Chamber, and Vocal Music

Before considering the music-dramas by which Boughton's achievement as a composer must stand or fall it is necessary to look into the qualities of his other music. He wrote in all forms, and these forms appear again as elements in his dramatic works. Thus what is strong or weak in them individually is likely to make a similar contribution to the music-dramas and for this reason alone would deserve serious attention.

It should be remembered what the world Boughton was born into was like. In 1878 there were no cars, no aeroplanes, no radio or television, no telephones or electricity in people's homes—scarcely any of the amenities that the more advanced societies now take for granted. It was a world of steam trains, horse-drawn carriages, gas light and oil lamps, unlimited servants for the rich, unlimited servitude for the poor. A world where Richard Wagner and Karl Marx were hard at work: Wagner on *Parsifal*, which he would complete; Marx on *Das Kapital*, which he would leave to Engels unfinished. Brahms, meantime, had just completed his Second Symphoy; Tchaikovsky was at work on *Eugene Onegin* and the Fourth Symhony; Sullivan was about to launch *HMS Pinafore*; and John Ruskin, who like Sullivan had twenty-two years to live, was at work on *Fors Clavigera*. Elgar was cautiously feeling his way to individuality by writing small pieces for wind ensembles; while Lenin, aged 8, still answered to the name Vladimir Ulyanov. The British Empire was at its apogee, for Disraeli, as Prime Minister, had just bestowed the title of Empress of India on Queen Victoria. In Europe the Congress of Berlin mulled over the eternal Balkan Question.

Though much would happen during his formative years, this was the world Boughton grew up in and the basis from which he operated both as a man and as a musician. He was therefore essentially a Victorian, with all the strengths and limitations that implied. His special misfortune was, however, to have been born into an intensely provincial milieu, without any vestige of the wider culture that enabled Vaughan Williams, his senior by six years, to rise above and beyond his nineteenth-century background and become, by British standards, a genuinely twentieth-century composer. Though he developed as a craftsman, Boughton's thought-patterns remained embedded in the society that first formed them, so that he lived to be something of an anachronism.

SONGS

Boughton's songs are scattered fairly evenly throughout his career, and just over half of them found a publisher. Though they cannot be said to rank with the finest English masters, his work in this field is both interesting and individual. The interest and individuality, however, lie less in the quality of the music than in the intention behind each song. For Boughton it was not enough to make a sensitive response to musically evocative words: the words had to say something, make some commentary on life. He took no interest in mere hey-nonny-no-ing. Though his purely literary perceptions were rather limited, his sensibility to the human application of the words he chose to set is beyond reproach. This led him to choose poems which most other composers were inclined to overlook: poems of socialism, of the relationship between men and women, religion and daily life. And when his musical inspiration matched his ethical intention the result is a song of peculiar force.

The influences present in his early songs are typical of the period: drawing-room Mendelssohn in the 1896 cycle *The Passing Year*, Brahms and patriotic ballads in the *Songs of the English* (1901). With the *Four Edward Carpenter Songs* of 1906–7 a more personal style began to emerge, but even after this date the influence of other men can be disconcertingly strong—some of the *Songs of Womanhood* (1911) are Schumann to a fault. Generally speaking, the stronger and more dramatic the emotional content of a poem the better the song. Simpler moods often miss the felicitous response of 'Little Boy Lost' and 'Little Boy Found' (*Songs of Childhood*, 1912) and fall into coy triviality, as in the remaining numbers of the cycle, or make do with Merry England banalities as in 'At Grafton' and 'The Feckenham Men' (1913). But given words that arouse his instinct for drama, Boughton can arrest the attention with a few powerful strokes—the grim strength of 'The Dead Christ' (the second of the *Four Edward Carpenter Songs*) is an outstanding example (Ex. 1).

In only a very few songs is the voice completely independent of the piano accompaniment: 'Immanence' and 'Sister Rain' are perhaps the best examples. It is far more usual for the piano to double the vocal line, thus robbing it of complete flexibility and creating a rather heavy effect. This heaviness is sometimes increased by full accompaniments—

Ex. 1

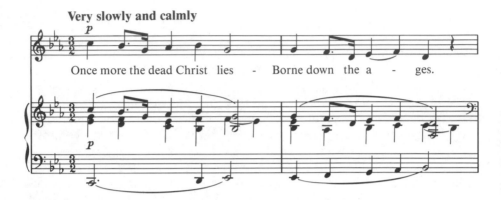

intricate figurations, wide-flung arpeggios, thick chords. The cold
sparseness of 'The Triumph of Civilization', the reticence of 'Child of
the Lonely Heart', and the Debussyish impressionism of 'The Lake of
Beauty' (*Three Edward Carpenter Songs*, 1914) make them exceptions
and places them high among his most effective songs.

For modern tastes Boughton's melodic lines are rather too dependent
on their harmonic underlay, and since the rate of harmonic change is
fairly constant the line is often impeded. They range in style from simple, self-sufficient tunes to dramatic declamation; the majority fall
somewhere in between these extremes as a series of related melodic
phrases held together by the guiding harmony. Boughton seldom set
words pictorially, but was content to aim at the overall mood of the
poem. Despite these strictures, a fair number of Boughton's songs win
through by sheer passion and sincerity. Like the man himself, they can
be awkward, gauche, and intractable, but they have an unerring
instinct for getting to the heart of the matter, and linger in the mind
long after criticism might think to have disposed of them.

PARTSONGS

Boughton's interest in the unaccompanied partsong lasted well into the
1920s: that is to say, as long as he had anything to do with choirs.
Nearly all of them were published, and many achieved a considerable
popularity.

Among the earlier pieces the sets of *Choral Variations on English
Folksongs* (1902, 1907, 1909) are the most interesting. As their title sug-

gests, the folksong becomes the basis for continuous variations designed to illustrate the unfolding drama. The melodic line is bent this way and that to accommodate expressive harmonies annd dramatic modulations, broken up to make points of imitation, passed from one voice to another, paraphrased and caricatured in the interests of the story and in a manner calculated to freeze the blood of any devotee of pure, unsullied folksong. Though they violate every canon of good taste in an exuberant display of academic part-writing and sentimental harmony, their vulgar good humour is hard to resist. They are, simply, very much the work of someone who had not yet recognized the value of folksong as a revitalizing agent in music and was content to exploit it in cavalier fashion. They must have been great fun to sing.

In the rough and tumble of the *Choral Variations* the inevitable period cliché scarcely matters, but in more serious pieces, such as the *Six Spiritual Songs* (1910) and the motet *The City* (1909), such lapses stand out more painfully. Patches of sticky chromaticism, sequences like decorative garlands, short-winded points of imitation that are worked mechanically, all speak of a time when the partsong was an industry and choirs had constantly to be fed with new material. In Boughton's case, the simpler the approach the better the outcome. Though his setting of Francis Thompson's 'The Kingdom of Heaven', the last of the *Six Spiritual Songs*, has considerable power and can rise to passages of great luminosity (Ex. 2), it is the tender melody of the 'St Bride's Cradle Song' and the folksong lilt of the 'St Bride's Milking Song' (settings of Fiona Macleod) that remain in the memory. Indeed, William Sharp's fey muse seems always to have struck a happy response in Rutland Boughton—as witness the *Six Celtic Songs* of 1914.

Many of Boughton's partsongs were written for particular occasions, or in compliment to certain individuals and groups of singers. Though they may not be fired by deep emotion, they are invariably well crafted and calculated to give pleasure to singers and audiences alike. The four delightful settings published in 1914 as Opus 39 are typical. The first of the group, 'Early Morn' (a setting of W. H. Davies) is particularly striking for its impressionistic delicacy. Only once, in 'Burglar Bill', a setting of a supposedly humorous poem by Frederick Anstey, did Boughton descend to a blatant pot-boiler. Quite what prompted it is difficult to imagine, but it may have been a misguided 'bright' idea of his publisher (Curwen), or simply a desperate bid to make a little money. Though he tackled Anstey's sickening tale bravely and with considerable ingenuity (a lisping moppet charms a burglar into repentance), he remained deeply ashamed at ever having put pen to paper.

Almost unimaginably, 'Burglar Bill' was written at exactly the same time as his finest cycle of partsongs: the sequence of six settings of

Ex. 2

poems by Henry Bryan Binns which he completed in December 1927 under the title *Child of Earth*. This celebration of his removal to the rural calm of Gloucestershire has a breadth and passion, a rhythmic vitality, a subtlety of melodic line, that cannot be matched in any of his earlier pieces (Ex. 3). On the strength of this cycle Boughton deserves a place among the foremost English partsong composers of his generation. In every sense it is a virtuoso piece; Boughton was proud of his *Child of Earth*, and rightly so.

Ex. 3

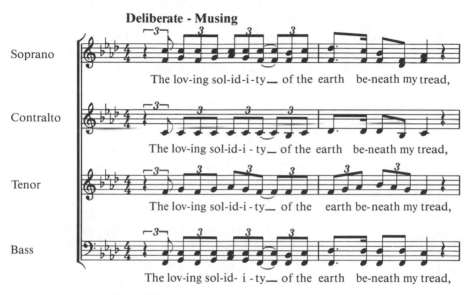

Copyright 1928, J. Curwen & Sons Ltd.

CHORAL MUSIC

Boughton's works for chorus and orchestra can only be judged against the background of choral festivals and choral competitions that played so vital a part in English musical life before the First World War. These gatherings were civic occasions and, as the greater part of the music they inspired now proves, not primarily the breeding grounds of great art. But they were enthusiastic and full-blooded, and offered the kind of opportunity that the more accomplished composer could turn to advantage. With occasions of this sort in mind Boughton wrote four major works: *The Skeleton in Armour* in 1898, *The Invincible Armada* in 1901, *Midnight* in 1907, and *Song of Liberty* in 1911. Two somewhat slighter works, *The Cloud* and *Pioneers*, followed in 1923 and 1925 respectively.

The Skeleton in Armour, a setting of words by Henry Wadsworth Longfellow, began life in February 1898 as a work for baritone and orchestra. Boughton arranged it for SATB chorus during the November and December of the same year. It is thus so exactly contemporary with Coleridge-Taylor's *Hiawatha's Wedding Feast*, which received its first performance at the Royal College of Music on 11 November 1898, that it is possible that the knowledge of a triumphant Longfellow choral setting had some bearing on Boughton's change of mind. He was a first-year student and, if he did not sing in the choir, would certainly have been aware of what was being rehearsed and the impact it was making. *The Skeleton in Armour* is an interesting work and has moments of real power and charm. Boughton is at his best when tackling the poem's grimmer aspects—the love element reduces him to teashop sentimentality. It is a bold, ambitious work, orchestrated with great aplomb, and an impressive achievement for a 20 year old.

The Invincible Armada marks a considerable advance in technique. It too seems to have begun life as a work for baritone solo and orchestra. Although the transformation to an SATB texture was carried out with great skill, the chromatic nature of its harmonic progressions has led to voice parts that are not always naturally vocal and certainly rather testing for the average choral society. Again it is the more powerful moments that make the deepest impression—as, for example, the growling triplets of the opening 'storm' theme, which Boughton made use of again in *The Birth of Arthur*. It is a great pity that the jingoistic text (Schiller translated by Bulwer Lytton) is now an embarrassment, for there are many fine moments in *The Invincible Armada*, including at least one theme (Ex. 4) worthy of Elgar.

Ex. 4

No such inhibitions attend the text of *Midnight*. Composed in 1907, in the first flush of Boughton's enthusiasm for socialism, it is a setting of lines from Edward Carpenter's *Towards Democracy*. The poem, separately entitled 'High in my Chamber' by Carpenter, traces the thoughts of one who hears the chimes of midnight and considers the sleeping city, its weariness and injustices for a moment at rest. But the tolling bells hint at a new dawn. Little by little the thoughts of men will surely change, for even now there are those who 'dream the impossible dream' of love and humanity, 'the sound of which is not yet come on earth'. The work ends on a note of tremulous hope.

The opening is masterly: high pianissimo strings over a deep tonic pedal, simple chords for the clear night air, augmented triads for the mystery of sleep (Ex. 5). Throughout the work the choral writing is beautifully judged and often thrilling in effect. It is the writing of a man who knows from experience what will come off with a choir and has a keen ear for vocal as well as instrumental colour. The passage dealing with the clanging and clatter of different bells startled the singers of 1909 as an expressionist innovation that was altogether too modern; but they found that, with practice, it too could be made effective. Only one section seems laboured: the setting of the words 'But the hour swings onward'. Here Boughton attempts to build a climax from a scrap of melody that is little more than an animated perfect cadence. Inevitably the passage stops and starts, short of breath and developing only mechanically. Fortunately such moments are rare. Far more typical is the passage that leads to the final climax. Here the music, simply by pivoting between C major and C sharp minor chords, blossoms with magical effect (Ex. 6).

Fired by the intense humanity of Carpenter's poem, *Midnight* is the first of Boughton's works to achieve his unique blend of mysticism, realism, and idealism. It is also the first of his large-scale works that could be revived without apology. Dedicated, somewhat alarmingly, to 'The Awakening Manhood of England', Boughton's setting of Helena Bantock's poem 'Song of Liberty' is probably beyond resurrection. This is a pity, for it has a number of good qualities: vigour for one, effective choral writing (not too difficult for the Comrades), and at least one passage of Elgarian nobility (Ex. 7). Should the time ever come when it is possible to give voice to such sentiments as:

> Now is the time, my brothers, to sing a battle song,
> To shame the cowards in the fight, the loiterers in the throng.

Boughton's *Song of Liberty* may have its day.

Probably nothing can save his setting, for SSA and piano, of Shelley's *The Cloud*. Beginning in the manner of Holst—a pendulum of

Ex. 5

Slow and Dreamy

Copyright 1909, Novello & Company Ltd.

Ex. 6

Ex. 7

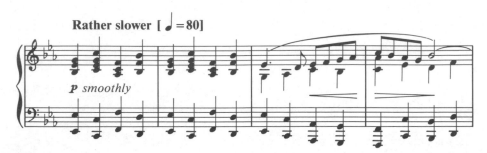

'mystic' chords supporting a wordless vocalization—it soon develops into whimsical prettiness that is not much helped by the poet's heavy-handed rhyme scheme. Nor is there likely to be much future for his setting of Walt Whitman's *Pioneers*, even thought it is vigorous and effective. The part-writing is relatively straightforward, as befits a work intended for amateurs, but somewhat too dependent on passages of predictable imitation. There are certainly moments of power and imagination: as, for example, the trumpet calls that herald the final climax. It was quite a favourite with the choirs of the London Labour Choral Union for whom it was written—though whether the singers entirely appreciated being hailed as 'tan-faced children' is a question whose answer is not beyond all conjecture.

CHAMBER MUSIC

Boughton wrote music for chamber ensembles throughout his life, but only two pieces of any consequence reached publication: the *Celtic Prelude* for violin, cello, and piano (1917), and the Sonata in D major for Violin and Piano (1921).

The *Celtic Prelude* is a short, rhapsodic piece in one movement, based on music he wrote for a Glastonbury production of W. B. Yeats's play *The Land of Heart's Desire*. Though tuneful and unpretentious, it is a slight work—its modal melodies, episodic formal structure, and unsurprising harmonic content creating an impression that is pleasant rather than powerful.

The Violin Sonata, on the other hand, is a much weightier work. It is also a very disconcerting one, for each movement begins boldly and with great effect, only to peter out in vapid, folk-dance tunes and an aimless bustle of semiquavers. Stylistically it is also uncertain. The first movement leans heavily on Brahms and César Franck, while the third movement builds its quasi-fugal opening on a theme that is almost dodecaphonic—anathema to Boughton, though the intention is presumably ironic, for the theme's tritonal basis can be explained away by the quotation from *Also sprach Zarathustra* that Boughton has placed at the head of the movement: 'I am the advocate of God in the presence of the devil . . .'. Of the three movements, the second ('Once having passions thou callest them evil. Now, however, thou hast nothing but thy virtues: they grow out of thy passions') is the most satisfying. It opens powerfully (Ex. 8).

Boughton's unpublished chamber music is of much greater interest than the two pieces that found a publisher. Without being of major importance, both Oboe Quartets (1932 and 1945) skilfully explore material that is never less than attractive. The first quartet has a particularly beautiful, Irish-sounding, melancholy tune as the basis of its last movement (a set of six variations). Boughton's own description of the work as 'small, Spring-like; with some of Spring's sadness in it' is singularly apt. With no piano to tempt him into muddy textures, Boughton's part-writing assumes greater clarity and a genuine sense of purpose.

The crown of his achievement is to be found in the two String Quartets composed in the months of June, July, and August 1923 in the excitement of his new-found happiness with Kathleen. The first, in A, is subtitled the 'Greek': its material being derived from certain modal studies he undertook before writing *Alkestis* and the unaccompanied choruses (based on Greek folksong) he contributed to the 1922 Glastonbury production of *The Trachiniae*. The second, in F major, bears the subtitle 'From the Welsh Hills' and was composed on holiday in Beddgelert. As the titles of its movements suggest ('Landscape from the Valley', 'Landscape from the Hilltops', 'Satire—Conversation', 'Song of the Hills') it is partly programmatic. Both quartets are fluently written, but pose a considerable challenge to performers. They have strong themes, a considerable harmonic range, and textures that are often adventurous and imaginative. While neither could be listed among the finest examples of quartet writing of the period, they have qualities that make their total neglect a matter of shame.

ORCHESTRAL MUSIC

So little has been published or, until recent years, performed that this side of Boughton's output has also been virtually ignored. The most accessible pieces have been the *Three Folk Dances* for string orchestra (1911), the Oboe Concerto No. 1 in C minor (1936), and the Flute Concerto (1937), all of which were published. The *Three Folk Dances*

Ex. 8

are simple and unassuming—worthy forerunners of Holst's *St Paul's Suite* and Warlock's *Capriol Suite*. The Oboe Concerto is instantly appealing. Its folkish themes are lively and memorable, the string orchestra is handled with vigour and imagination, and the writing for the soloist shows a very complete understanding of the instrument. The slow movement is particularly beautiful. A second Oboe Concerto, in G minor, was completed in the summer of 1937 and dedicated to Leon Goossens, who gave the first performance on 11 September 1943 at the Wigmore Hall with Kathleen Riddick's string orchestra. Like its predecessor it is a virtuoso work. Its thematic material is, for Boughton, unusually chromatic (at least in the first movement), but rather less compelling than that of the first concerto. Much the same criticism might be levelled at the Flute Concerto which, despite its virtuoso qualities and moments of imagination, drifts all too easily into folkish clichés. The slow movement is the strongest: a simple, sixteen-bar melody framed by an introduction and epilogue built on the sustained chords of a rising, nine-note scale decorated with delicate cascading arpeggios for the soloist.

Virtuoso demands delayed the Trumpet Concerto's first performance for forty-six years. Completed in September 1943, no player felt equal to the challenge until John Wallace tackled it in September 1989. Cast in two movements and scored for a medium-sized orchestra, rather than the strings that had served the earlier concertos, it proved to be an effective work in which passages of dark, brooding power (as in the slow opening of the first movement), of solemn majesty (as in the chorale that appears in both movements), and ghostly solemnity (as in the disembodied, Holstian march that opens the second movement) are effectively offset by lively fanfares and cheerful folk-dance rhythms. The 1937 Concerto for String Orchestra seems not to have been performed—possibly because certain divisi passages proved too great a virtuoso challenge for the orchestras of the period. The second of the four movements (another melancholy, folk-like tune of devastating simplicity) is particularly effective and curiously anticipates the slow movement of Tippett's *Concerto for Double String Orchestra*. The outer movements are vigorous, rhythmically positive, and very effective.

Much of Boughton's purely orchestral music belongs to his early days and came to be discarded by him, even though it had been well received at the time. Such works as the symphonic poem *A Summer Night*, the *Imperial Elegy: Into the Everlasting*, and the 'Oliver Cromwell' Symphony (all of which were raided for later music-dramas) have undoubted strengths—bold thematic work, exceptionally skilled orchestration—even if the material is somewhat derivative and their whole ethos that of a vanished age. But they might well stand revival as curiosities. Nothing, however, could revive the score of the ballet *May Day*: a thoroughly misbegotten piece of political claptrap. Nor is there much to be said for the ballet *Snow White and the Seven Little Dwarfs*, save that it is a competent piece of utility music that served its purpose at the time.

Of the orchestral works Boughton wrote in connection with his music-dramas, the overture to *The Queen of Cornwall* is probably the most important. He composed it in 1926, two years after the music-drama had been successfully performed, and probably at the suggestion of his publisher as a means of bringing the stage work to a more practical length. Being derived from its parent work, the thematic material is striking, but it is also ill-adapted to symphonic development and does not therefore gel into a convincing movement. A much sounder proposition is to be found in the short symphonic sketch *Tintagel*—an orchestral version of the opening of the second act of *The Birth of Arthur*. The final scene of the same work also exists in an effective orchestral arrangement.

By far and away the most important contributions Boughton made to British orchestral music are the two symphonies: the 'Celtic' Symphony (subtitled 'Deirdre'), composed in 1926–7, and the Symphony No. 3 in B minor, completed in the autumn of 1937. Though both were performed during the composer's lifetime, it is only in recent years (since 1986, in fact) that broadcasts and recordings have revealed their very considerable qualities.

Boughton's 'Celtic' Symphony began life as a ballet. When it became clear that no one was interested in it in that form, he decided to cut his losses and turn it into a purely orchestral work which, if not exactly a symphony in the usual sense of the word, might be regarded as 'a music-drama without action' and pass muster as a cross between a three-movement symphony and a sequence of three symphonic poems. To do this he simply removed 67 bars: a 49-bar introduction in which a Bard had set the scene in song, a 3-bar link between what is now the first and second movements, and a 15-bar sung interjection in what is now the first movement. Nothing else needed to be altered, for the ballet had been conceived in symphonic terms. The programme now took the following shape:

First Movement: Allegro vivace

The Young Girl—The Old King—The Young Lover

Deirdre is as wild as the mountains where she has her home. She cares much for the beasts, the winds, the skies, the flowers, and the mystery of the stars and the tarns. She is indifferent to the fate which, she has been told, is in store for her. For King Conochar has willed that Deirdre shall become his Queen. Her music, as the expression of free, virginal, hill-life is stated and fully developed, chiefly by strings and woodwind [Ex. 9]. Conochar's music is introduced by trumpets. With the coming of the King, come also the three young sons of Usna. When Deirdre's eyes fall upon Naisi it is he who enters her heart. The first timid phrase of what becomes the love-tune is heard on oboe and violas, with a dark background of trombones. The conflict arising in Deirdre's heart through her fear of the King and her passion for Naisi consti-tutes the middle section of the movement. She makes her choice; and the last section is devoted to a happy dance-rhythm, developed chiefly in transformed themes of Deirdre and her love. The closing bars sound a new theme—as of a shadow in the background, stalking their happiness.

Ex. 9

Allegro vivace

Second Movement: Adagio molto

Moonlit Idyll

The lovers run away together, and this is the music of their consummated love [Ex. 10]. A smaller orchestra is used. The strings are much sub-divided, into solo quartet with two lines of tutti in each of the four upper parts. The cloud which shadowed the joy of the first movement passes also over the peace of this movement.

Ex. 10

Third Movement: Allegro moderato

Love and Death: A Dance of Death-defiance

This is the longest movement, and is less simple in its dramatico-emotional development. Deirdre may love Naisi, but Conochar has power over their lives. The movement opens with the suggestion of that adverse power [Ex. 11]. A feeling persists that the power may be used tyrannically. (This being a 'Deirdre' symphony, its moods are developed only from her point of view—until the last few bars, when she no longer has any point of view!) When the tyrannous music has been fully stated, it is followed by Deirdre's counter-music: first, a soft coaxing appeal, which fails; then, conscious of the certainty of death, a weak sobbing reaction. But an innate sense of her own right to life and to Naisi reawakens the flood of her love and the assertion of her own full womanhood. That having been asserted, it strives with the tyranny which is more crudely and emphatically proclaimed. With ever-fiercer exaltation, Deirdre faces her doom by the side of her lover—finally leaving the tyrant only the clay of the beauty he was unable to win.

Ex. 11

Though the work may have limitations as logical symphonic argument, there is no denying its power and beauty. The themes are bold and evocative, the orchestration brilliant and imaginative, the sense of dramatic direction full of purpose. Influences are to be found, of course—Elgar, inevitably, and also Richard Strauss (the moment of Deirdre's coaxing appeal in the third movement suddenly and rather comically taps a *Salome* vein!)—but the overall effect is that of a positive, individual voice that knows exactly where it is going.

If the voice that informs the material of Boughton's third and last symphony is rather more eclectic, the sense of true symphonic development is altogether more positive. The symphony astonished listeners to the BBC's 1986 broadcasts (the first geniunely 'public' performances). Boughton, it was discovered, had written a real symphony: powerful, exuberant, and memorable. Completed in the autumn of 1937, the symphony would then have been considered wildly old-fashioned: contemporary with Brahms and Tchaikovsky or, so far as this country is concerned, with the Elgar of 1911. Fifty years later this was of little significance. What mattered to the 1986 audiences was that it worked as a symphony, that it was full-blooded and vigorous, that its thematic material was admirably suited to symphonic development, and that, above all, it was extremely beautiful. The spirit of Elgar may indeed hover over the entire work; Dvořák's *Slavonic Dances* may intrude briefly in the working out of the third movement; the orchestration may have the *fin de siècle* opulence of Richard Strauss—but what of that? Boughton's Third Symphony was the real thing: music that had guts, heart, soul—a splendid affirmation of all the finest values of a bygone age.

Structurally the four movements follow a fairly traditional plan: sonata forms for the first, third, and fourth movements, and an elaborate ternary form for the slow movement. The fourth movement's sonata form is prefaced by an adagio section which returns, briefly, to herald the beginning of the development section, and is later expanded into the glorious peroration that brings the work to its triumphant close (Ex. 12). The tonal patterns involved in all four movements are also entirely traditional, as is the nature of the thematic material—beginning in the first movement with the staccato four-note germ cell from which everything else arises (Ex. 13). Especially impressive is the sheer fertility of Boughton's invention. Ideas generate ideas in remarkable profusion. For once, he has no recourse to simplistic folk rhythms; the thematic material is fluid and unrestrained. Boughton's Third Symphony is the work of a master.

Ex. 12

Ex. 13

23

The Music-Dramas

Of the ten full-length music-dramas Boughton wrote in his maturity, five belong to the Arthurian cycle. Four of the remaining five reached publication—two, *The Immortal Hour* and *Bethlehem*, in handsome full score editions as well as the usual vocal score. Nothing of the Arthurian cycle has been published; Boughton himself damped Curwen's interest in *The Round Table* by insisting that to publish it would automatically bind them to undertake the complete cycle, three parts of which were still to be written. It therefore seems appropriate to ignore chronology and divide his stage works into Arthurian and non-Arthurian categories.

THE IMMORTAL HOUR

In setting Fiona Macleod's 'psychic drama' to music Boughton breathed life into a text which, for all its poetic beauty, can scarcely be said to stand up as a play in its own right. William Sharp had unwittingly written a libretto—a text that cries out for music and seems incomplete without it. Part of Boughton's genius lay in the fact that he recognized the play's potential. Even so, he did not set it exactly as Sharp had written it. He made a number of substantial cuts, including a mimed scene in which Eochaidh and Midir play chess as a preliminary to Midir's fateful request. He also added eight of the Fiona Macleod poems at points which required lyrical expansion. Four appear as set-piece arias: Dalua's 'I have heard you calling', at the end of the first scene; and in Act 2, Eochaidh's apprehensive 'Where the Waters Whisper', Midir's ecstatic 'In the Days of the Great Fires', and the Old Bard's song, 'I have seen all things pass'. The remaining four, together with eight lines transposed from Act 1, became the basis of the elaborate choral ceremony that opens the second act. Moreover, instead of giving the so-called 'Faery Song' to Etain at the end of Act 1, as Sharp did, Boughton hit on the brilliantly theatrical device of having it sung chorally, off-stage and 'heard' only by her. In many respects Boughton's libretto for *The Immortal Hour* is one of the most theatrically effective and literate books any British composer had so far set to music.

Like all his music-dramas, *The Immortal Hour* stands apart from the average British opera of the period by its deeply serious nature. Indeed, its nature is so 'inward', so 'religious' in its intensity, that a worthy stage presentation is not an easy challenge to meet in the average opera-house, or be understood by the average producer. The temptation is to fill it with 'business', for there is little dramatic action as such. The drama unfolds on a spiritual plane and at levels of mystical awareness. *The Immortal Hour* is a parable of the human spirit's eternal quest for beauty and perfection, brought to inevitable destruction by the very nature of mortality. Briefly, the story is this:

Act 1, Scene 1

Dalua, Lord of Shadow, the agent of all dark and unknown powers, whose touch brings madness and death, wanders alone in the forest. He is observed by invisible spirits who mock him as the outcast of gods and men. At first he does not know why he has come to so remote a spot, but when he senses the approach of others he understands the workings of a destiny beyond his control.

Etain enters, wandering as in a dream. She no longer remembers that she is a Princess of the faery Land of Youth. She knows only her name. Dalua tells her that a king draws near, searching for the Immortal Hour, the Fountain of all Beauty, and that he will believe that in her he has found what he desires, not knowing that death is the one true end to his quest. Etain goes on her way, while Dalua waits for Eochaidh, the king.

When at last he arrives, Dalua reveals the Fountain of all Beauty and despite his apprehension Eochaidh allows himself to be led further into the forest by Dalua's mocking voice.

Act 1, Scene 2

A storm has arisen and Etain has taken shelter in the hut of two peasants, Manus and Maive. Led by Dalua, Eochaidh comes to the same place and at first glance knows that his search is at an end. As night draws on he sinks into a deep sleep, his head in Etain's lap, while she, half in dream, hears in the far distance the echoing faery songs of her own people.

Act 2

At a great feast Eochaidh celebrates his 'year of joy' with Etain. But both are troubled by strange dreams and forebodings. Etain leaves the gathering, but no sooner than she has gone than a stranger enters and makes a simple request of the king: that he may 'touch the white hand of the Queen' and sing to her 'a little echoing song'. Sadly, compelled by a fate he cannot alter, the king sends for Etain. While he and the stranger wait an Old Bard sings of the transience of mortal life. Etain comes in, dressed in the strange garments she wore when Eochaidh first saw her. At the touch of the stranger's lips and the first notes of his song she recognizes that he is Midir, Prince of the Land of Youth, her rightful lord. Joyfully she goes with him. Eochaidh falls stricken, his dreams shattered. Dalua's shadow, the shadow of death, covers him completely.

As William Sharp well understood, Etain's story is a Celtic version of Euridice won back to life by Midir/Orpheus, into which are woven elements of the Persephone legend—spring won from winter's grasp. Parallels may also be found in the Sanskrit Vedas: Eochaidh the Dreamer, mocked by Dalua/Varuna the shadow God of Night, longs for Etain, the Dawn. He enjoys a brief moment of happiness in her beauty, only to lose her when Midir/Mithra the Sun God appears in all his splendour. Sharp's story is thus firmly rooted in archetypes, hence its potency even for those who, at first glance, find it merely whimsical. As for the world of faery, it goes without saying that it does not involve creatures with gossamer wings sitting on toadstools and prancing in magic circles. Rather, it offers a mirror-image of the mortal world: a proud, fierce race to whom the comings and goings of mortals are of no more importance than the peregrinations of ants. The words of the song that lures Etain back to faery reality set the tone: 'They laugh and are glad and are *terrible*'. Through the persona of Fiona Macleod, William Sharp invented a story that could touch the deepest levels of the psyche. It gripped Boughton and brought into focus all his powers as a creative artist, so that he was able to match it with music that would carry its message across the footlights and into the hearts and minds of everyone that heard it.

The first of Boughton's strokes of genius lay, then, in his ability to recognize the play's musical possibilities and mould them to his own ends. The second is to be found in his ability to blend three potentially conflicting musical ingredients: the purely choral utterance, inherited from oratorio and therefore an important part of the British musical heritage; the system of representative themes derived from Wagner though carried out in a distinctly un-Wagnerian way; and the self-contained song, stemming from the traditional British tendency towards ballad-opera methods. These three elements he combined into a convincing and individual style, thereby creating a sound-world that has no real parallel in the music of his contemporaries. Overtones of other composers there may be (and what composer is ever free of them?) but the sum total is highly individual and particular only to Rutland Boughton.

The network of representative themes is carried through in terms of a subtle mosaic rather than a true symphonic development. It is a method more akin to Puccini than Wagner. The themes themselves are bold and clear-cut and always the exact measure of the mood or thought to be conveyed. Take, for example, the ideas with which the drama opens: solemn, pulsating string chords, out of whose depths rises a shadowy, yearning figure on the clarinet—pentatonic, as are most of the important themes in the score. As an image of the 'Immortal Hour', the 'Joy that is more great than Joy', the 'Beauty of all Beauty', it is singularly evocative and memorable, exactly capturing the necessary mood of mystery and wonder. Equally precise is the theme that follows: a painful, weary sequence of chords, scarcely able to rise out of the depths, which tells us all we need to know about Dalua (Ex. 14).

Throughout the score such themes are transformed according to the needs of the moment. Thus at the words 'For Lu and Oengus laugh not', Dalua's theme is allowed to expand, thereby revealing his tortured relationship to the other gods (Ex. 15). Closely related to this idea, particularly through its second phrase, is the theme that announces Etain's appearance—she too belongs to the Immortal Clan (Ex. 16). While a further transformation occurs as Dalua salutes her with playful, courtly mockery (Ex. 17). Similar thematic transformations occur throughout the score—one need only compare the solemn, Parsifal-like chords that accompany the first mention of Midir's name (Ex. 18) with the lilting, deceptively lighthearted tune that accompanies his fateful appearance in the second act (Ex. 19), or note how the anxious three-note phrase that rounds off Eochaidh's otherwise bold, kingly theme (Ex. 20) becomes the basis of the theme that represents Midir (Ex. 21)—Eochaidh unconsciously recognizing the sorrow that lies ahead. It may thus be seen that *The Immortal Hour* is not, as it may first appear, merely a stream of agreeable melody, but a subtle, highly integrated score of considerable psychological depth.

Melody nevertheless plays a crucial part, and Boughton is prodigal when it comes to its invention. The lyrical set-pieces mainly take the form of metrical songs, some short and self-contained, others a sequence of several tunes—as, for example, Midir's confrontation with Eochaidh. For the most part the melodic contours are gentle, owing much to folksong, and give great scope to the lyric type of voice. Only Midir, as suits his nature, is required to express himself in anything approaching the heroics of grand opera; though even in his case the element of vocal display is muted. The entire work swims on a sea of magnificent tunes, whose only weakness is a tendency to be four-square and pegged down by too many cadences.

Ex. 14

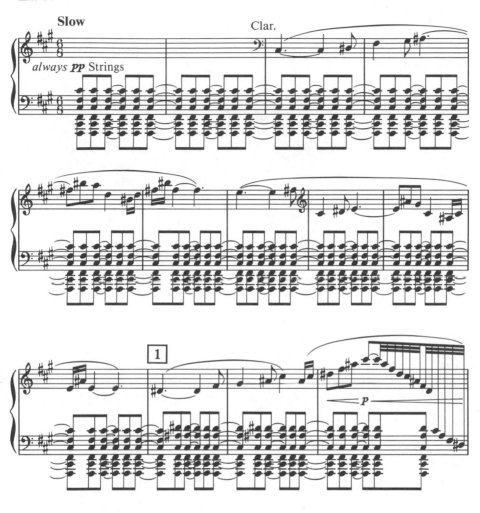

Dalua is seen -
weary and painful, a strange creature of faery - The Lord of Shadow.

Ex. 15

Ex. 16

Ex. 17

Ex. 18

Ex. 19

Ex. 20

Ex. 21

Ex. 22

The melodic material for the entire work is based mainly on pentatonic scale formations. Certain tunes, such as the 'Faery Song', are wholly pentatonic, while others stray only momentarily outside its confines. A short example, the Old Bard's song from Act 2, may give some indication of the power of Boughton's invention without the limitations of the scale (Ex. 22). By insisting on a pentatonic basis, Boughton imposes an impressive unity on his thematic material and, incidentally, provides the work with its extraordinary moonlit atmosphere.

Considerable importance is given to the chorus, as both a seen and unseen protagonist in the drama. In particular its presence at the beginning and end of each act helps to give shape to the work. In both acts the opening choral section functions as a largely static prelude to the action, thereby presenting considerable problems to any producer. In Act 1 the lengthy dialogue between Dalua and the unseen chorus is partly helped by the 'continuous ballet of tree spirits' that Boughton requests. The choral procession that opens Act 2 presents an almost insuperable problem—and would have done even as originally envisaged in an open-air setting. However, they have their compensations: Dalua's monologue builds up from a murmured, chant-like opening to a passionate climax that provides the perfect foil for Etain's entrance; while the climax of the Act 2 processional scene (three quite distinct tunes combined, Sullivan-like, in a glorious free-for-all) is a moment of ingenuity that provides an exciting preface to the broad melody of 'Green Fire of Life'. Both acts end with an unseen chorus singing the 'Faery Song'—the first time as a distant, sad memory; the second as a joyous fulfilment. The parallel is extremely effective.

To all this must be added Boughton's considerable orchestral skills: these show him to be the equal of any of his contemporaries. He deploys an orchestra of modest size with great imagination and, moreover, with an economy that enables the singers to make very word tell. It is, in short, a singularly accomplished score.

BETHLEHEM

The choral drama *Bethlehem* was the first of Boughton's stage works to be written with the knowledge of the precise limitations and advantages of a Glastonbury performance. Knowing what he could and could not depend upon, Boughton clarified his style accordingly. Gone are the last vestiges of Wagner, gone are the production problems. *Bethlehem* was tailor-made for Glastonbury.

For his libretto Boughton turned to the fourteenth-century Coventry cycle of mystery plays as acted by the Shearmen and Taylors Company, using Thomas Sharp's 1825 transcription of the surviving fifteenth- and sixteenth-century manuscript versions. From this he extracted that part of the pageant that deals with the Annunciation, the Shepherds, the Magi, and King Herod, but omitting Isaiah's prophetic prologue, a dialogue between two unnamed prophets, and the Massacre of the Innocents. He also worked lines from some of the interludes between the 'Seeker' and the Herald into the Herod scene. The main scenes he set almost word for word, adding only a number of well-known fifteenth-century carols for Mary and the Angels to sing. In place of the original prologue and interludes he introduced a series of popular carols. This was a stroke of genius, for not only did they provide a framework for the entire drama, but they formed a powerful link between the stage and the audience. The personal application of each event is thus underlined by words and music that were already part of the fabric of daily life. Bach's use of Chorales in his Passion music is the obvious prototype—Tippett's use of Spirituals in *A Child of Our Time* and Britten's recourse to well-known hymns in *Noyes Fludde* are more recent manifestations. Boughton also introduced two popular carols into the 'Street Scene' in the second act, thus involving the audience's emotions even more deeply in the action. Originally he intended the interlude carols to be sung in unison by the audience—as was done at the first production—but second thoughts prompted him to make extended arrangements in the manner of his *Folksong Variations*. The drama now stands within an elaborate choral framework:

ACT 1:

Choral Prelude: Earth his day rejoices

Scene i: The home of Mary and Joseph

The Angel Gabriel tells Mary that she is chosen above all women to conceive and bear God's only son. On learning of her condition, Joseph chides Mary for deceiving him. He prepares to leave her, but the Angel Gabriel reassures him of her innocence. Together, Joseph and Mary set out for Bethlehem.

Choral Interlude: In the ending of the year

Scene ii: A Moor at Night

The Shepherds, Jem, Dave, and Sym, see a great Star—the promised sign. Angel voices tell of the birth of a Saviour. Joyfully they set out to find the Holy Child, the Star their guide.

Scene iii: The Stable at Bethlehem

Mary sings a lullaby to the new-born child. The Shepherds enter and present their simple gifts—a penny whistle, a hat, a pair of warm mittens.

Choral Interlude: The Holly and the Ivy

ACT 2:

Scene i: Jerusalem: outside Herod's palace

Three Wise Men, Zarathustra, Nubar, and Merlin, meet. Each has followed the Star and will now set out together for Bethlehem.

A woman brings news of the birth of a Saviour. The tidings spread among the people and there is great rejoicing. This is interrupted by the sudden and terrifying appearance of Herod who demands that the Wise Men be brought before him to explain their mission. As he waits, dancers and a singer entertain him. The Wise Men present themselves and explain that a Child has been born who will be King over all the world—not Herod's own son, but a Heavenly Child, born in a stable. Herod rages against the Child and vows to kill him. The scene ends with the crowd singing mockingly of Herod's wrath.

Choral Interlude: The Seven Joys of Mary

Scene ii: The Stable at Bethlehem

The Wise Men present their gifts of Gold, Frankincense, and Myrrh. The Angel Gabriel warns them of Herod's evil intentions: Mary and Joseph must take the Holy Child into Egypt. As they set out on their journey the chorus tell of the Passion and Resurrection, while angels sing a pean of praise.

The most immediately striking thing about the music of *Bethlehem* is its extreme simplicity. The dominant influence is that of folksong, although, apart from the carols, only one authentic example is used, to characterize Zarathustra at the beginning of Act 2. Everything else is Boughton's invention in the folksong style, carried out with such purity of purpose that several of the tunes have been taken to be authentic— even, on one occasion, by so eminent an authority as Vaughan Williams. Three of the carols are also by Boughton: the prologue, 'Earth this day rejoices', 'King Herod and the Cock' (a direct quotation from the first act of *The Round Table*), and the angel's carol 'The Stars of the Morning shall Dance and Sing'. Their style matches that of the traditional carols.

But without some element of contrast Boughton's folksong style would have been too bland. That contrast is supplied by the music

allotted to Herod and his court. Herod's own song is a brilliant virtuoso character sketch of empty pomp and vanity—a challenge to any
tenor. Herodias is given a curious, mock oriental setting of lines from
the Song of Solomon, complete with monotonous drum accompaniment, a sinuous, whining cor anglais counterpoint, and a plentiful supply of augmented seconds. The ballet music that follows is less happy.
Boughton's intention was to provide a deliberately meretricious music,
so that the hollow sham of court life might be made clear, especially
when contrasted with the simple, truthful music of the Holy Family.
Unfortunately the parody is not clever enough. Too many passages
seem to have strayed from the pages of *Chu Chin Chow* or the weaker
moments of Albert Ketelby, to be really effective. Boughton recognized
the fact and was quick to sanction cuts if he felt that the stage presentation would not be lavish enough to offset the music's inadequacies.

Although he continued to structure his choral-drama from a number
of representative themes, these are now much more fluid—varying in
length from the short motivic cell to the extended, virtually self-contained melody. The result is a light, uncomplicated score that flows
without effort. Considerable ingenuity is shown in providing each
melody with slight changes of accompaniment and orchestration, so
that the dramatic development can be followed closely, yet without any
distortion of the melodic thread. Here and there, melodies are combined with surprising felicity—as when the three tunes that represent
the Wise Men are combined, with the four-note 'Star' motive thrown
in for good measure; or when the lilting 6/8 of 'I saw three ships come
sailing by' is made to combine, magically and unexpectedly, with the
sturdy 3/4 of 'The First Nowell' as the crowd rejoice before Herod's
palace. The Wise Men's ensemble is made even more interesting by
Merlin's theme being a quotation from *The Birth of Arthur*, while
Zarathustra's is an authentic Somerset folksong, and Nubar's an amusing snatch of ragtime (Ex. 23).

Like *The Immortal Hour*, *Bethlehem* is rich in melody—melodies of
child-like simplicity, heartbreakingly tender. Of the larger set-piece
tunes both the lullabies of the Virgin Mary are outstandingly beautiful
(Ex. 24). They epitomize the loving pain of motherhood: 'I believe',
wrote Boughton, 'that Christ returns each time a child is born.'

The choral variations which make up the carol interludes are carried
out with an assurance and taste that raises them far above his earlier
experiments with English folksong. When the carol tells a story, as in
'The Seven Joys of Mary' or 'The Holly and the Ivy', the variations
illustrate each point, momentarily leaving or modifying the tune where
necessary. When it is simply a carol of praise, as for example 'O come,
all ye faithful', the variations are jubilant displays of virtuosity. In

Ex. 23

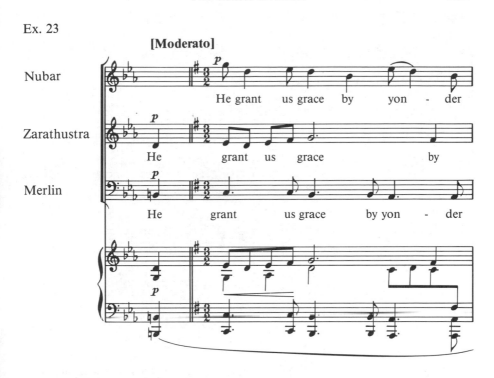

cont.

Ex. 23 contd.

Copyright 1920, J. Curwen & Sons Ltd.

neither case, however, is the integrity of the carol submerged in mere technical ingenuity. In *Bethlehem* there is no striving for effect. What happens, happens naturally, and with the inevitability that characterizes all great art.

In striking contrast to the efforts of most of his contemporaries, Boughton reduces his harmonic vocabulary to bare essentials. The texture of the music is equally simple, and the orchestration appropriately lucid. In such circumstances sham is not possible. Boughton's simplicity is the simplicity of an uncomplicated nature responding heart and soul to a universal truth. The dedication he placed at the head of the score says it all: 'To my children and to all children'.

ALKESTIS

By the time *Alkestis* came to be written Boughton's resources at Glastonbury had established a pattern which differed from his original intentions. Although the chorus still drew on local talent, the principal singers were now young professionals who gave their services because they believed in the validity of the work. *Alkestis* was written with these conditions in mind and, as a result, blossomed with music of

Ex. 24

Allegretto

Mary

Lul-lay, lul-lay, lul - lay,_____ So soft - ly sleep-ing there,_____ It

aches my heart to   see thee lie With cov-'ring thin   and   spare,_____

Copyright 1920, J. Curwen & Sons Ltd.

greater breadth and subtlety. The materials he chose to use do not dif-
fer greatly from those he employed in *Bethlehem*. The harmonic vocabulary is still audaciously simple, though a certain amount of
chromaticism is called in to deal with the wider range of emotions.
The melodies remain serenely diatonic, though they are broader in outline and less obviously metrical in pattern. The system of representative
themes is deployed in much the same way as before: a mosaic of characteristic melodies which yield fragments suitable for motivic development. In short, *Alkestis* demonstrates Boughton's unshakeable
conviction that traditional materials and methods could convey most
emotional or dramatic situations, and that the discoveries of such composers as Stravinsky and Schoenberg were no more than unhealthy
aberrations in the development of music.

In certain respects the word-setting in *Alkestis* is more flexible and

naturalistic than that of its predecessors. This is because more of the
melodic argument devolves upon the orchestra: the vocal line being left
free to join it, or venture upon some brief quasi-recitative of its own,
according to the demands of the text. The chorus is more important
than ever—plumbing depths of feeling far beyond the scope of his earlier works, and assuming a key role in the drama. The story unfolds in
a series of tableaux, separated by choruses which comment on each
new situation. The pattern established in *Bethlehem* is thus carried a
stage further: the chorus is deeply involved in the action, while serving,
at the same time, as mediator between the audience and the mounting
tensions of the drama. In the conventions of classical Greek drama
Boughton had found the logical solution to his idea of what an operatic chorus should be.

For his libretto he turned to Gilbert Murray's rhymed-couplet translation of the Euripides play, first published in August 1915. Nothing
had to be added. All that was needed was the removal of passages that
were either obscure in their reference, or unnecessarily digressive—a
matter of some 280 lines. Boughton was then left with a shapely
libretto: powerful, poetic, and, above all, literate. The story is as follows:

Act 1

Apollo, the God of Life and Light, encounters Thanatos, servant of the God
of Death and Darkness. Thanatos has come to claim Alkestis, the wife of
Admetus, for she has volunteered to die in place of her husband. Apollo cannot remain in the presence of Death, but foretells that even now someone may
come to rescue her. Thanatos and Apollo depart, and citizens assemble to hear
news of their beloved Alkestis. Their sorrow and apprehension is increased as
her Maidservant describes how Alklestis has calmly accepted her fate. At last
she emerges from the palace and bids farewell to her children, her husband,
and all that she holds dear. The grief-stricken citizens take her body away to
prepare it for burial.

Act 2

The citizens discuss the situation, marvelling that, alone of all his friends and
relatives, only Alkestis loved Admetus more than life itself. They are interrupted by the tumultuous arrival of Herakles. Rather than dishonour his
house Ademtus offers him hospitality, pretending that it is not a day of special
mourning. Herakles enters the palace and the funeral procession assembles.
Admetus bids farewell to his wife and, stricken with grief, rails against his
father for feigning sorrow when he has refused to take her sacrifice upon him-

self, even though he is old and near to death. Pheres is indignant and calls his son a coward. Admetus is left alone with his grief and shame. Slowly the procession moves towards burial.

A servant girl now enters, weeping tears of shame at the drunken behaviour of Admetus's guest. Herakles appears and begins to make merry at her expense. Angrily she tells him the real cause of their mourning. Overcome with shame, Herakles resolves to rescue Alkestis from the jaws of Death itself. Admetus and the citizens return. He is inconsolable: ashamed of his own weakness, and crushed by the knowledge that he has lost the noblest of wives. But Herakles appears, dishevelled and with a veiled woman. He is excited and in a mood for jokes. He offers Admetus a new wife. Admetus is horrified and angry, but at last yields to Herakles and touches the veiled woman. It is Alkestis.

Boughton's music for *Alkestis* has a classic nobility and calm. A mood of complete serenity runs throughout the work, giving measure and control even to the moments of greatest anguish. This is achieved by the simplest thematic ideas, of which Apollo's theme (Ex. 25) is the simplest of all. It took courage to risk a bare tonic-subdominant structure in 1922, but scored for clarinets, horns, and harps, with a high sustained pedal G in the upper strings, it has a radiance that is truly Apollonian. Thanatos, grim and implacable, is pinpointed with equal economy (Ex. 26), as is Alkestis's calm acceptance of her fate (Ex. 27). Moments of greater tension require only a slightly larger harmonic vocabulary. The relentless diminished fifths and the despairing chromaticism of Admetus's sorrow (Ex. 28) are all the more effective when set against the plain harmonies that have served the other characters and situations.

Ex. 25

Ex. 26

Ex. 27

Ex. 28

The all-important contribution made by the chorus is equally simple. Textures range from solemn unisons, through syllabic four-part writing, hymn-like in its economy, to the independence of accompanied and unaccompanied partsong. Nothing is allowed to stand in the way of a clear presentation of the words and the full impact of their raw emotion. With the barest of means, Boughton meets the drama's supreme challenge: the threnody that precedes the return of Alkestis at the moment when Admetus realizes the full measure of his grief (Ex. 30). It is music of devastating simplicity, superb and inevitable.

Ex. 29

If there are weaknesses in *Alkestis* they are to be found in Boughton's handling of Herakles. Seldom at ease with comedy, either in life or art, he clearly found it difficult to characterize the satyr-play element with which Euripides invested his trickster-hero. Inspiration apparently failed him and he fell back on an earlier work, the 'Oliver Cromwell' Symphony of 1904. From this he borrowed a theme that had served to portray Cromwell's strength of purpose, modifying its rhythm and somewhat foreshortening it so that it might now serve to suggest the heroic determination of Herakles (Ex. 29). He might have done better to retain the original form. As it is, the idea seems too short-breathed to be really convincing. A second theme from the same symphony worked to better effect as an expression of Herakles's sympathy for Admetus, while a theme borrowed from the *Imperial Elegy* that Stanford had so disliked proved a very precise metaphor for the anguish that Admetus feels (Ex. 28). But borrowed or newly minted, the thematic content of Boughton's *Alkestis* is of a standard that matches the profound emotions he was called upon to deal with, and the result is a work that is utterly convincing.

THE QUEEN OF CORNWALL

For his contribution to the 1924 Glastonbury Summer Festival Boughton turned to the one act 'Mummers' Play' that Thomas Hardy had published in 1923 under the title: *The Famous Tragedy of the Queen of Cornwall at Tintagel in Lyonesse*. It appealed to him on aesthetic grounds, in that Hardy had framed the story of Tristram and the two Iseults with a series of choruses ('Shades of Dead Old Cornishmen and Women') that comment on the action in the manner of Greek drama; and on purely personal grounds in that the story mirrored his

Ex. 30

[Slow]

Ex. 30 contd.

own emotional problems—torn, like Tristram, between the love of two women—much as *Alkestis* had echoed Christina's self-sacrifice.

In order to shape his libretto Boughton consulted Hardy over the cuts he required, the additional words he needed, and the six poems he thought might be inserted at those moments that called for greater lyrical expansion. Hardy agreed to each suggestion, including a number of word changes that clarify the meaning, and even wrote several short additional passages, declaring that Boughton's libretto was an improvement on the original play. He was particularly surprised to see how appropriate Boughton's choice of additional poems were: 'Bereft, she thinks she dreams', 'When I set out for Lyonesse', 'Beeny Cliff', 'If its ever Spring again', 'The End of the Episode', and 'A Spot' fitting each dramatic situation as if made for it. The only words they could not find were for King Mark's drunken knights to sing off-stage at the end of Boughton's proposed first act. Hardy finally suggested that the knights were probably so drunk that they would only be capable of a burbling 'la-la', for which Boughton was happy to find suitably tipsy music.

Despite Hardy's eccentric archaisms, the result was another first-rate libretto: powerfully poetic, economical, and crying out for music. Once again it was to be the chorus, rising out of the sea and wind and the very stones of Tintagel, that provided the framework of the drama. And once again it was the chorus that would coax from Boughton some of his finest music. The story is as follows:

Act 1

Iseult, Queen of Cornwall, has returned to Tintagel after a voyage to Brittany, made in the hope of seeing Tristram who has been wounded unto death. Her husband, King Mark, is deeply suspicious, but accepts her word that Tristram is dead. Indeed, this is what she believes; for Tristram's wife, Iseult of the White Hands, has told her so when she tried to land in Brittany. At the moment of her greatest despair Iseult hears the song of a harper outside the castle walls. It is Tristram.

Act 2

Iseult of the White Hands has followed Tristram to Cornwall and now pleads in vain that he return to her. King Mark's suspicions are confirmed. He traps the lovers and stabs Tristram. Snatching up the dagger, Iseult kills King Mark and hurls herself onto the rocks below Tintagel. Iseult of the White Hands is left to mourn the unhappy dead. As the chorus brings the work to a close the ghostly voices of the lovers are heard singing of their eternal love.

Musically, the drama is presented as if it were a dream—a legend borne down the ages, enacted by ghosts. It begins and ends with magnificent 'storm' choruses, out of which the action gradually unfolds and into which everything ultimately dissolves (Ex. 31). The whole conception is both original and startlingly beautiful: the wordless chorus becoming a extension of the orchestral palette.

In almost every respect *The Queen of Cornwall* inhabits a very different sound-world from Boughton's earlier music-dramas. The harmonies are much more astringent, raw, and exciting, and the orchestration noticeably harsher. Lyrical moments now appear as momentary lulls in a storm of passion, heavy with brooding tragedy. The system of representative themes is used, as before, to pinpoint the main characters and ideas, and they are skilfully woven into the general fabric which is more consistently symphonic. The more highly charged the emotion of a scene, the less formal are its lyrical patterns. Thus Iseult's main arias are really dramatic *scenas* which, though they may incorporate stretches of metrical song, constantly melt into lyrical speech rhythms. The degree of flexibility thus available gives pace and breadth to the entire work and marks a distinct advance in Boughton's approach to the problem of opera. On the other hand, Tristram's songs, which occur at moments of quietude and are in any case expressive of his assumed role as minstrel, return to the metrical patterns of *The Immortal Hour* and *Bethlehem*. Like them, they draw upon pentatonic scale formations, but are somewhat broader in concept (Ex. 32).

Ex. 31

cont.

Ex. 31 contd.

Ex. 32

Quick and happy

When I set out for Ly - o-nesse a hun - dred miles a-
way The rime was on the spray And star - light lit my
lone-some-ness When I set out for Ly - o-nesse a hun-dred miles
a - way.

In the twelve years that separate *The Queen of Cornwall* from *The Immortal Hour* Boughton had travelled a long way. But nothing had been lost. His inspiration ran as steadily as ever, strengthened by a greater command over his materials, enriched in emotional depth, broader and yet more subtle. One final example may serve to underline the extent of this development. Two themes are presented, the one growing effortlessly out of the other. Both are powerful characterizations: King Mark, regal and arrogant; Queen Iseult, tragic and in despair (Ex. 33). On any count, *The Queen of Cornwall* is a remarkable work.

AGINCOURT AND THE MOON MAIDEN

Boughton's published stage works include two short pieces: *Agincourt* and *The Moon Maiden*. *Agincourt* is a setting of words cannibalized from Act 4, Scenes 1 and 3, of Shakespeare's *Henry V*. It is, frankly, a mess—Shakespeare trivialized by trivial tunes. Boughton's heart simply cannot have been in it. *The Moon Maiden*, on the other hand, has charm. Boughton described it as a 'choral ballet', using as his libretto a translation by Dr Marie Stopes of the Japanese Noh play *Hogaromo*.

Ex. 33

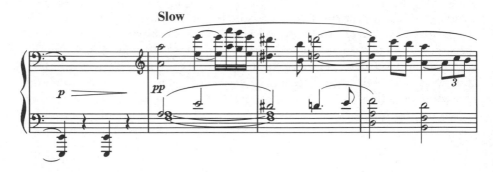

Even though dancing is called for, it is, in effect, a one-act opera. The story is simplicity itself:

The Moon Maiden drops her cloak as she dances through the world. A fisherman finds it, and at first will not give it back to her. When he sees her sorrow (for she cannot return to her home in the sky without it) he relents, but asks her to dance for him. Overjoyed, she agrees. As she dances, night turns into day.

The music is tuneful and delicately imagined, with nicely judged writing for a female chorus, and an unusual orchestration for flutes, trumpet, harp, and strings. It is minor Boughton, and perhaps a little too cloying in its overall sweetness, but does not deserve to be entirely forgotten.

THE EVER YOUNG

Boughton began work on *The Ever Young* in the summer of 1928 and completed the score in July 1929. He wrote his own libretto, evolving the plot from a paragraph he had come across in a book of Irish legends which tells the story of Aengus, the Celtic god of love, and his passion for a mortal maid, Caer. It occurred to Boughton to wonder what would happen when the mortal woman grew older and wiser, while the immortal god remained forever young and impetuous. Something that on one level would merely be an attractive legend might, on a deeper level, become a parable of human development—how men and women can outgrow each other, as Boughton himself had reason to know. In the programme notes for the first, and only, production (at Bath in September 1935) he wrote: 'There is no plot, and little incident in this work. The drama moves in a sphere wherein music, colour, and symbolic movement seem to be natural mediums of expression—that is to say, in the feelings of the audience.' He regarded it as the most complete expression of the type of music-drama he had sought throughout his life: the drama which unites the realities of this world with the realities of the mystical world—a drama that is both flesh and spirit, and moral in every sense of the word. The story, as he saw it, unfolds thus:

Scene 1: A mountain lake in Spring.

By way of overture: an elaborate Choral Dance—a bridge between the world of reality and the world of faery, through which passes Caeria, the mortal maid. She has dreamed of Aengus, the God of Youth and Love, and now will search until she finds him.

Orchestral Interlude: Caeria's Dream Journey.

Scene 2: The Hall of Dagda, Father of the Gods.

Aengus has dreamed of a mortal maid and longs to win her love. His father, the Dagda, is scornful of his passion, and his brother, Lugh, indifferent to such matters. But his mother, Boyanna the Consoler, is more sympathetic and tells him to consult Bride, the Moon Goddess. She warns him that a mortal maid must grow old and die, but he believes he has the power to change the order of things. Braving his father's wrath, Aengus sets out on his quest.

Scene 3: The Dome of Bride, Goddess of the Moon.

To Bride come all the dreamers of the world: children, full of life and hope; old men and women, weary and full of sorrow. Caeria enters, wandering as in a dream, her heart full of longing. As if in answer to her desire, Aengus appears. But in these cold realms there can be no union for a mortal and a god. Caeria and Aengus must seek happiness in the realm of Dana.

Scene 4: The Cave of Dana, Goddess of the Earth.

Dana, the mother of gods and men, has waited through all eternity for the day when she may be released from the never-ending cycle of Spring and Winter, Day and Night, Joy and Sorrow, Birth and Death. She believes that the coming of Aengus may bring this to pass. But when, led by Caeria, he arrives, she realizes that in believing he can win eternal youth for the woman he loves he dreams an impossible dream. The circle of change can never be broken.

Orchestral Interlude: Caeria's Awakening.

Scene 5: A mountain lake in Autumn.

By way of finale: an elaborate Choral Dance. Daeria has grown old. She knows that the cycle of life and death is the only reality. Aengus, locked in eternal youth, must leave her and she must die. The Gods realize that their power over humankind is waning and that the false hopes they offer will no longer serve. Henceforth men and women must take responsibility for their own destiny.

Although Boughton's libretto contains many passages of great beauty, it is far from easy to understand. As in all mystical matters, clarity of expression seems to be inimical to revelation. At best we are allowed to see through the glass but darkly. Since it is difficult enough to tease out a meaning in the calm of the study, it is hard to imagine what sort of sense would come over the footlights in even the most dedicated performance. Add to this the problems in creating a stage picture for Celtic gods and goddesses who do very little and talk far too much, and it seems that the only future *The Ever Young* may have is on disk and in the imagination of the listener. One thing is certain: a work whose meaning is so elusive calls for music of exceptional intensity if it is to make any sort of impact on any sort of audience.

The dominant impression left by the music of *The Ever Young* is not very encouraging. Apart from Caeria's main songs, the only moments of real distinction come in the third scene, where Bride is revealed in the cold splendour of night. Here the music takes on a delicate, silvery quality (Ex. 34). The chorus of old men and women has a dignified pathos (Ex. 35), and the children are given appropriately innocent

Ex. 34

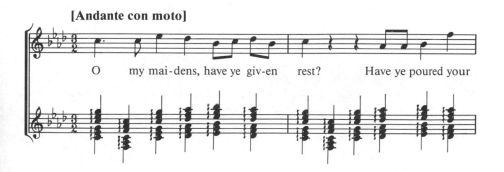

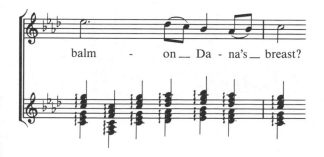

Ex. 35

Ex. 36

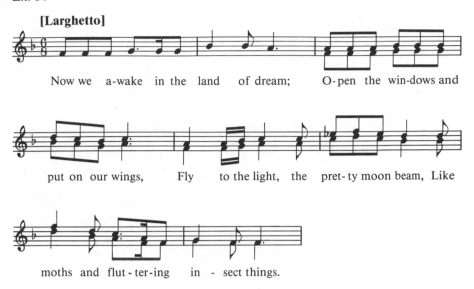

[Larghetto]

Now we a-wake in the land of dream; O-pen the win-dows and put on our wings, Fly to the light, the pret-ty moon beam, Like moths and flut-ter-ing in - sect things.

music (Ex. 36). The whole scene has the character of a masque, and critics were quick to single out its beauties.

Less happily, Dagda and Boyanna are sketched in semi-comic terms: a middle-aged, middle-class couple, bickering with each other in an irritating 'stage' brogue:

DAGDA. Pray thee, wife, put no more words upon me. It's not myself that weaves my beard into my loom. And it's with me thou must be laughing the way Aengus has been wildered by his own birds.

BOYANNA. Little laughter has he of the sickness upon him: and we must help him to the quiet of his heart.

Presumably intended to reduce the 'gods' to a domestic twilight in preparation for their ultimate dismissal, it is scarcely convincing even as pastoral comedy.

Aengus borrows from the early symphonic poem, *A Summer Night* for his main love song, and in so doing immediately plunges the work into the nineteenth century (Ex. 37). Only Caeria has music of beauty and character (Ex. 38), but even her themes lose their impact through lack of any vigorous contrast. In *The Ever Young* Boughton's penchant for litling 6/8 rhythms sometimes degenerates into spinelessness.

Superficially, the music of *The Ever Young* has the same characteristics as Boughton's published music-dramas. It has the same ingredients, the same fingerprints. It is orchestrated with the same delicacy and imagination—indeed, its orchestration is exceptionally beautiful. The

Ex. 37

idea behind the story is an important one: a parable well worth pon-
dering. What is missing is that special quality his finest works possess
in abundance: the power to crystallize each emotion in phrases that are
not only apt but inevitable.

Ex. 38

24

The Arthurian Cycle

Rutland Boughton worked at the five dramas of his Arthurian cycle on and off over a period of nearly forty years. It began as a Wagnerian epic, developed through a conventional tale of kings and queens and unhappy lovers, sidetracked brilliantly into the moving personal story of Elaine the Lily Maid, and ended as a political tract with strong religious overtones. In the sense that it embraces every aspect of his personal development it is the consummation of Boughton's life-work.

When he began work, in January 1908, he had every intention of setting Reginald Ramsden Buckley's sequence of four dramas exactly as their author intended. And, save for a few minor adjustments, he did indeed set the first part, *The Birth of Arthur*, word for word. But in 1908 his experience of opera was entirely theoretical. When he came to tackle the second drama, *The Round Table* (1915-16), he had gained immeasurably in practical experience through writing and producing *The Immortal Hour* and *Bethlehem*. His approach was now that of a practical man of the theatre and, to Buckley's dismay, he took it upon himself to reshape the libretto. Thereafter he seems to have cooled towards the entire cycle as Buckley saw it. His working copy of the complete cycle shows changes only to the first act of *The Holy Grail*, Buckley's third drama; the remainder of the text is untouched. Moreover, in 1917 he completed the first draft of a libretto of his own on the subject of Lancelot and the Lily Maid of Astolat. Though this has no counterpart in Buckley's scheme it was to become the third part of Boughton's complete cycle. It therefore seems obvious that Buckley's untimely death in 1919 gave Boughton the freedom to do openly what he had privately decided to do anyway.

On the whole, Buckley's cycle, published in 1914 under the general title *Arthur of Britain*, is reasonably faithful to legend. The first drama, *The Birth of Arthur*, departs from the traditional story in only two respects: Igraine is shown to be eager for Uther's embrace, and Merlin brings Uther to her by cunning rather than magic. Arthur, the Hero-King, is therefore born of a passionate, consenting union and we are spared the element of deceit and rape that make the original story somewhat distasteful. Buckley's second drama, *The Round Table*, also follows tradition. It begins with the boy Arthur releasing the sword from the stone and, vouched for by Merlin, being accepted as king by the knights and the people. Guenevere is introduced, rather precipitately, as someone Arthur has already met and admired. She tells of a Round Table, owned by her father, that has the power to bind the kingdom and bring peace and happiness. The people decide she shall be Arthur's queen and the Round Table her dower. In the second act Arthur's reign has prospered, and Lancelot's support is shown to be crucial. Though there is as yet no doubt about Guenevere's love for the King, it is clear that she is sympathetic to Lancelot and that he is in love with her. In the third and last act Arthur visits the Lake of Wonder and is given the sword Excalibur which will enable him to overcome his enemies. Merlin's task is now done and he can die in peace, even though he knows that only a stainless knight (which Arthur no longer is) will find the Holy Grail.

Boughton found Buckley's scheme inadequate. Not only did the protagonists seem to lack character and motivation, but the dramatic shape was awkward and moved, if anywhere, only to an anticlimax. He therefore demanded a radical reorganization. Merlin was made to set the action in motion by visibly planting the sword in the stone. Arthur's lowly state was made very clear, so that his winning of the sword became more dramatic, and the knights' initial dismay more understandable. What was Buckley's third act became the second act: Arthur learning of his origins through Merlin and the Lady of the Lake, and being advised to win Lancelot's allegiance by marrying Guenevere and claiming the Round Table as her dowry. Buckley's second act then became the natural climax to the drama: Guenevere being shown to be in love with Lancelot, whom she persuades to support Arthur's bid to bring peace to the land by means of the Quest for the Holy Grail.

Buckley's version of the cycle is completed in two further dramas: *The Holy Grail* and *The Death of Arthur*. In *The Holy Grail* he prepares for Arthur's downfall by introducing Morgan le Fay, his evil sister. She stirs up the ambition and discontent of Mordred, Arthur's bastard son, and sows doubt in his mind about Guenevere's faithfulness. The decline and disorder in Arthur's kingdom is temporarily averted by the arrival of Galahad, who proposes the Quest of the Grail. Though Arthur remains aloof, the knights take up the challenge. Lancelot reaches the threshold of the Grail, but Galahad, the only spotless knight, finally attains it.

In the fourth and final part of the cycle Buckley shows a kingdom wracked by discontent, Mordred now openly plotting King Arthur's downfall. Lancelot and Guenevere admit the hopelessness of their love and go their separate ways. Led on by Mordred's promise of a time of freedom when there will be 'No God, no King, no Grail', many of Arthur's knights rebel and he is defeated. Sir Bedivere returns Excalibur to the Lake of Wonder, and the Lady of the Lake bears the dying King away.

It was to be twenty-six years before Boughton felt able to complete his version of the Arthurian cycle—by which time his understanding of life and himself had grown immeasurably. His interpretation of the legend underwent a correspondingly radical change. Though he was eventually to decide that *The Birth of Arthur* no longer fitted the style of the other four dramas, and would therefore have to be dropped from the cycle, the complete scheme unfolds in the following manner:

THE BIRTH OF ARTHUR

Act 1

Uther, king of Britain, has seen Igraine at a banquet and has fallen in love with her. He now holds Gorlois, her husband, prisoner and is about to lay siege to Tintagel where Igraine is in hiding. Merlin, who dreams of a Hero-King who will lead Britain to greatness, decides to use Uther's passion to his own ends.

Act 2

Igraine, attended by the faithful Sir Brastias, remembers Uther's passion and contrasts it with her husband's coldness. She longs for a child. When Uther is brought secretly into Tintagel by Merlin she welcomes him with a love as fierce as his own. She agrees that the child of their union shall be given into Merlin's safekeeping.

THE ROUND TABLE

Act 1

Merlin plants the Sword in the Stone outside the Cathedral, to await the coming of the Hero-King. Arthur's lowly state is made clear (he is shown working alongside the cooks in Sir Ector's kitchens), but he is allowed to accompany Sir Kay and Sir Ector to church. Kay and Arthur find the Sword, and where Kay fails Arthur succeeds in removing it from the Stone (he alone has the wit to release a hidden catch). Arthur gives the Sword to Kay, who claims he has forgotten to bring his own. Kay is hailed as King, but Sir Dagonet throws doubt on his claim and replaces the Sword. Kay fails to remove it, but Arthur, in full view of the knights, succeeds. The knights are dismayed, for Arthur's origins are unknown to them. Sir Ector vouches for his noble birth and tells Arthur to seek out Merlin for an explanation.

Act 2

Led by the seductive Nimue, Merlin comes to the Lake of Wonder to await Arthur. Merlin tells him of his parentage and advises him to marry Guenevere and ask for the Round Table as her dowry. Because she loves and is loved by Lancelot, the most powerful knight in the kingdom, Lancelot will be bound to Arthur's cause. Thus united, the knights will bring peace to the kingdom. Arthur accepts his destiny, and Merlin is led to his tomb by Nimue, the avenging spirit of his renunciation of human love for the sake of wisdom.

Act 3

The knights gather for the great Pentecostal Feast. Servants gossip about the power of the Seat Perilous, while Sir Dagonet proceeds to get tipsy. Queen Guenevere tells the Lady Bettris of her longing for her native Wales. The reason for her melancholy becomes clear with the arrival of Lancelot: they love each other, but are separated by circumstances they cannot, dare not, challenge. Guenevere begs Lancelot to support King Arthur's plan to bring peace to the land by sending the knights on the Quest of the Grail. But Lancelot has no faith in the Quest and refuses to do as she asks. The knights assemble to hear Arthur's words. He is supported by the Queen and the Bishop (who appreciates the mystical value of such a Quest), but the knights are unimpressed. Arthur makes a final appeal, citing his proven valour and his Christian hatred of bloodshed. This gains him the support of the women. With a look, Guenevere makes a last appeal to Lancelot. He gives his support and the knights follow his example: they take the oath and the Quest of the Grail begins.

THE LILY MAID

Act 1

Sir Bernard, old and disillusioned, lives with his sons, Sir Terry and Sir Lavaine, and his daughter Elaine, the Lily Maid. His wife is long since dead and he will not speak her name, for she left his house to follow another, stronger love. Into this sad and dreamlike atmosphere comes Lancelot, on his way to joust at Winchester. He is in disguise, but Elaine recognizes him as the knight whose fame she and her younger brother, Sir Lavaine, have worshipped from afar. Innocently, scarcely realizing what she is doing, Elaine falls in love with him and gives him a Lily and a Rose to wear as emblems in the joust.

Act 2

Lancelot returns from Winchester. Tormented by his love for Guenevere, and troubled by the Quest he follows but cannot believe in, he allows himself to return Elaine's love.

Act 3

Sir Gawaine arrives with a message from the Queen. Her suspicions have been aroused by the favours Lancelot wore at the joust and she demands that he return. Lancelot is forced to tell Elaine the truth. He leaves, and in despair she drowns herself. Only Hod, the old serving man, understands her tragedy.

GALAHAD

Scene 1

Galahad lives in Lancelot's castle, unaware that he is his son. He is filled with wonder at the beauty and truth of Christ's teachings, but cannot understand the Quest of the Grail. To him it is simply a magic dish—mere superstition. He questions three travellers: a knight, a monk, and a pedlar. Their attitudes to religion puzzle him: either they scorn it, believe only in its outward trappings, or ignore it altogether. None of them seem able to understand the true message that Christ brought to the world.

Scene 2

Galahad comes to a strange chapel. Before its altar there lies a knight, wounded and dying—grieving, it seems, over some lost, unhappy love. From him Galahad takes the Silver Shield of Innocence and the Golden Sword of Truth.

Scene 3

The poor folk wait outside the doors of the cathedral in which Arthur and his knights are at prayer. They hope for a few pence and the excitement of a splendid procession. Galahad joins them. He is perturbed that the poor may not enter a building that is intended for all men. The King and his knights emerge, followed by the priests and choir. The choir sing the Beatitudes, using St Luke's version of the gospel. The priests answer with St Mark's version. Galahad is struck by the subtle differences:

> Blessed are ye poor, for yours if the Kingdom of Heaven.
> Blessed are the *poor in spirit*, for theirs is the Kingdom of Heaven.

He questions the priests, but receives evasive answers—they regard him as subversive of good order. The nobles and the clergy depart, singing a hymn in praise of military might. Galahad is left alone with Will, the leader of the poor. Will offers to take him to friends who may be able to answer his questions.

Scene 4

Will leads Galahad to a wood where live two strange figures: Penimel and Melicora. By means of a cinematograph they show Galahad how life evolves, and how Nature both protects and kills her creatures. Men and women will only rise above such conditions when they learn to love and help one another. Will hints that a new order is coming into existence. He accepts Galahad's ring as a token of friendship, promising to send it as a signal for Galahad to join him when the time is ripe. Left alone, Galahad ponders what he has seen and heard. He believes he has found a solution to mankind's problems in the laws that govern music and make the sound of many voices into one harmonious song. Full of joy, he decides to tell the King of his discovery.

AVALON

Act 1

Galahad arrives at Camelot. As he waits for an audience with the King he learns that Lancelot is his father. The knights assemble and King Arthur tells them that the Quest of the Grail has been unsuccessful, but asks them to give some account of their adventures. Sir Gawaine hints that Lancelot's journeys that not been entirely without incident. A quarrel breaks out and is only brought to an end when Galahad asks to be allowed to sit in the Seat Perilous. He does so and comes to no harm, but immediately disproves the legend by tumbling Sir Dagonet in. Lancelot is delighted, but the priests are angered at the loss of a useful myth. When Arthur asks Galahad to speak his mind he tells him that the poor must be set free. Only the women understand and sympathize. Encouraged by Guenevere, Galahad reveals his faith in the power of music. But he is interrupted by a messenger who speaks of a barge that bears the body of a dead lady, tended by an old man. The court depart and Galahad is left alone with Lancelot. They talk of the unrest that stirs in the land and what may come of it. A messenger arrives with the ring that Galahad gave to Will. With Lancelot's blessing Galahad leaves to join his friend and the uprising.

The knights now return with the body of the dead lady. It is Elaine. Hod has told her story, and Lancelot must do penance. Left alone, Arthur and Guenevere consider their lives. He thanks her for her loyalty, and comforts her with the thought that their sacrifice has been made for the good of the country. When Lancelot returns, Arthur leaves him alone with Guenevere. She tells him they must part. She wronged him when she married the King, just as he wronged the Lily Maid and Galahad's mother. Lancelot protests his innocence. But the damage has been done: Arthur, Guenevere, Lancelot have failed—failed themselves, failed each other, failed the country. All that is left to Guenevere is to do penance in some remote nunnery. Lancelot curses the Church and vows never to fight for Christianity again.

Act 2

Scene 1

An army of peasants, led by Mordred and Mordaunt, has gathered in Merlin's Cave beneath Tintagel Castle. Mordaunt leads the people because he believes in their cause, Mordred because he is ambitious. Galahad arrives, but the peasants will have none of him. They tie him up and leave him to the mercy of the rising tide. At the last moment Will returns in a desperate bid to save his friend.

Scene 2

The rebellion has been crushed. Mordaunt is dead, but Mordred still leads the remnants of the peasant army. He is joined by Sir Gawaine who cynically offers to make a pact with him so that together they may seize power. Mordred readily agrees.

Scene 3

Sir Bedivere carries the dying King to the Lake of Wonder. Arthur is in despair: his kingdom is divided into rich and poor, bitterly opposed to one another. Weapons have failed him as surely as corrupted Faith. But he is granted three visions: of the Past, the Present, and the Future—the Star that rose over Bethlehem, the White Star of Hope that rose with Arthur himself, and the Red Star that will one day rise to unite all men in liberty and equality. Joyful at this promise of better things, Arthur dies in peace.

Though world events have made it impossible to take the climax of *Avalon* seriously—and thus cast doubt over the possibility of ever performing the cycle as a whole—credit must be given for the ingenuity with which Boughton shaped the Arthurian legend to his own ends. Whatever shortcomings there may be in its arguments, the cycle is a fascinating document of its author's political odyssey. Significantly, it is Galahad and his touchingly naïve faith in the sanity of art that Boughton seems most to favour. The peasants are shown to be impotent and easily misled; the knights equally gullible to the claims of fascists (Mordred) and cynical opportunists (Gawaine); while Arthur's vision of a rose-red future merely substitutes a new and equally dubious act of faith for those that have been overthrown by experience. When in 1956 Boughton had finally to measure his hopes against reality he must indeed have felt that he had been whistling in the dark.

THE MUSIC OF THE ARTHURIAN CYCLE

Despite its network of representative themes, the musical style of the five Arthurian dramas is far from consistent. Nor, indeed, could complete consistency be expected. Both as a man and as a musician the Boughton that wrote *The Birth of Arthur* in 1908 and 1909 was very different from the Boughton that completed *Avalon* in November 1945. A greater degree of consistency might have been secured had Boughton remained faithful to Buckley's dramatic scheme, which, after all, had been put together in a relatively short space of time: 1905–6 in its original form. But he did not, and forty years were to elapse between Buckley's first draft for *The Birth of Arthur* (*Uther and Igraine*, as it was then called) and Boughton's final notes. On such a time-scale, consistency could hardly be expected. Leaving aside the immeasurable difference in stature between the two men, not even Wagner had to contend with quite so formidable a challenge—the librettos for the *Ring* being completed in a matter of five full years (1848–52), and the music in two monumental stages: 1853–7, and 1869–74.

What can be expected is that each of the five dramas should give satisfaction on its own terms. This they do in varying degrees, though only *The Lily Maid* possesses the unity and strength of purpose that is likely to ensure it any degree of genuine viability as a stage work. A simple explanation suggests itself: *The Lily Maid* is the only episode in the cycle which deals with characters and situations that Boughton could identify with on the deepest personal level. He could sympathize with many of the characters and situations in the other episodes—some (Galahad and Guenevere, for example) more than others. But his interest in them was more a matter of intellectual conviction than emotional identification. It is not too much to say that he fell in love with his Lily Maid, just as he had fallen in love with Etain, Mary, Alkestis, and the two Iseults. Like them, she is a aspect of the idealized womanhood that seems always to have been the mainspring of his creativity. Other characters and situations might inspire music that is powerful, apt, and effective: but only Etain and her sisters could draw from him the unforgettable magic.

The Birth of Arthur

Though Boughton described the first part of his Arthurian cycle as a 'choral drama', the method he employed to articulate its musical structure is that of Wagnerian music-drama. A system of representative themes is used and, save for moments of Debussyish, tritonal ambiguity to illustrate the mystical aspects of the drama, these themes are unmistakably Wagnerian in character. They are, however, not subject to the same degree of subtle transformation that gives the *Ring* its organic unity. The effect is less of continuous symphonic development than of a mosaic of relatively static themes, shifting according to the demands of the text but not growing, as it were, from within.

The themes themselves are mostly striking in conception, and convincing as musical imagery. They may be presented chorally, as is the theme (Ex. 39) that ultimately comes to stand for the Grail, or orchestrally, as with the 'Tintagel' theme (Ex. 40) that opens Act 2—on this occasion combined (bars 11–12) with one of Merlin's themes. They range in character from the heroic (Ex. 41), as befits the first mention of Arthur, to the lyrically tender (Ex. 42), as in the love duet between Uther and Igraine.

The un-Wagnerian feature is, of course, the use of the chorus. This,

Ex. 39

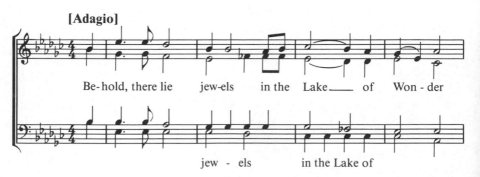

Ex. 40

[Moderato]

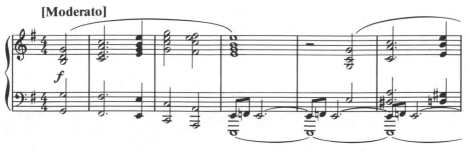

Ex. 41

[Andante]

Ex. 42

Ex. 42 contd.

as we have seen, can be motivic, or simply a commentary on the dramatic situation. For example, the choral commentary (Ex. 43) that follows the love duet acts as a tactful drawing of veils over a moment of
intense sexual passion. The texture, as in all the choral passages in this
work, is complex and far removed from the average operatic chorus—
we are, in fact, in the world of oratorio rather than opera.

The overall impression left by *The Birth of Arthur* is that it is a
work of considerable power and imagination. Had Boughton felt
sufficiently encouraged to complete the orchestration (he abandoned it
a few bars before the end of the first act), it might well have proved
itself an effective piece—assuming that a way could be found to present the choruses in a convincing manner. Though Buckley's libretto
may creak, the sweep of Boughton's music is more than enough to
paper over the cracks and lend conviction to the whole enterprise.

The Round Table

Of all Boughton's stage works *The Round Table* comes closest to conventional opera. The chorus is now part of the action, representing
nothing more symbolic than cooks and courtiers. The main characters
have set-piece arias in true operatic fashion, and though the network of
representative themes remains it is deployed in a more relaxed manner.
As for the drama itself, it is played out almost wholly in terms of
action. By thus adopting a more pragmatic approach Boughton allowed
himself a greater variety of operatic ingredients and produced a much
more dramatic and finely paced score. It is not surprising that, next to
The Immortal Hour and *Bethlehem*, *The Round Table* drew the most
enthusiastic audiences at Glastonbury.

Where the work falters is in the variable quality of Boughton's inspiration. A passage of startling beauty and originality, such as that which

Ex. 43

[Sostenuto]

Ex. 43 contd.

depicts the cathedral as the knights assemble for prayers—bells and
hymn-tune combining in a way that anticipates Britten (Ex. 44)—is
hard to reconcile with the scene that has just gone before: cooks
preparing the food in Sir Ector's kitchens, in a manner that is merry
enough but fatally derived from the example of Sir Edward German
(Ex. 45). The two styles simply do not match.

   Despite these discrepancies the dominant impression left by the
music of *The Round Table* is one of richness and variety. Passages of
folk-like simplicity, such as Guenevere's song of homesick longing (Ex.
46), sit comfortably alongside the Brahmsian harmonies that accompany Lancelot's declaration of love (Ex. 47), and the melting,
Tristanesque chromatics of Nimue's blandishments (Ex. 48)); not to
mention the Debussy-inspired whole tones that give the Lady of the
Lake's delphic utterances their air of mystery and magic (Ex. 49). It is
a remarkable mixture—a heady brew that speaks volumes for
Boughton's powers of invention. A work of such vitality does not
deserve the oblivion that has descended upon it since its last staged
production in 1925. And although extracts have been presented with
orchestral accompaniment there appears never to have been a complete

Ex. 44

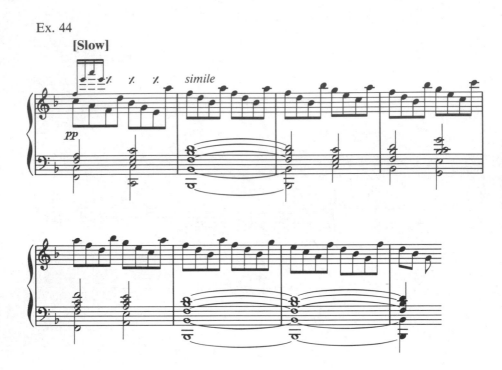

Ex. 45

It's a lord-ly life to be a knight And to wear a sur-coat red

orchestral performance. What this would add to the work's effectiveness can well be imagined. Of all the historical operas by British composers to that date it is by far the most impressive; massively
out-classing even Sullivan's *Ivanhoe*, and certainly the various turgid
efforts of such composers as Sir Alexander Campbell Mackenzie and
Sir Frederick Hymen Cowen.

Ex. 46

In the homeland of my fa-thers There's ne-ver a wind that blows

Ex. 47

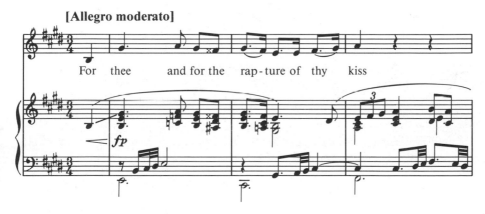

For thee and for the rap-ture of thy kiss

I would go quest.

Ex. 48

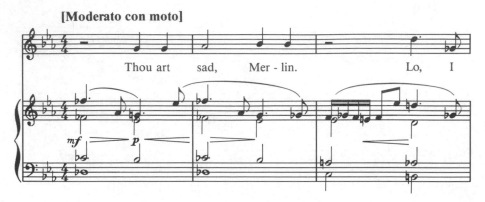

Thou art sad, Mer-lin. Lo, I

bring thee love be-lat-ed.

Ex. 49

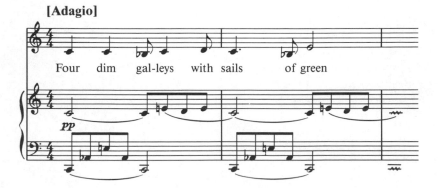

Four dim gal-leys with sails of green

The Lily Maid

The third drama in the cycle stands apart from the others on two counts. Firstly, it is virtually a self-sufficient story, its links with the rest of the cycle being tenuous and built round the slender fact that they involve Sir Lancelot. Little would be lost to the main thrust of the cycle had it never existed. Secondly, by virtue of the reduced orchestra it requires (single woodwind, instead of double and triple; two horns and a trumpet, instead of the full brass complement), and the intimate nature of the story, it is nearer chamber opera than full-blown music-drama. Though it uses a chorus both to explain events that are not being presented on the stage, and comment on the action as it unfolds, Boughton was happy to have this sung by a group of four soloists— thus underlining the music's intimate nature. It is also, as we have seen, the only part of the cycle he could identify with on the deepest level. But if it is the odd man out, it is also by far the most impressive of the five dramas.

Musically it has all Boughton's virtues: a strong choral framework, an unfailing spring of melody that is characteristic of its composer's mature style, a sumptuous orchestration (achieved with impressive economy of means), a powerful and moving story, and, above all, that quality of indefinable magic that is only to be found in the authentic Boughton music-drama.

As compared with those of *The Birth of Arthur*, the choral sections are much simpler in style: homophonic in texture, with words set syllabically and therefore that much easier to hear. The lengthy 'choral preludes' which open the first and second acts are cast as strophic ballads (with some variation according to the way the verses unfold as narrative), and as such are perhaps a shade too long to be completely satisfactory. The chorus that opens the third act is altogether more fluid, and thus more effective. But by far the most remarkable choral episodes are the brief commentaries that punctuate the action and act both as a bridge between characters and audience, and as a means of catharsis at moments of extreme emotional tension—as in Act 2 when Elaine is brought face to face with the implications of her love for Lancelot (Ex. 50).

Each character is neatly caught in musical terms. Elaine's cool, timid beauty (Ex. 51) as decisively as her brother Lavaine's youthful impetuosity (Ex. 52). So also is the general atmosphere that pervades the tragedy—a dreamlike foreboding, as of some ancient tale lost in the mists of time (Ex. 53). The love music is particularly beautiful. It somehow contrives to be both voluptuous and chaste, and is certainly

Ex. 50

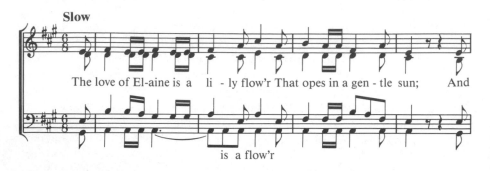

The love of El-aine is a li - ly flow'r That opes in a gen - tle sun; And

is a flow'r

he who may take her love for dower The joy of peace hath won.

he——— who may take this dower

among the most passionate examples that any British composer had so
far allowed himself in an opera (Ex. 54). Even Hod, the old serving-
man, who sings in unaccompanied, unmeasured recitative as befits his
garrulous nature, is more effective than most of Boughton's attempts at
humour—though his Mummerset dialect and all-too realistic bumbling
goes on too long for complete comfort.

Despite one or two longeurs, easily remedied by judicious cutting,

Ex. 51

Ex. 52

[Quick]

Ex. 53

[Slow]

Ex. 54

Ex. 55

The Lily Maid can be counted among Boughton's most successful operatic essays. It creates a world of its own and meets every emotional challenge with music that is both apt and beautiful. By 1934 standards it was, of course, hopelessly out of date; but by the timeless standards of art it is completely successful and convincing. And by the standards of 1934 it was also remarkably courageous, for how many other composers would have dared to clothe a moment of supreme pathos, such as Elaine's heartbreaking message to Lancelot, her betrayer, with music of such simplicity (Ex. 55)? It is this capacity to state his case with child-like directness that made Boughton what he indisputably was: the foremost British opera composer of his time.

Galahad *and* Avalon

The last two parts of the Arthurian cycle present special difficulties to any commentator. Neither have yet been performed, and it is notorious that an opera which looks doubtful on paper may, in the theatre, prove to have strength and validity. Nevertheless, if a reading of the score can be said to give any indication of a stage work's potential it has to be admitted that both *Galahad* and *Avalon* show something of a decline in Boughton's powers. For while the general technique is, if anything, even more fluent and assured, the thematic material seldom rises above the commonplace.

Galahad begins well. A short orchestral prelude based on themes already established in the earlier episodes of the cycle ends with a new theme which introduces Galahad himself (Ex. 56). His first song establishes his innocent vision of life. Its words are based on a poem that Boughton wrote in 1901, and the music is a variant on a theme already associated with the Holy Grail and the idea of lasting peace (Ex. 57). Short motives associated with the old nurse, a stableman, a knight, a

Ex. 56

Ex. 57

Andante con moto

O that I could sing   now the song my soul— is sing - ing!

monk, and a pedlar are lively and apt, but moments when Galahad
reveals his innermost thoughts come dangerously near to being banal
(Ex. 58). A small off-stage chorus, representing Galahad's unconscious,
adds appropriate comments from time to time.

The second scene consists of the short cantata 'The Chapel in
Lyonesse' which Boughton had composed in 1904 to words by William
Morris. Rescored for full orchestra, and with some minor adjustments
to the text, it sits uneasily in a work conceived forty years later. The
style is that of Brahms and Tchaikovsky and therefore belongs to
another sound-world. It is astonishing that Boughton did not consider
it incongruous, especially since it has very little to do with the main
thrust of the drama. But he seems always to have believed that one
facet of his life's work was automatically on the same level as every
other.

If the second scene leans heavily on Brahms, the third owes more
than a little to Mussorgsky, as the theme which represents the angry
peasantry shows (Ex. 59). It is into this scene that some of the more
unfortunate lapses of judgement begin to creep. The hymn sung by the
knights and clergy in praise of military glory, even if intended as satire
(which one supposes it must be), ought to be something more than a
flat-footed reminiscence of 'Onward, Christian Soldiers' (Ex. 60). And
the pivotal moment when Will offers to take Galahad to his friends
Penimel and Melicora, confident that they will be able to impart
important truths to him, is ill served by a four-square tune that mistakes bathos for simplicity (Ex. 61). The same tune is later made to
serve Galahad's vision of music as the solution to life's problems!

Ex. 58

Andante con moto

Dream - ing must tru-ly a - bide My an - gel near, Sure - ly no ill can be- tide

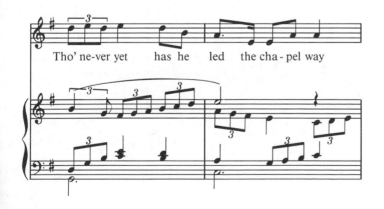

Tho' ne-ver yet has he led the cha - pel way

Ex. 59

Fortunately the music of the fourth and last scene, the visit to Penimel and Melicora, is on a somewhat higher plane and contains many passages of simple but effective lyricism. The final 'Magnificat' unfolds with a certain splendour: its bold, diatonic tune offset by cascades of pealing bells, which finally give way to Galahad's naïve vision and then return to accompany the choral peroration that brings the drama to its end. In its own simple terms the ending is effective enough, though not exactly awe-inspiring. Like all the thematic material in *Galahad* it lacks that quality of inevitability—of being exactly the right musical image—that alone carries complete conviction. The music of the final drama, *Avalon*, is similarly afflicted.

Ex. 60

Ex. 61

Allegro

My friend has a trick of _ mak-ing you free In the things you hear and the

things you see.

Nor is there much to be said in favour of Boughton's librettos. Whereas Buckley's contribution, for all its hyperbole and affected archaisms, is at least consistent, Boughton's attempts to forge a language that will bring contemporary meaning to an heroic tale set in mythic times are far from convincing. Only when he is dealing with emotions he can relate to on a personal level, as in *The Lily Maid*, does he achieve what he presumably set out to find—the simplicity of folk-poetry. One such moment is Elaine's plaintive:

> I have found the shadowy deer
> That you were hunting when you came here.
> Unto all ladies I make my cry,
> But you must offer my mass-penny.

In *Galahad* and *Avalon* we are faced all too often with passages which combine doggerel with platitude in the most reckless manner:

MORDRED:	I wit you know the King's dis-ease,
	No longer manly to hold the throne.
	So we are met, Sir Galahad,
	to 'stablish a braver rule of our own.
GALAHAD:	Witted I not things were so bad:
	But we do need Christian government.

Hard enough to disentangle on the page, it would be almost impossible to understand when sung.

Saddled with a text that reads like the worst of nineteenth-century operatic translations, it is amazing that the music of *Galahad* and *Avalon* is as interesting as it is. In *Avalon* particularly, Boughton works his representative themes to considerable purpose—virtually all of them reappearing in appropriate summation of the various dramatic threads. But as the text is largely concerned with dialogue, the thematic underlay is obliged to shift ground to accommodate what amounts to each character's visiting-card. The result is a rather nervous, unsettled structure: everything proceeding in short bursts, and without that sense of symphonic spaciousness that alone can give credence to the method. Nor is the situation much helped by Boughton's penchant for an animated flow of quasi-contrapuntal part-writing. The ear longs for the accompaniment to sink into the background and allow the voice to take charge of the situation in some purely lyrical outburst. Boughton's disdain for mere operatic 'entertainment', his puritanical reservations against sensuous vocal display, is self-defeating. A passage of vocal bravura, as Wagner well knew, can sometimes say more than any amount of logical thematic argument.

Nevertheless, the score of *Avalon*, like that of *Galahad*, contains much that is impressive. It is vigorous where it needs to be, and embraces a wider range of harmonic ground than any of his previous works (with the possible exception of *The Queen of Cornwall*). The scenes of the peasants' rebellion are carried out with considerable panache—the orchestra quivering with semiquaver figurations as the angry revolutionaries chant their song of defiance. In the last act the Debussyesque music of the Lake of Wonder is reintroduced in much the same form as it first appeared in *The Round Table*; and there are moments of affecting recall, such as the detested Puccini himself might have resorted to, as the dying King remembers fragments of the carol he sang as Sir Ector's kitchen boy. Only the choral climax that follows King Arthur's death seems less than adequate. It is built, appropriately enough, on the peasants' 'Song of Revolution'—a bold, simplistic hymn that would certainly satisfy the proletarian recipes of any latter-day Zhdanov (Ex. 62). Somehow it does not sound like the culmination of an epic.

Ex. 62

New   hope is__ ris-ing   not__built on____ a__ throne,

New__life that   men may__make__ fast__ for their   own.__

A Personal Postscript

Biographers ought not to insert themselves into their accounts of other people's lives, but in the case of Rutland Boughton an exception may perhaps be permitted. I knew him during the last ten years of his life, and since his death I have become closely involved not only with his music, but also with his family: so closely that my interest in him has almost become an obsession, and for this reason it seems only proper to state my position as clearly and objectively as possible.

I first met Rutland Boughton in 1949. I had known and admired *The Immortal Hour* and *Bethlehem* ever since I had borrowed the vocal scores from the Gloucester City Public Library some five or six years earlier. But it was not until 1949 that I discovered he lived within relatively easy reach of my home, and eventually plucked up the courage to write to him and ask advice about my own music. His reply, on my self-addressed postcard, came promptly:

Would Oct 25 suit you? Ross bus leaves Westgate St at 3.15. Get off at Gorsley Church. Then a mile's walk. Ask for Bevan's Hill. Return bus gets you back to Gloster at 7 o/c.

R. Boughton.

From Gorsley Church take first turning to left. Signpost directs to Little Gorsley.

I followed his directions and found myself deep in the Gloucestershire countryside, apparently miles from anywhere. The house, set well back from the lane, looked warm and inviting. He was there to greet me.

He was smaller than I expected: bird-like and neat in a pair of corduroy knickerbockers, unusual for the period but eminently practical and comfortable. His white hair was worn much longer than was then the fashion, but it suited him. He looked a little like Wagner, I thought. His voice was rather high-pitched, but pleasant and very much the countryman's. He was brimming with vitality, bombarding me with questions about myself and my music, willing to answer my own questions without any of the reserve that I, as a 20 year old, had come to expect of men fifty years my senior. I was completely captivated, and captivated too by Kathleen when I met her.

From that day I visited him several times a year, showing him my music and gradually learning more and more about his—though pumping him about it demanded perseverance and not a little cunning. I began to talk optimistically about writing a biography, but he always shook his head. And then, one day early in 1959, he suddenly agreed that it might be a possibility. It soon became obvious, to me at least, that if I was to tackle the job with any hope of success I would have to give up my position on the staff of the Royal Marines School of Music. For a while I dithered—until a friend pointed out that I would only have one life in which to do what I wanted to do. I handed in my resignation, and in October 1959 began to visit Kilcot several times a week in order to question Rutland Boughton in detail about his life. By Christmas I had covered almost every aspect that a verbal questioning might decently encompass, though there were a number of personal matters that our difference in age and my inexperience made it impossible for me to pursue. We had also begun to go through the letters we had squirrelled away—only to have him smile if he thought a letter too personal and then tuck it out of sight.

And then, of course, he died.

The last time I saw him was the day before he left for London to celebrate his eighty-second birthday at the home of his daughter Joy. I had decided to leave on an earlier bus that afternoon, for it was beginning to snow and I did not want to risk a difficult journey back to the equally remote spot in which I was living at the time. Boughton was not a tactile person but on this occasion, as he came to the door with me and as I turned to say goodbye, he took a step forward as if to embrace me. He withdrew again almost as soon as he had made the gesture and we parted like the good Englishmen we were. Even so, I was aware that something important had happened, though at the time I could not say what.

In fact it was his death that made the biography possible. It soon became obvious that I already knew more about his life and work than most of his family and could therefore help with all the sorting out and ordering, weeding and preserving, that arises on such occasions. I went to live at Kilcot, sleeping in the bed that had been placed in his downstairs study when it had become something of an effort for him to climb the stairs, surrounded by his books and music, playing his piano and immersing myself in the vast amount of material he had left behind. Kathleen, I believe, was glad to have me as a companion: we had always got on well, and she seemed content to hand over everything to me.

It was strange, trawling through his life. Letters which he thought had been lost (a large box of Shaw's, for instance) began to turn up. So also did the various attempts he had made at writing an autobiography.

Vast quantities of newspaper cuttings fell from forgotten cupboards, and a tea-chest full of old programmes and posters emerged from one of the apple sheds—mostly intact, though the bottom layer had been turned to pulp by a persistent trickle of rainwater.

Gradually a coherent picture began to appear. Kathleen was always willing to answer my questions, as were all his children. I began to meet such friends and associates as had survived him—Roger and Sarah Clark, Steuart Wilson and Frederick Woodhouse, Gwen Ffrangcon-Davies, Desirée Ames, Kathleen Beer and Kathleen Dillon, Adolph and Emil Borsdorf—and slowly I began to learn things from their point of view. I had met Christina once, years before; but that was at a time when I did not know I would write about their lives. Flo I could have met, but the family thought it best I should not do so and I was foolish enough to take their advice. Nor would his brother Ted agree to see me—fearing, I think, that I would be too interested in the politics that had been a bone of contention between them for so many years. But the necessary material was accumulating none the less.

Boughton's funeral took place at the St Marylebone Crematorium in Finchley. There was no religious service, but Frederick Woodhouse spoke tenderly of him and a group of musicians played Beethoven's *Three Equali* as a solemn farewell. A week or so later, as I was working in his study, alone in the house, the postman brought a registered parcel and I realized I was holding all that remained of the man I had most admired. I placed it on his piano and waited for Kathleen to return. That evening, towards sunset, we scattered the ashes over the rose beds in front of the house. They made little grey patches in the thin covering of snow, but soon sank into the dark earth beneath.

I completed my biography and, to my surprise, immediately found a publisher. Though I was never very happy with the title I was advised to give it, I was delighted and relieved by the way it was received. It sold reasonably well, but in the 1960s Rutland Boughton was no longer news—any more than were a dozen of equally individual composers of his generation. There was room, apparently, only for Elgar and Vaughan Williams, with Holst and perhaps Bax on the sidelines. Lesser men were a mere footnote in the history of British music, judged, more often than not, on hearsay, and seldom on any living experience of their work. It was for this reason that I had adopted a rather defensive tone in writing about him. I allowed myself to be ironic (I was suffering from an undue admiration for Lytton Strachey at the time), and tried not to go overboard about a career that had often been quixotic and misguided. I desperately wanted other people to see what I saw in the man and his music, but it was clear that I was fighting a cause that had long since gone out of fashion.

The trouble was that my own feelings were also rather mixed, and always had been. I was about as different in temperament from my subject as it was possible to be. I loved Verdi and Puccini, whereas he thought them purveyors of mere 'entertainment'. He allowed their genius (at least Verdi's), but thought it had been used for ignoble ends. What he would have said of my passion for the music of Erich Wolfgang Korngold I tremble to think, but fortunately he had never heard of him. Nor could I persuade him that Mozart was an operatic genius of Shakespearean depths, or that there was enormous delight to be had from Gilbert and Sullivan. As for the music of the twentieth century, he considered my interest in it to be little more than proof of a sensibility that had been corrupted. Perceptive irony—that touchstone of twentieth-century art—was totally foreign to his view of the world. And though my natural sympathies were left of centre, I had to hide the fact that I found communism humourless and inhuman. Whereas he was always in earnest, Oxford had undermined whatever capacity I had to burn with a hard, gem-like flame, and National Service in Vienna had added the finishing touch. I could see everything that was wrong with his music-dramas, and still can—and yet I found them admirable, and still do.

For a few years after his death I served, with Joy Boughton, her husband Christopher Ede, and Frederick Woodhouse, on a 'Boughton Trust' that Adolph Borsdorf had brought into existence. It mounted performances of *Bethlehem*, *Alkestis*, and *The Queen of Cornwall*, but they inevitably had too much of a provisional quality about them to be good advertisements for his music. Borsdorf, who was, after all, a professional promoter of live concerts, would not contemplate anything so useful as a commercial recording, and in the end the Trust was wound up.

And then Boughton's friends and family began to die. First Joy, in 1963, then Estelle, Maire, and Ted in 1972. Kathleen followed in June 1973, and Flo, perverse to the end, a few months later. Roger and Sarah Clark had long since gone, and Frederick Woodhouse and Steuart Wilson were soon to follow. In 1978 it would be a hundred years since Rutland Boughton's birth. It now occured to me that the centenary might serve as an excuse to persuade the BBC to take an interest, and that this in turn might bring about an increase in the royalties that still came in from time to time. If I could persuade his four surviving children to forgo their claims, we could set up a charitable Trust whose main aim would be to sponsor the finest possible commercial recording of his music so that, little by little, we might be able to convince the musical world that there was something more to him than was generally allowed. I shall never cease to be grateful for the generosity that led them to agree to my suggestion, or the support that they, and their children, have given the Trust ever since.

Thanks to Elaine Padmore, the BBC allowed me to present a two-hour programme about Boughton's life and works in January 1978, specially recording substantial extracts from *The Immortal Hour*, *Bethlehem*, *Alkestis*, *The Queen of Cornwall*, and *The Lily Maid*. From that moment Boughton began to be included in the list of those composers who were being reassessed by the younger, less prejudiced music-lovers. In the following year the BBC broadcast studio performances of *The Immortal Hour*, and this added to the interest that was beginning to make itself felt. No longer was it necessary to rely on the misleading 'potted' version that Columbia Records had issued in 1932—I can remember one critic claiming that Boughton simply did not know how to orchestrate, so inadequate had that recording seemed!

It took time for the new 'Boughton Trust' to gather together sufficient money to sponsor its first recording, and almost as long to persuade any reputable record company that it was a project worth pursuing. In the end, of course, it was Ted Perry of Hyperion Records who saw the point, and in June 1983 he issued the first complete recording of *The Immortal Hour*. It proved to be a turning point. Boughton, it seemed, was not the misguided crank that legend liked to suggest. BBC performances of his Second and Third Symphonies helped to clinch the matter, as did Hyperion's splendid recording of the Third Symphony and Oboe Concerto No. 1. The Trust now began to receive anxious requests for news of its next recording projects.

By this time, my biography had long been out of print and I had retrieved the copyright in the hope that I might one day make another attempt. I no longer felt I had to defend eccentric interest in an eccentric composer. Moreover, I realized there was much more to be said. It was not that I wanted to disown what I had published in 1962, but simply that I had now experienced at first hand something of what he had faced and overcome. I had written an opera of my own (a black comedy which would certainly have caused him pain) and, even more instructively, I had produced it, conducted it, and taken it on tour. As a conductor I had by now not only worked on innumerable operettas and musicals, but also directed performances of *Bethlehem* and *The Lily Maid*. If I had admired Rutland Boughton's Glastonbury achievements in the abstract, I could now admire them with a genuine appreciation of the immense amount of work they had involved.

It was also fascinating to observe how many aspects of the Glastonbury Festivals had come to the surface in new guises. Leaving aside the relative quality of their genius, what difference was there between Boughton and Britten? Both had found it necessary to create their own operatic conditions. Aldeburgh was nothing but Glastonbury writ large and with the greater sophistication of a more sophisticated age. And what was the increasing interest in 'Community Arts' but an echo of Boughton's intentions? Or, for that matter, the resurgence of the commune as an alternative way of life, and the recognition that a failed relationship might be better ended? It seemed that the time might be ripe to recount Boughton's story in greater detail and without apology.

Even so, my obsession with the man and his music has never been so all-embracing as to blind me to his weaknesses. He was not an Elgar or a Vaughan Williams, a Britten or a Tippett. His music-dramas present many problems and are quite unsuited to the ethos of the commercial opera-house. And like most composers he wrote a great deal that is better forgotten. His views were often wrong-headed, and he was very much the architect of his own misfortunes. And yet . . .

I have devoted much of my life to Rutland Boughton and his music, for it seems to me that his finest works have a genuine originality and inhabit a world that is uniquely their own. I do not know quite why they move me so, but they do, and I know that I have been right to stand up for them.

NOTES AND SOURCES

This biography has been based on the notes made by the author in conversation with Rutland Boughton in a series of interviews beginning in October 1959 and ending in late January 1960. These notes have been further supplemented by the various attempts Boughton made to write his autobiography: a sheaf of typed, incomplete chapters dating from the 1930s, and fourteen hand-written note books dating from 1945. The material in the autobiographical attempts often duplicate, word for word, Boughton's conversations with the author. It is as if he had a complete autobiography in mind but was unable to shape it into a convincing book. Any quotation in the present biography that is not attributed in the text or the notes may be assumed to come from one or other of Boughton's autobiographical attempts, or the notes made in conversation with the author.

Many of the letters written to Boughton have been placed in the British Library. Those from other composers, and such friends and colleagues as Edith, Lady Londonderry, are to be found under Add. 52364–6. The Edward Clark sequence is to be found under Add. 52256, and the George Bernard Shaw collection under Add. 50529. Other letters (including carbon copies of Boughton's own letters) are at present still in the possession of the author and the Rutland Boughton Music Trust, but it is anticipated that these will eventually be lodged with the British Library.

Details of the Glastonbury Festivals have come from the author's collection of original programmes, heavily supplemented by the regular reports that appeared in the *Central Somerset Gazette* at the time. Being a provincial newspaper, the *Central Somerset Gazette* was accustomed to include lengthy cast lists and elaborately detailed accounts of each event—an invaluable means of checking what actually happened against what the printed programme hoped would happen.

All other sources are included in the following notes.

CHAPTER 1: 1878–1899

1. Details of Boughton ancestry supplied by various members of his family, supplemented by Terence Mclaughlin, *Journal of the Buckinghamshire Family History Society: Worthies of Bucks*, No. 4.
2. Boughton claimed that the advert appeared in the *Musical Times* but I have not been able to trace anything approximating to it.

3. In an early, handwritten catalogue of compositions Boughton lists six settings of poems by Heine: 'Dream Tears', 'All Night in Vision', 'Autumn', 'The earth is so fair', 'Prayer', and 'Presentiment'. They are dated July 1896 and Feb. 1897. Of these only 'Presentiment' has survived. His claim to 'seventy' such settings seems rather extreme and may, perhaps, be due to a lapse of memory.

4. Source: *The Speaker*, 29 Mar. 1902. Ernest Newman's article was entitled 'The New School of British Music'.

5. This appears to have been destroyed, unless it is an earlier version of the 1901 'March of the British'.

CHAPTER 2: 1899–1905

1. From *The Times* and *Daily Telegraph*, 23 Sept. 1903, quoted in *The Self-Advertisement of Rutland Boughton*.

2. David Ffrangcon-Davies, 1855–1918. Baritone, celebrated for his oratorio and operatic roles. Created the part of Cedric in Sullivan's *Ivanhoe* (1891) and was much admired by his friend Sir Edward Elgar. His 1906 treatise *Singing in the Future* was reissued in 1938 by his daughter Marjorie, together with a brief but useful biography under the title *David Ffrangcon-Davies, His Life and Book*. Both Marjorie and her more famous sister, Dame Gwen Ffrangcon-Davies (1891-1992), sang for Rutland Boughton.

CHAPTER 3: 1905–1907

1. Sir Granville Bantock, 1868–1946. Composer, conductor, and tireless champion of British music. Principal of the Birmingham and Midland Institute's School of Music 1900–34. Professor of Music, Birmingham Univ. from 1908.

2. Source: George Painter's reminiscences in *Fanfare*, the bulletin of the Birmingham School of Music, 1/9, Apr. 1948, on the occasion Boughton being made an Honorary Fellow.

3. Whether Elgar met Boughton on this occasion, is not known, or if he remembered the letter he had written on 30 Nov. 1899 in answer to Boughton's intention of writing about his music. See Add. 50850.

4. John Drinkwater, 1884–1937. Poet, playwright, producer, and actor. General Manager of the Birmingham Repertory Theatre 1913–19, 1929–31. Plays include *Abraham Lincoln*, 1918, and *Bird in Hand*, 1928.

CHAPTER 4: 1907–1911

1. Quoted in *Music Drama of the Future*.
2. Ibid.
3. Lady Isabel Margesson, d. 1946. Born Lady Isabel Hobart-Hampden, daughter of the 6th Earl of Buckingham. Wife of Sir Mortimer Margesson, 1861–1947.

CHAPTER 5: 1911–1912

1. *Ann Veronica* (1909), novel by H. G. Wells.
2. Frederick Jackson, 1832–1915. Solicitor and music-lover. Wrote articles on socialism under the pseudonym 'Vox Clamans'. A letter written by Shaw on 17 Nov. 1912 reveals that Jackson had been very upset when Boughton explained the nature of his relationship with Christina. Shaw was highly amused: 'I should have thought that one glance at that dear Chris would have been enough for any man on ten minutes experience. An engaged couple, perhaps. A several-year-married one, impossible.' (Add. 50539) The cottage that Boughton lived in on Jackson's estate was called, appropriately enough, 'The Limit'.
3. 27 Rosenau Road, Battersea, London SW11.
4. Charles Kennedy Scott, 1876–1965. Conductor and choir trainer. He founded the Oriana Madrigal Society (1904) and the Philharmonic Choir (1919). Editor of 16th- and 17th-cent. vocal music in the Euterpe Edition, 1905–14.
5. Philip Tom Oyler, 1879–1974. Agriculturalist and educationalist. Author of *An Invitation to the Woods* (1910), and *How to Bring About a Social Revolution* (1914). His school at Hindehead was probably that same Mount Arlington that Boughton used for his first Summer School in 1912.

CHAPTER 6: 1912–1913

1. Margaret Morris, *My Galsworthy Story* (London, 1867).
2. Jessie Kennedy, d. 1937, sister of Marjory Kennedy-Fraser, and wife of Tobias Matthay, 1858–1945. Better known as a reciter than as a singer.

CHAPTER 7: 1914

1. Professor Adshead, 1868–1946. Professor of Town Planning, Liverpool Univ. 1907–1914. Designer of the Liverpool Repertory Theatre.
2. Add 52364.

3. Holst set Alice M. Buckton's poems 'The Heart Worships', and 'A Vigil of Pentecost', and wrote incidental music for her play *Nabou*.
4. Archives of the Margaret Morris Movement, Manchester.
5. John Rodker, 1894–1955. Poet and dramatist. Involved in Modernist literary movements. Foreign editor of *The Little Review* in succession to Ezra Pound.
6. Letter dated 'Oct 1914, Friday evening'. J & W Clark Archive. When *The Immortal Hour* became a commercial success, Clark insisted that Boughton forget the 'debt'.

CHAPTER 8: 1915

1. Margaret Morris, *My Life in Movement* (London, 1969).
2. Rutland Boughton, *The Glastonbury Festival Movement* (Somerset Folk Press, London, 1922).
3. Hugh Carey, *Duet for Two Voices* (Cambridge Univ. Press, Cambridge, 1979), an informal biography of Edward J. Dent.
4. Clarence Raybould, 1886–1972. Conductor and composer. Joined the BBC in 1936. Assistant conductor BBC Symphony Orchestra, 1936–45. His letter is undated, but postmarked 5 Sept. 1915. Frederic Austin, 1872–1952. Singer and composer. Sang Gunther in Richter's *Ring* cycle at Covent Garden, 1908. Edited and arranged *The Beggar's Opera* for Nigel Playfair's famous 1920 production. His letter is undated, but internal evidence suggests Sept. 1916.
5. Miss Buckton's lecture was advertised in the *Central Somerset Gazette* for 20 Aug. 1915, but not reported in subsequent issues. The 'Angels', who were said to have hovered protectively over the British retreat from Mons (Aug. 1915) were in fact a popular fantasy embroidered on a story that Arthur Machan had published in Sept. 1914.

CHAPTER 9: 1916–1918

1. *Central Somerset Gazette*, 4 Aug. 1916.
2. *Birmingham Post*, 11 Sept. 1916.
3. *Central Somerset Gazette*, 1 Sept. 1916.

CHAPTER 10: 1919

1. *Central Somerset Gazette*, 2 May 1919.
2. By courtesy of Diana McVeagh and the Finzi Trust.

3. *Central Somerset Gazette*, 5 Sept. 1919.
4. Philip Napier Miles, 1865–1935. Composer and wealthy landowner. He studied under Parry and Dannreuther and was mainly concerned with opera, of which the three-act *Westward Ho!* (1913) and the one-act *Markheim* (1919) are the most important.
5. Boughton in conversation with the author, 1959.
6. Reported in the *Central Somerset Gazette*, 5 Dec. 1919.

CHAPTER 11: 1920–1921

1. Margaret Kennedy, *The Constant Nymph*, a novel published in 1924 and later successfully dramatized.
2. Geoffrey Shaw, 1879–1943. Composer and organist. Brother of Martin Shaw, and like him an ardent champion of better church music.
3. *Daily Mail*, 21 Aug. 1920.

CHAPTER 12: 1921–1922

1. The reference is to the Wills Tobacco Company.
2. Sir Barry Jackson, 1879–1961. Founder and Governing Director of the Birmingham Repertory Theatre, which grew out of a semi-amateur group, The Pilgrim Players. His father had been the successful proprietor of the Maypole Dairies, and Jackson used his fortune to finance some of the most interesting theatrical ventures of the day. He was an enthusiastic promoter of Shaw's plays, and launched the careers of a remarkable number of the century's finest actors.
3. The idea of a 'comic opera', or its equivalent, occurred to Boughton at various moments of crisis—as in May 1921 when Maurice Baring gave him permission to make a libretto of *King Alfred and the Neat Herd* (see letter: Add 52364). Even as late as 1949 he approached Sean O'Casey with a proposal to collaborate on a 'satirical' opera. But O'Casey turned down the idea (see letter: Add 52364).
4. This and other personal letters are currently in the possession of Boughton's daughter Jennifer and are not likely to be placed in any public collection.

CHAPTER 13: 1922–1923

1. Quoted in the *Sunday Times*, 29 Apr. 1923.
2. Bax's letter is undated. Add 52364.

3. Holst's letter is undated. Add 52364.

4. Add 52364.

5. Add 52365.

6. Quoted in Bevis Hillier's *Young Betjeman* (London, 1988).

7. Paul Shelving, 1888–1968. Principal designer for the Birmingham Repertory Theatre.

8. Exhibited in 'Paul Shelving (1888–1968) Stage Designer'. Birmingham City Museum and Art Gallery, 7 June–17 July 1986.

9. The passage, a modification of Herod's aria, is rhythmically but not melodically similar to the 'march' theme in the third movement of Tchaikovsky's Sixth Symphony.

10. Undated letter in possession of Jennifer Boughton. 'Tommy' is Miss Agnes Thomas, friend of the Boughton family.

11. Undated letter, ibid.

12. Undated letter, ibid.

13. Undated letter, ibid.

14. Letter, 19 Mar. 1923, sent from St Paul's Rectory, Covent Garden; in possession of Jennifer Boughton.

15. Undated letter, ibid.

CHAPTER 14: 1923–1924

1. *Musical News and Herald*, 15 Feb. 1928.

2. Laurence Housman, 1865–1959. Novelist and dramatist. Brother of A. E. Housman, best known for two sequences of short plays: *Victoria Regina* and *The Little Plays of St Francis*. Many of his plays fell foul of the censor, for reasons that now seem absurd.

3. According to Boughton the dream occurred during his childhood. He recorded it in an interview given to the *Daily Express* which was reprinted in the *Central Somerset Gazette*, 18 Jan. 1924.

CHAPTER 15: 1924–1927

1. Among the other contributors were Edgar Bainton, Felix White, and Alan Bush.

2. 'Song of War': words by Langdon Everard, music by Boughton. No. 4 of 'Twelve Labour Choruses'.

3. Elizabeth von Arnim, 1866–1941.

4. Letter to Miss Agnes Thomas, 32 Feb. 1926. Add 52364.

5. Letter from The Opera Players Inc., 25 May 1926. The Grove Street Theater was located at 22–4 Grove Street, New York City.

6. 18 Jan. 1927.
7. Kenneth Curwen, 1881–1935. Director of J. Curwen & Sons, Ltd.—the most supportive of Boughton's publishers.

CHAPTER 16: 1927–1929

1. A complete list of Boughton's addresses is virutally impossible to compile, but the following may give some idea of his nomadic existence. The dates are taken from letters and contracts emanating from these addresses at the time.

December 1911	38 Charleville Road, West Kensington.
Early 1912	154 York Road, London SE.
June 1912	'The Limit', Grayshott, Hampshire.
December 1912	29 Anhalt Road, Battersea Park, London SW.
April 1913	15 Waterloo Bridge, London SE.
July 1914	27 Rosenau Road, Battersea, London SW11.
Summer 1914	3 Bere Lane, Glastonbury.
	'Bona Vista', Street Road, Glastonbury.
Summer 1919	'Mount Avalon', Glastonbury.
Spring 1922	18 Emperor's Gate, London SW7.
July 1923	19 (Flat 4) Arundel Gardens, London W11.
September 1923	Plas Colwyn, Beddgelert, Wales.
November 1923	14 Cremorne Road, London SW10.
Spring 1924	'Shirley Cottage', Shirley Holmes, Lymington, Hampshire.
	'The Bungalow', Bledlow, Buckinghamshire.
November 1925	33 Parliament Hill, Hampstead.
Spring 1926	'Tanglands Castle', Tatsfield, Surrey.
September 1927	'Windmill Farm', Rusden, Hertfordshire.
November 1927	Kilcot, Newent, Gloucestershire. The house was originally called 'Glen More', but this was never used by the Boughton family. The Post Office accepted 'Kilcot' as sufficient.

2. He also saw Tchaikovsky's *Eugene Onegin*, and Rimsky-Korsakov's *Sadko* and *A May Night*.
3. Boughton took advice from the local MP, socialist landowner Morgan Phillips Price, whose family became close friends and enthusiastic support-ers of Boughton's family and enterprises. A letter dated 25 July 1927 describes the land as 'Not bad for fruit growing', but considers the price, £1,250 for 10 acres, as 'outrageous, unless, of course there is a very good house attached to it'. It would seem that Boughton was able to beat down the asking price.

4. Ursula Greville, 1894–1991. Singer and editor of the *Sackbut*. She was intimately associated with Kenneth Curwen.
5. The composer Isidore de Lara (1858–1925) had launched a scheme for an Imperial Opera House in 1924, but Beecham's project overtook it. Boughton had many meetings with de Lara in 1926.
6. 'A British Composer's View of the Opera Question'. Unpublished MS article by Boughton, *c.*1929.
7. Letter to Sir John Reith as it exists in Boughton's final draft.
8. Edward Clark's letters: Add 52256. Shaw's draft letter: Add 50529.

CHAPTER 17: 1929–1933

1. Choral versions eventually published in 'Twelve Labour Choruses', 1930.
2. Terence James Stannus Gray, b.1892. Founder of the Cambridge Festival Theatre. Author of *Dance Drama* (1926), *Cuchulain* (1925).
3. Boughton's appointments diary records a meeting with Ninette de Valois on 24 Sept. 1927, but Dame Ninette could not recall the occasion.
4. Letter from Lord Berners, 12 Feb. 1930. Add 52364.
5. Letter dated 4 Mar. 1923. Add 52364.
6. Letter dated 8 Jan. 1934.
7. Letter undated, but received 9 July 1926. Add 52364.
8. Letter dated 7 May 1930. Add 52364.
9 . Boughton had adjudicated at a choral festival in Mountain Ash on 1 Apr. 1929. His diary includes an entry, 'Mountain Ash? opera', on 4 May 1929, which is presumably a reference to the meeting with Holst and Vaughan Williams. The draft schedule is quoted in Ursula Vaughan Williams, *R.V.W.: A Biography of Vaughan Williams* (Oxford, Oxford Univ. press, 1964).
10. Sir Barry Jackson's first Malvern Festival was held in Aug. 1929.

CHAPTER 18: 1934–1937

1. Christopher Gordon Ede, 1914–88. Producer and Pageant Master. He was the grandson of Elgar's doctor, and great-grandson of the Dean of Worcester. Though he trained as an organ builder, he turned to stage direction while an undergraduate at Cambridge. He won a considerable reputation as a director of Son et Lumière productions.
2. Boughton's MS libretto was extant in 1960, but cannot now be traced. This extract was copied by the author in 1960. Words deleted by Shaw are enclosed in angled brackets and his additions are in **bold** type.
3. Boughton's income was recorded in the Civil List Petition as:

1922	£511. 15s. 3d.
1923	£1946. 0s. 3½d.
1924	£378. 3s. 7d.
1925	£481. 17s. 0d.
1926	£983. 13s. 6d.
1927	£268. 4s. 3d.
1928	£325. 9s. 1½d.
1929	£382. 15s. 0d.
1930	£120. 17s. 2½d.
1931	£441. 7s. 0½d.
1932	£553. 8s. 0d.
1933	£196. 0s. 4d.
1934	£117. 10s. 1½d.
1935	£95. 4s. 5d.
1936	£509. 6s. 0d.
1937	To end of May, under £50.

4. The Elgar Estate attempted to repay the loan after his death, but Shaw, characteristically, insisted that the money be credited to the composer's daughter, Carice, as a 'birthday present'.

CHAPTER 19: 1937–1944

1. Quoted in Margaret Wilson, *English Singer: The Life of Steuart Wilson* (London, 1970).
2. Letter dated 10 Sept. 1944.

CHAPTER 20: 1945–1960

1. Letter to Desirée Ames (Mrs Arthur Newell).
2. Letter dated 8 Apr. 1949.
3. Letter to George and William Lloyd, dated 1 July 1935. Quoted by courtesy of George Lloyd.
4. A draft letter to Norman Fulton, dated 22 Apr. 1954, providing information for a programme, 'Apollo in the West', to be broadcast from BBC Bristol.
5. Included in *Letters from a Life: Selected Letters and Dairies of Benjamin Britten*, ed. D. Mitchell and P. Reed (1991), ii add. 4 of notes. Reproduced by kind permission of the trustees of the Britten-Pears Foundation.
6. Postcard, undated, acknowledging (50th?) birthday congratulations from Brian Boughton. Original with the Rutland Boughton Music Trust, copy of the Britten-Pears Archive. Reproduced by kind permission of the Trustees of the Britten-Pears Foundation.

7. Add 52364.
8. Printed in *Music and Life*, Summer 1960.
9. Add 52364.
10. Christina Walshe died, in a car accident, on 24 Dec. 1959 in Vence, where she had been teaching at the École Freinet. Kathleen died, of cancer, on 12 June 1973, and Flo on 27 Oct. 1973.
11. Boughton is recorded as having died at approximately 5.30 a.m. on 25 Jan. 1960. The cause of death is given as 'Ruptured aorta due to Hyaline Medial Necrosis'.

APPENDIX A

Boughton's Musical Compositions

A.	Augener	L.	Larway
B.	Cecil Barth	N.	Novello & Company
B&H	Bossey & Hawkes	OUP	Oxford University Press
C.	J. & W. Curwen	P.	Keith Prowse
G&T	Goodwin & Tabb	R.	William Reeves
ILP	Independent Labour Party	Rob.	Roberton
IM	International Music Company	S&B	Stainer & Bell
JW	Joseph Williams	S.	Swan Sonnenschein
K.	Klene	W.	Weekes

In certain instances the copyrights of published works have been returned to the Rutland Boughton Music Trust. This is indicated by an asterisk (*) after the publisher's name. Where Boughton withdrew and destroyed unpublished full scores, but did not destroy the individual orchestral parts (which therefore still exist), the usual MS indication is replaced by MSP.

The bulk of Boughton's manuscripts have been placed in the British Library, where they are to be found under Add 50960-51012. It is ancitipated that the collection will be added to from time to time.

MUSIC-DRAMA

Title	Libretto	Composed	Performed	Published
The Birth of Arthur	Reginald Buckley	1908–9	Glastonbury 16 Aug. 1920	MS
The Immortal Hour	Fiona Macleod	1912–13	Glastonbury 16 Aug. 1914	S&B 1920
Bethlehem	Coventry Nativity Play	1915	Street 28 Dec. 1915	C. 1920
The Round Table	Reginald Buckley and Rutland Boughton	1915–16	Glastonbury 14 Aug. 1916	MS
Alkestis	Euripides/Murray edn.	1920–2	Glastonbury 26 Aug. 1922	G&T* 1923
The Queen of Cornwall	Thomas Hardy	1923–4	Glastonbury 21 Aug. 1924	J.W.* 1926
The Ever Young	Rutland Boughton	1928–9	Bath 9 Sept. 1935	MS

Title	Libretto	Composed	Performed	Published
The Lily Maid	Rutland Boughton	1933–4	Stroud 10 Sept. 1934	MS
Galahad	Rutland Boughton	1943–4	Unperformed	MS
Avalon	Rutland Boughton	1944–5	Unperformed	MS

SHORT DRAMATIC WORKS

Title	Source for Libretto	Composed	Performed	Published
The Chapel in Lyonesse	William Morris	1904	Glastonbury 15 Aug. 1914	MS
Agincourt	Shakespeare	1918	Glastonbury 25 Aug. 1924	JW* 1926
The Moon Maiden	Noh play/Stopes	1918	Glastonbury 23 Apr. 1919	JW* 1926

BALLET

Title	Scenario	Composed	Performed	Published
Death Dance of Grania	Margaret Morris	1912	London, 1912	MSP
Mystic Dance of the Grail	Margaret Morris	1913	London, 1913	Lost
Snow White	Margaret Morris	1914	London, 1914	MS
Pandora's Box	Margaret Morris	1914	London, 1914	Lost
The Death of Columbine	John Bostock	1918	Glastonbury, 1 Sept. 1921	MS
May Day	Rutland Boughton	1926–7	Bournemouth, 2 Mar. 1929	C.* 1929

INCIDENTAL MUSIC

Title	Author	Composed	Performed	Published
Dante and Beatrice	Emily Underdown	1902?	1902?	S. 1903
The Land of Heart's Desire	W. B. Yeats	1917	Glastonbury, 24 Jan. 1917	MS
Sacrifice	Rabindranath Tagore	1918	Glastonbury, 11 Sept. 1918.	Lost

Title	Author	Composed	Performed	Published
The Robin, the Mouse, and the Sausage	John Bostock	1918	Glastonbury, 11 Sept. 1918	Lost
Wee Men	Brenda Girvin and Monica Cosens	1920	Birmingham, 26 Dec. 1923	Lost
Everyman	Mystery Play	1922	Bath, 23 Dec. 1923	Lost
Little Plays of St Francis	Laurence Housman	1924–5	Glastonbury, 19 Aug. 1924	MS
'Our Lady of Poverty' 'The Builders', 'Sister Clare'				
'Brother Sun', 'Brother Juniper', 'The Seraphic Vision'			20 Aug. 1924	
'The Revellers', 'Fellow Prisoners', 'The Bride Feast'			5 Aug. 1925	
'Sister Death' 'Brief Life', 'Blind Eyes', 'Brother Wolf', 'The Lepers', Sister Gold', 'The Chapter', 'Brother Elias'			6 Aug. 1925 1926–7	
Passion Play	Antonia Bevan Williams	1933	Unknown	Lost
Isolt	Antonia Bevan Williams	1935	Unknown	MS

Lost or Incomplete Stage Works

The Bride of Messina (music–drama)	Schiller/Bulwer Lytton	1898	Unperformed	Destroyed
Eolf (music–drama)	Rutland Boughton	1903	Unperformed	MS
Butterflies and Wasps (musical comedy)	Eimar O'Duffy	1931	Incomplete	MS
The Hunchback (opera)	Gladys Morton	1933	Incomplete	MS

ORCHESTRAL MUSIC

Title	Composed	Published
Symphonies		
Symphony No. 1, 'Oliver Cromwell'	1904–5	MS
Symphony No. 2, 'Deirdre'	1926–7	MS

Title	Composed	Published
Symphony No. 3, in B minor	1937	MS

Symphonic Poems

Lucifer	1898	Destroyed
A Summer Night	1899, rev. 1903	MS
Imperial Elegy: Into the Everlasting	1901	MSP
Troilus and Cressida ('Thou and I')	1902	MS
Love in Spring ('A Song in Spring')	1906	MS

Overtures

School for Scandal	1903	MSP
The Round Table	1916	MS
The Queen of Cornwall	1926	JW 1926
Overture to the Arthurian Cycle	1936	MS

Concertos

Piano Concerto in A flat	1897–8	Destroyed
Concerto for Oboe and Strings, No. 1 in C	1936	B&H 1937
Concerto for Oboe and Strings, No. 2 in G	1937	MS
Concerto for Flute and Strings	1937	B&H 1937
Concerto for Strings ('Four English Pieces')	1937	MS
Concerto for Trumpet and Orchestra	1943	MS
Concertante for Cello and Orchestra	1955	Sketches

Miscellaneous

Symphonic Suite: *The Chilterns*	1900	MSP
Symphonic March: *Britannia*	1901	R. 1910
Variations on a Theme of Purcell	1901	MS
Prelude and Finale: *The Birth of Arthur*	1910	MS
Three Folk Dances for Strings	1912	C. 1913
Love Duet, Luring Scene: *The Immortal Hour*	1924	S&B 1924
Three Interludes: *The Ever Young*	1929	MS
Three Flights for Orchestra	1929	MS
Winter Sun (*see also* Chamber Music)	1932	MS
Rondo in Wartime	1941	MS
Orchestral Prelude on a Christmas Hymn	1941	MS
Reunion Variations	1945	MS
Suite for Strings: *Aylesbury Games*	1952	Rob. 1991

CHORUS AND ORCHESTRA

Title	Text	Composed	Published
The Skeleton in Armour	Longfellow	1898, rev. 1903	N. 1909
Sir Galahad	Tennyson	1898	C. 1910

Title	*Text*	*Composed*	*Published*
The Invincible Armada	Schiller/Lytton	1901	N. 1909
Midnight	Edward Carpenter	1907	N. 1909
Song of Liberty	Helena Bantock	1911	C. 1911
Three Festival Choruses	John Drinkwater	1911	S&B 1919
(1. 'Song of our Fathers';			
2. 'Song of Summer';			
3. 'Song of Evening')			
'The Cloud'	Shelley	1923	S&B 1925
'Pioneers'	Walt Whitman	1925	C. 1925
'The Raggle Taggle Gypsies'	Traditional	1931	C. 1931
'Bridal Chorus'	Anon.	1932	MS
Prelude and Fugue	Psalm 46	1941	MS

CHAMBER MUSIC

Title	*Composed*	*Published*
Japanese Suite (piano solo)	1893	P. 1927
'Hungarian Fantasia' (piano duet)	1894	W. 1897
'Sentry's March' (piano solo)	1895?	K. 1898
'Gaelic Romance' (string quartet)	1912?	Lost
Celtic Prelude (violin, cello, piano)	1917	A. 1923
Variations on a Ground (violin and piano)	1917	S&B 1922
Dorian Study for String Quartet	1920	Lost
Sonata for Violin and Piano	1921	C. 1922
String Quartet No. 1 in A, 'The Greek'	1923	MS
String Quartet No. 2 in F, 'From the Welsh Hills'	1923	MS
Trio for Flute, Oboe, and Piano	1925	MS
'A Tale, a Game, a Trouble, and a Romp' (piano)	1926	IM 1926
Oboe Quartet, No. 1	1932	MS
Winter Sun (violin and piano)	1932	MS
Two Pieces for Oboe and Strings		
(1. 'Somerset Pastoral'; 2. 'The Passing of the Faery')	1937	MS
Three Songs without Words (Oboe Quartet)	1937	MS
(1. 'Whence!'; 2. 'Faery Flout'; 3. 'Barcarolle')		
String Trio	1944	MS
Oboe Quartet No. 2	1945	MS
Trio for Piano, Violin, and Cello	1948	MS
Sonata for Cello and Piano	1948	MS
Kilcoteri 1 and 2 (cello and piano)	1955	MS

PARTSONGS

Title	*Author*	*Composed*	*Published*
Choral Variations: 1st set	Traditional	1905	R. 1909
(1. 'The Berkshire Tragedy';			
2. 'King Arthur')			

Title	Author	Composed	Published
Choral Variations: 2nd set (1. 'William and Margaret'; 2. 'Widdicombe Fair')	Traditional	1905	R. 1909
'Sunset'	George Eliot	1906	JW 1928
'The Devon Maid'	Anon.	1906	MS
'In Arcadia'	Milton	1906	JW 1928
Choral Variation: 'Young Herchard'	Traditional	1907	N. 1909
'Meg Merrilees'	Keats	1908	N. 1909
Two Folksongs (1. 'Men of Harlech'; 2. 'The Black Monk')	Traditional	1909	N. 1909
The City (Motet)	Henry Bryan Binns	1909	S&B 1913
'The Wind'	Edward Carpenter	1909	MS
Two Folksongs (1. 'Early One Morning'; 2. 'Bronwen')	Traditional	1910	C. 1909
Six Spiritual Songs		1910	R. 1911
1. 'A Prayer'	John Drinkwater		
2. 'St Bride's Milking Song'	Fiona Macleod		
3. 'St Bride's Cradle Song'	Fiona Macleod		
4. 'The Bird of Christ'	Fiona Macleod		
5. 'Song of Easter'	Fiona Macleod		
6. 'The Kingdom of Heaven'	Francis Thompson		
Five Choruses:			
The Immortal Hour (1. 'Faery Song'; 2. 'Druid's Chorus'; 3. 'The Bells of Youth'; 4. 'Warrior's Chorus'; 5. 'Green Fire of Life')	Fiona Macleod	1914	S&B 1920
Six Celtic Choruses		1914	S&B 1924
1. 'Dalua'	Fiona Macleod		
2. 'Avalon'	Fiona Macleod		
3. 'A Sea Rune'	Fiona Macleod		
4. 'A Celtic Lullaby'	Fiona Macleod		
5. 'Niahm'	Ethna Carberry		
6. 'Angus the Lover'	Ethna Carberry		
Five Partsongs		1914	C. 1914
1. 'Early Morn'	W. H. Davies		
2. 'Pan'	Beaumont and Fletcher		
3. 'Spring'	Lewis Spence		
4. 'Lady of May'	Nicholas Breton		
5. 'Little Billee'	W. M. Thackeray		
Four Partsongs		1924	C. 1924
1. 'Prospice'	Browning		
2. 'Contentment'	Thomas Traherne		
3. 'The Blacksmiths'	Henry Bryan Binns		
4. 'Quick March'	Thomas Hardy		
'The Dreamers'	Robert Buchanan	1924	C. 1926

Title	*Author*	*Composed*	*Published*
'A Song of Graves'	G. K. Chesterton	1926	C. 1926
Child of Earth	Henry Bryan Binns	1927	C. 1929
(1. 'Earth's Fosterling';			
2. 'The Return';			
3. 'The Cage'; 4. 'The Storm';			
5. 'The Sword'; 6. 'Insecurity')			
'Burglar Bill'	F. H. Anstey	1927	C. 1929
Carol: 'I Sing the Birth'	Ben Jonson	1928	OUP 1928
Three Partsongs		1928	JW 1929
1. 'The Tyger'	William Blake		
2. 'The Gentle Heart'	D. G. Rossetti		
3. 'The Donkey'	G. K. Chesterton		
'Song of War'	Langdon Everard	1930	ILP 1930
'The Day of Days'	William Morris	1930	ILP 1930
'A Dance at Parson's Green'	Stuart Fletcher	1931	C. 1931

SONGS

Title	*Author*	*Composed*	*Published*
'Passing Joys'	Unknown	1893	B. 1893
'Thou and I'	Edward King	1895	MS
'The Midnight Wind'	F. Motherwell	1895	MS
'The Fox'	Gerald Griffin	1896	MS
The Passing Year (cycle)	Lizzie M. Pengelly	1896	W. 1896
'The Lark'	Unknown	1898	Lost
'Lament of Ethleun'	Unknown	1900	Lost
Songs of the English	Rudyard Kipling	1901	MS
(1. 'Fair is our Lot';			
2. 'The Coastwise Lights';			
3. 'Song of the Dead';			
4. 'The Price of Admiralty';			
5. 'The Deep-Sea Cables';			
6. 'Song of the Sons')			
Four Faery Songs	Keats	1901	MS
(1. 'Shed no tear';			
2. 'Ah, woe is me';			
3. 'Unfelt, unheard, unseen';			
4. 'The Witching Hour')			
'May and Death'	Browning	1902	MS
Three Baby Songs	Swinburne	1902	MS
(1. 'In a Garden';			
2. 'A Cycle of Roundels';			
3. 'Baby-Bird')			
'Go, Little Lay'	Emily Underdown	1902	S. 1903
Six Songs of Manhood		1903	MS
1. 'The Great Grey Mother'	Rutland Boughton		
2. 'Sea Grave'	W. E. Henley		

Title	*Author*	*Composed*	*Published*
3. 'Song of the Labourer'	Ellwyn Hoffman		
4. 'Man and Men'	George Meredith		
5. 'In Prison'	William Morris		
6. 'The Love of Comrades'	Edward Carpenter		
Four Songs	Edward Carpenter	1906–7	N. 1909
(1. 'To Freedom';			
2. 'The Dead Christ';			
3. 'Fly, Messenger';			
4. 'Standing Beyond Time')			
'Rhapsody'	Song of Solomon	1908	Lost
'Sweet Evenings'	George Eliot	1908	MS
'Joy is Fleet'	George Meredith	1908	MS
'A Sight in Camp'	Walt Whitman	1908	MS
Five Celtic Songs	Fiona Macleod	1910	S&B, 1923
(1. 'Green Branches';			
2. 'Daughter of the Sun';			
3. 'Tragic Lullaby';			
4. 'Shule Agrah';			
5. 'My Grief')			
Songs of Womanhood	Christina Walshe	1911	L. 1912
(1. 'Prayer to Isis';			
2. 'A Woman to her Lover';			
3. 'A Song of Giving';			
4. 'A Song of Taking';			
5. 'Woman's Song of			
Creation')			
Songs of Childhood	William Blake	1912	N. 1912
(1. 'The Blossom';			
2. 'Infant Joy';			
3. 'Spring'; 4. 'Little Boy			
Lost'; 5. 'Little Boy Found';			
6. 'The Lamb')			
Two Duets	William Blake	1912	N. 1912
(1. 'Piper's Song';			
2. 'Holy Thursday')			
Two Songs	John Drinkwater	1913	B&H 1919
(1. 'At Grafton';			
2. 'The Feckenham Men')			
Three Songs	Edward Carpenter	1914	C. 1919
(1. 'The Triumph of			
Civilization'; 2. 'The			
Lake of Beauty';			
3. 'Child of the Lonely			
Heart')			
Songs from The Immortal			
Hour	Fiona Macleod	1914	S&B 1920
(1. 'The Old Bard's Song';			
2. 'Song of Creation';			
3. 'Great Lords of Shadow';			

Title	Author	Composed	Published
4. 'Faery Song';			
5. 'Luring Song';			
'Immanence'	John Rodker	1914	C. 1919
'A Newe Year (*Bethlehem*)	Anon., 15th century	1915	C. 1920
'Hangman's Song'	Thomas Hardy	1916	Lost
'Into the Twilight'	W. B. Yeats	1917	MS
'The Wind'	W. B. Yeats	1917	MS
Two Folksong Duets	Traditional	1918	MS
(1. 'The Bailiff's Daughter';			
2. 'Come Lassies and Lads')			
'Apollo'	Henry Bryan Binns	1919	MS
Symbol Songs	Mary Richardson	1920	C. 1921
(1. 'Mother Mary';			
2. 'Honeysuckle';			
3. 'Blue in the Woods';			
4. 'Fierce Love Song';			
5. 'The New Madonna')			
'Sister Rain'	Henry Bryan Binns	1922	C. 1923
Four Everyman Songs	Anon., 15th century	1922	MS
(1. 'Dance of Death';			
2. 'Inconstancy';			
3. 'Angel's Song';			
4. 'Soul Rest')			
'Holiness'	John Drinkwater	1923	C. 1923
'Foam Song'	Ursula Greville	1923	C. 1923
Three Hardy Songs	Thomas Hardy	1924	JW 1924
(1. 'Song of Lyonesse';			
2. 'Evensong';			
3. 'Foreboding')			
'Sweet Ass'	Eleanor Fargeon	1928	C. 1928
'Maiden's Song'	William Blake	1928	C. 1928
Two Duets	W. H. Davies	1928	C. 1928
(1. 'Clouds';			
2. 'The Green Tent')			
'Song of Cyder'	G. K. Chesteron	1930	MS
'Morning Song'	Anon	1930	N. 1930
Five Songs	Joe Corrie	1931	MS
(1. 'God'; 2. 'The Captive';			
3. 'Pedlar's Song';			
4. 'The Traitor';			
5. 'Treasures')			
'Eros'	Mary Webb	1931	MS
'The Chester Carol'	Anon., 15th century	1933	C. 1933
'Lorna's Song'	R. D. Blackmore	1934	B&H 1934
'Aylesbury Grammar School Song'	Andrew Hurst	1938	C. 1938
'The Faery People'	Mary Webb	1940	MS
'The Street'	A. E. Housman	1940	MS

Title	Author	Composed	Published
Five Songs		1944	MS
1. 'Sunset'	John Masefield		
2. 'The Emigrant'	John Masefield		
3. 'Alone'	Thomas Hardy		
4. 'By the Blackthorn'	Mary Webb		
5. 'Laugh and be Merry'	John Masefield		
'Clown's Congé'	Montague Slater	1948	MS
'Antibombastical Shanty'	Shakespeare/Boughton	1948	MS

APPENDIX B

Boughton's Prose Writings

BOOKS

Bach (Music of the Masters, ed. Wakeling Dry; John Lane, 1907)
Music Drama of the Future, with R. R. Buckley (William Reeves, 1911)
Bach (Masters of Music, ed. Landon Ronald; Kegan Paul, Trench, Trubner, 1930)
The Reality of Music (Kegan Paul, Trench, Trubner, 1934)

BOOKLETS

The Self-Advertisement of Rutland Boughton (William Reeves, 1909)
The Death and Resurrection of the Music Festival (William Reeves, 1913)
A National Music-Drama: The Glastonbury Festival (Royal Musical Association, 1917)
Parsifal (Musical Opinion Booklets, 1920)
The Glastonbury Festival Movement (Somerset Folk Press, 1922)
Music and the Co-operative Movement (Co-operative Union, Ltd., 1929)

MONTHLY MAGAZINES

Music

1903   Mar.–Apr.: 'Richard Strauss'
       Aug. 'British Music'
       Dec.–1904 Sept. 'The Prostitution of Music'

Music Opinion and Music Trades Review

1898   July: 'The Songs of Rubinstein'
       Nov. 'Brahms's Variations for Piano Solo'
1899   Oct.: 'Liszt's Songs'
       Nov.: 'Beethoven's "Mount of Olives"'
1900   Aug.–Sept.: 'Gottschalk's Pianoforte Works'

1901   June: 'Schubert's Piano Sonatas'
1902   Jan.: 'The Masterpiece of Brahms'
1905   Oct.–Dec.: 'Loewe's Ballads'
1906   Dec.–1907 Mar.: 'The Music of Algernon Ashton'
1912   Feb.–Nov.: 'The Pictorial Element in Music'
1913   Mar.: 'Der Rosenkavalier'
        July: 'Ariadne in Naxos'
1919   Oct.: 'The Glastonbury Experiment'
1925   Mar.–May: 'English Singers and Voice Production'

Musical Times

1910   July: 'English Folk Song and English Music'
1912   Sept.: 'The Future of Chamber Music'
1916   July: 'The Glastonbury Festival School'
1921   Aug.–Sept.: 'Early English Chamber Music'
1922   Apr.: 'Modern Music and a Way Out'
        July: 'Our Decadence'
1926   Mar.: 'A Musical Impressionist: William Baines'
1928   Oct.–Nov.: 'Schubert and Melodic Design'
1929   June–July: 'In Mutual Contempt'
        Oct.–1930 Feb.: 'The Arts in Revolt'
1930   Sept.: 'From the Street Corner'
1931   Jan.: 'A Postscript to my Bach Book'

Music Student

1908   Nov.: 'Wagner'
        Dec.: 'Bach'
1909   Jan.: 'A Foreword to the Study of Sonata Form'
        May: 'A Foreword to the Study of Wagner'
        Sept.–1911 May: 'Wagner: Studies of the Music Dramas'
1911   Nov.: 'The Failure of the Symphony'
1912   June: 'The New Woman in Music'
        Sept.–1913 Apr.: 'A Study of Folk Music'
1916   Oct.: 'G.B.S.'

Sackbut

1924   Apr.: 'Stories for Opera'
1927   Dec.: 'Clear Out the Reds and Give me 2d.'
1928   Feb.: 'Composers and Executive Artists'
        Mar.: 'Criticism, Life and Laughter'

1928   Apr.: 'Grunts from a Pigstye'
        June: 'Is it Life or Death for the Arts (i)'
        Aug.: 'Is it Life or Death for the Arts (ii)'
        Sept.: 'Professors and Pickpockets'
        Oct.: 'A Supressed Speech'
        Dec.: 'Carols of Antichrist (i)'
1929   Jan.: 'Carols of Antichrist (ii)'
        Feb.: 'The Music of Soviet Russia (i)'
        Mar.: 'The Music of Soviet Russia (ii)'
        A.pr.: 'The Delicate Decadent'
        May: 'Heaven and Hell in Wales'
        June: 'The Lifting of a False Hood'
        Aug.: 'On Smacking the Other Cheek'
        Sept.: 'The Fall of Bernard Shaw'
        Oct.: 'Common Persons and Great Art'
        Dec.: 'More Words for [Aylmer] Maude'
1930   Feb.: 'The Age of Monkey Gland'
        June: 'Bach on the Gramophone'
        Aug.: 'Art and Publicity (i)'
        Sept.: 'Art and Publicity (ii)'
        Oct.: 'Festivalediction (i)'
        Nov.: 'Festivalediction (ii)'
        Dec.: 'Festivalediction (iii)'
1931   Jan.: 'The Opera Subsidy Scandal'
        July: 'The Triumph and Tragedy of Wagner'
        Oct.: 'The Recurrent Revolution'
        Nov.: 'What the Devil!'

NEWSPAPERS AND WEEKLY PUBLICATIONS

Central Somerset Gazette and County Advertiser

21 Feb.–18 Apr. 1918, 'Homes or Hovels: The Future Houses of Somerset'

Clarion

24 Nov. 1911, 'Musical Matters'
 1 Dec. 1911, 'The Blowing of Clarions'
 8 Dec. 1911, 'How the cold musician may warm himself'
22 Dec. 1911, 'A Note on Crators' [*sic*]
29 Dec. 1911, 'Sanctified Sensuality at Olympia'
 5 Jan. 1912, 'Stale Plum Pudding'

19 Jan. 1912, 'The Tragedy of Kingship'
26 Jan. 1912, 'The Vindication of Wagner'
16 Feb. 1912, 'Socialist Music Festivals'
 1 Mar. 1912, 'Concert Room Christianity'
19 Mar. 1912, 'A Singing Lesson'
20 Feb. 1914, 'The Confession of the Blackleg'

Daily Citizen

 8 Oct. 1912, 'The New British Music'
10 Oct. 1912, 'Better Music: The Effect of the Festivals'
12 Oct. 1912, 'Music and the Child: The Future of English Opera'
14 Oct. 1912, 'Choral Singing: The North Leading the Way'
25 Oct. 1912, 'Jingle Music: The Passing of a Vogue'
 6 Nov. 1912, 'Music Notes'
30 Nov. 1912, 'The Music Machine: Its Faults and Future'
 4 Dec. 1912, 'English and Maori Folk Music'
 9 Dec. 1912, 'Song and Labour: The Magic Influence of Rhythm'
20 Dec. 1912, 'The Messiah Habit: Handel Worship and Hypocrisy'

Daily Herald

 3 May 1923, 'A Forecast of Revolution'
10 Oct. 1923, 'Shall We Sing?'
17 Oct. 1923, 'Music in Socialist Civilization'
 7 Nov. 1923, 'The Best Music for Everybody'
13 Nov. 1923, 'The Healing Art'
21 Nov. 1923, 'Music that Makes us Love and Hate'
28 Nov. 1923, 'How Not to Praise God'
19 Dec. 1923, 'When Emotion Soars into Music'
 2 Jan. 1924, 'British Opera or Viennese?'
24 Jan. 1924, 'Mozart's Don Giovanni'
30 Jan. 1924, 'Wagner on the Political Situation'
8 Feb. 1924, 'Are we a Musical People?'
13 Feb. 1924, 'The Labour Movement and Art Criticism'
19 Mar. 1924, 'Choral Music and the Socialist Ideal'

Musical News and Herald (published monthly from 15 January 1927)

23 June 1923, 'The Future of Opera in Great britain'
29 Nov. 1924, 'Living Music and Dead Modernism'
 4 July 1925, 'A Book of Judgement'
15 April 1927, 'The Present Crisis in British Music'

15 June 1927, 'From Romance to Reality'
15 July 1927, 'A Sale of Our Birthright'
15 Feb. 1928, 'A Musical Association with Thomas Hardy'
15 May 1928, 'Authors, Composers, Publishers and the Public'

Musical Standard

12 Mar. 1904, 'Wiggery: Self-satisfaction in Art'
18 June 1904, 'The Co-operative Orchestra'
 2 July 1904, 'Infant Prodigies'
23 July 1904, 'Music-Piracy'
 6 Aug. 1904, 'Socialism'
17 Sept. 1924, 'The Time Spirit'
 1, 8, 15, 22 Oct. 1924, 'Man and Music: A Belated Review'
19 Nov. 1924, 'Man and Music (cont.)'
 3 Dece. 1924, 'Man and Music (concl.)'
14, 21, 28 Oct. 1905, 'Music and the People'
 4, 11, 18, 25 Nov. 1905, 'Music and the People (cont.)'
 2, 9, 16 Dec. 1905, 'Music and the People (concl.)'
 9 Mar. 1907, 'Anti-Wagnerianism'
16, 23 Mar. 1907, 'Studies in Anti-Wagnerianism'
13 Apr. 1907, 'Studies in Anti-Wagnerianism (concl.)'
 7 Nov. 1907, 'Vaughan Williams: Toward the Unknown Region'
 9 Nov. 1907, Holbrooke's Romantic Songs'
16 Nov. 1907, 'Bainton's The Blessed Damozel'
30 Nov. 1907, 'Ashton's Cello Sonata in B flat'
 4 Jan. 1908, 'Bethoven's Orchestration'
18 Jan. 1908, 'Musical Criticism'
 9 Jan. 1909, 'Mind in Music: The Element of Intellect'
16, 23 Jan. 1909, 'Mind in Music: Music Titles'
30 Jan. 1909, 'Mind in Music: Of Scientific Titles'
13 Feb. 1909, 'Mind in Music: Poetic Ideas in Music'
13 Feb. 1909, 'Mind in Music: Of Realism of Music'
20 Feb. 1909, 'Mind in Music: Programme Music'
27 Feb. 1909, 'Mind in Music: Vocal Music'
 6 Mar. 1909, 'Mind in Music: Music and Drama'
13, 20 Mar. 1909, 'The Musician's Bookshelf: Review of "The Rhythmic
   Conception of Music" by M. Glyn'
10, 17, 24 Apr. 1909, 'The Aesthetic of Song'
30 Sept. 1910, 'The Music of Gordon Craig'
19 Nov. 1910, 'The Need of Music'
 6 Nov. 1911, 'The Failure of the Symphony'

Railway Review

21 July 1933, 'Music as Fertilizer and Dope'
28 July 1933, 'Lies for Choral Singing'
 4 Aug. 1933, 'Sugar-Music'
11 Aug. 1933, 'Brass and Military Bands'
18 Aug. 1933, 'Low Brow Music'
25 Aug. 1933, 'Beethoven the Revolutionary'
 1 Sept. 1933, 'Musical Soloists and Political Leaders'
 8 Sept. 1933, 'A Cure for the Pip'
15 Sept. 1933, 'Conductors and the Interpretation of Music'
22 Sept. 1933, 'Switching off Beethoven'
29 Sept. 1933, 'What is Music Making?'
 6 Oct. 1933, 'The Massed Bands of Britain'
13 Oct. 1933, 'Defects in Musical Broadcasting'
20 Oct. 1933, 'More about Brass Bands'
27 Oct. 1933, 'Mastery in Music'
 3 Nov. 1933, 'Welsh Music, War Music and Fancy Stuff'
10 Nov. 1933, 'Musical Criticism; British Genius; Hungarian Music'
17 Nov. 1933, 'A Popular Composer of Bad Operas'
24 Nov. 1933, 'Sir Thomas Beecham goes to Heaven'
 1 Dec. 1933, 'An Introduction to Bach'
 8 Dec. 1933, 'Bach as Broadcasted'
15 Dec. 1933, 'A Joy Forever'
22 Dec. 1933, 'Music in a Mining Village'
29 Dec. 1933, 'Ways of Enjoying Music'
 5 Jan. 1934, 'A Fairy Opera for Grown Ups'
12 Jan. 1934, 'Tannhäuser at the Vic'
16 Jan. 1934, 'Hearing and Seeing in Opera'
26 Jan. 1934, 'Gluck's Orpheus at the Old Vic'
 2 Feb. 1934, 'Merriment and Joy in Music'
 9 Feb. 1934, 'Modern Ghosts on the Wireless'
16 Feb. 1934, 'Dances Dead and Living'
23 Feb. 1934, 'Satan Gets a Fall'
 2 Mar. 1934, 'Elgar'
16 Mar. 1934, 'Music and Workers'
23 Mar. 1934, 'A Musical Organization for Railwaymen?'
 6 Apr. 1934, 'Poison-music'
28 June 1935, 'Hands Out, Comrades, a Genius!'

Sunday Worker

 2 Sept. 1928, 'Gramophone Review'
 9 Sept. 1928, 'Gramophone Review'

23 Sept. 1928, 'A Singer you will Love'
28 Oct. 1928, 'The Right and Wrong Sort of Sweetheart'
 4 Nov. 1928, 'Mozart: The Demon-Angel'
18 Nov. 1928, 'International Music Triumph'
25 Nov. 1928, 'Records which Make Good Christmas Gifts'
 2 Dec. 1928, 'A Great Symphony'
 6 Jan. 1929, 'Jazz for Mayfair and Music for the Workers'
20 Jan. 1929, 'Great Composer as Passionate Rebel'
10 Mar. 1929, 'Chaliapin's Decadence'
24 Mar. 1929, 'Colour that is Beautiful'
14 Apr. 1929, 'Gramophone Notes'
12 May 1929, 'Rebellion in Works of Genius'
16 June 1929, 'Getting Drunk on Beethoven'
 7 July 1929, 'Boughton Breaks with Morrison'; 'Offering Children Musical Dope'
11 Aug. 1929, 'Pandering to what They Think is Popular Taste'
22 Sept. 1929, 'Music that might Go On For Ever'
29 Sept. 1929, 'Jazz Justifies itself: The Music of Civilization Crumbling to Pieces'
20 Oct. 1929, 'Modern Noises—and Some Music'
27 Oct. 1929, 'A Fine Record—and Some Others'
 3 Nov. 1929, 'Revolution Reaches the Newer Music: Creative Effort in Soviet Russia'
24 Nov. 1929, 'Gramophone Notes'

Workers' Weekly

11 June 1926, 'Wanted: A Proletarian Editor for the Daily Herald'

MISCELLANEOUS PUBLICATIONS

Scallop Shell

1911 Feb.: 'Music and Drama'

World's Work

1912 July: 'The Pioneer of British Choralism'
1913 Nov.: 'The Rights and Wrongs of Stagecraft'

TP's Weekly

1913 14 Mar.: 'How I Began'

Musical Quarterly

1915 Oct.: 'The Soul of Music'
1916 Oct.: 'Shakespeare's Ariel'

Musician

1919 Oct.: 'Glastonbury's Musical Gospel'

Theatre Craft

1920 Summer: 'Dramatic Art and the Voice'

Music Bulletin

1923 Feb.: 'Dame Ethel Smyth'

Labour Magazine

1923 Nov.: 'The Arts as Part of the Worker's Life'

Labour Monthly

1926 Aug.: 'How Come these Traitors?'

Communist Review

1926 June: 'D. H. Lawrence: Fascist or Communist?'

Aria

1929 Mar.: 'The Dusk of Opera and a New Dawn'

Millgate

1929 Dec.: 'Christmas and Music'

Author

1933 Summer: 'The Isolation of Composers'

New Britain

1934 30 May, 13 June: 'The Anti-British Opera'

Fanfare

1946 Apr.: 'Music Study and Social Life'
1947 Mar.: 'A Note to the Opera Class'

Philharmonic Post

1947 Dec.: 'British Opera—What Hopes?'
1949 Apr.: 'The Immortal Hour'

Modern Quality

1948 Summer: 'The Science and Art of Music'

Our Time

1948 Oct.: 'Jerico Jazz'

Radio Times

1950 17 Feb.: 'The Queen of Cornwall'

APPENDIX C

The Glastonbury Festivals

CAST LISTS OF PRINCIPAL PRODUCTIONS 1914–1926

Summer Festival, 5–30 August 1914

There were seventeen events: ten concerts mixing songs, partsongs, dances, and short plays, three performances of Alice M. Buckton's pageant play *The Coming of Bride*, and one of her *Beauty and the Beast*, and three performances of *The Immortal Hour*. Cast lists for the Buckton events have not survived.

The Night Shift: Wilfred W. Gibson (5 August)

Performed by 'members of Margaret Morris's Company of the Criterion and Court Theatre, London'. Individual names not given.

The Mystic Dance of the Grail: Boughton (5, 8 August)

Celestial Being	Kathleen Dillon
Creatures of Earth	Margery Drew, Beatrice Filmer, Flossie Jolley

Parsifal—the 'Grail' Scene: Wagner (15, 28 August)

Amfortas	George Painter/Arthur Trowbridge
Parsifal	Rutland Boughton
Guernemanz & Titurel	Gerald Warre Cornish

The Travelling Man: Lady Gregory (19 August)

The Mother	Christina Walshe
The Child	Kathleen Dillon
The Man	George Painter

Alice in Wonderland (19 August): impromptu scene directed by John Rodker

Giant	Gerald Warre Cornish
Queen of Hearts	Sonia Perovskaia

Soldiers   Flossie Jolley, Kathleen Dillon, Barbara Kennedy Scott, Charles Kennedy Scott jun.

Soul Sight: Walter Merry (22 August)

Mary	Irene Lemon
Bridget	Agnes Thomas
Joan	May Walden
Robert	R. Neville Strutt
Thomas	Bernard Lemon

'The Chapel in Lyonesse': Boughton (25 August)

Sir Ozana	Rutland Boughton
Sir Bors	R. Neville Strutt
Sir Galahad	Arthur Jordan

The Death Dance of Grania: Boughton (19, 25 August)

Grania	Margery Drew
Spirits	Flossie Jolley, Kathleen Dillon
Wakewomen	May Lemon, Agnes Thomas, Frances Chase, Muriel Boughton, Irene Lemon, Gladys Fisher, Sonia Perovskaia, Nora Searson

The Heracleidae, scene: Euripides (25, 28 August)

Demophon	Robert Billingham
Iolaus	Gerald Warre Cornish
Macaria	Kathleen Dillon
Citizen	R. Neville Strutt
Sons of Hercules	Charles Kennedy Scott, jun., Ruby Boughton

'Nimue's Dance' (The Quest of the Grail): Vincent Thomas (21 August)

Merlin	Gerald Warre Cornish
Nimue	Beatrice Filmer

The Immortal Hour (26, 27, 28 August)

Dalua	Rutland Boughton
Etain	Irene Lemon/Gladys Fisher
Eochaidh	Frederic Austin

Manus	R. Neville Strutt
Maive	Agnes Thomas
Spirit Voice	Muriel Boughton
Old Bard	Arthur Trowbridge
Midir	Arthur Jordan
Druids, Soldiers	The Wookey Hole Male Voice Choir
Women	Members of the Summer School
Tree Spirits (dancers)	Flossie Jolley, Margery Drew, Beatrice Filmer, Kathleen Dillon, Gwendolen Foulke, May Lemon, Sonia Perovskaia, Barbara Kennedy Scott
Pianist/Conductor	Charles Kennedy Scott
Costumes and Design	Margaret Morris, Gerda Giobel, Christina Walshe

Bournemouth, 7–9 January 1915

The Immortal Hour (four performances)

Dalua	Herbert Langley
Etain	Gladys Fisher/Marjorie Ffrangcon-Davies
Eochaidh	Frederic Austin
Manus	Rutland Boughton
Maive	Agnes Thomas
Spirit Voice	Muriel Boughton
Old Bard	Arthur Trowbridge
Midir	Arthur Jordan
Chorus and dancers	*as in August 1914*
Conductor	Charles Kennedy Scott
Costumes	Margaret Morris, Christina Walshe

Easter Festival, 7 April 1915

Dido and Aeneas: Purcell (two performances)

Dido	Gladys Fisher
Belinda	Daisy Boyd
Attendant	Edith Phipps
Sorceress	Herbert Anderton
Evil Spirit	Carrie Windmill
Witches	Agnes Thomas, Helen Skinner
Sailor	Kathleen Dillon
Aeneas	Ralph Chivers
Aeneas's companion	Tom Gilbert

Conductor   David Scott
Costumes and design   Christina Walshe

The Immortal Hour, Act 1, Scene 2 (two performances)

Etain   Gladys Fisher
Eochaidh   George Painter
Manus   R. H. Green
Maive   Agnes Thomas
Pianist/conductor   Rutland Boughton
Costumes   Christina Walshe

Whitsuntide Festival, 25–26 May 1915

The Birth of Arthur, Act 2, Scene 1 (three performances)

Igraine   Gladys Fisher
Brastias   Arthur Jordan
Choral dances led by   Kathleen Dillon, Flossie Jolley
Pianist/conductor   Rutland Boughton
Costumes   Christina Walshe

Dido and Aeneas (three performances)

Dido   Gladys Fisher
Belinda   Daisy Boyd
Attendant   Edith Phipps
Sorceress   Herbert Anderton
Evil Spirit   Judy Norman
Witches   Agnes Thomas, Helen Skinner
Sailor   Kathleen Dillon
Aeneas   Ralph Chivers
Aeneas's companion   Tom Gilbert
Pianist/conductor   Rutland Boughton
Costumes and design   Christina Walshe

Summer Festival, 11–28 August 1915

There were twelve events, including a singing competition and a final concert (at Street) for the competitors. Some of the shorter operatic programmes included songs and dances as a second half.

Oithona: Edgar Bainton (11, 12 August, four performances)

Oithona	Marjorie Ffrangcon-Davies
Gaul	Frank Mullings
Dunrommath	Herbert Langley
Pianist/conductor	Rutland Boughton
Costumes and design	Christina Walshe

Tristan und Isolde, Act 2 (11, 21 August, four performances)

Tristan	Frank Mullings
Isolde	Ciciley Gleeson White
Brangaena	Jessie Norman
Melot	David Scott
Kurvenal	Bernard Lemon
King Marke	George Painter
Conductor	Clarence Raybould
Costumes and design	Christina Walshe

The Birth of Arthur, Act 2, Scenes 1 and 3 (18, 23, 25 August)

Igraine	Gladys Fisher
Brastias	Arthur Jordan
Merlin	Herbert Langley
Choral dances led by	Kathleen Dillon, Flossie Jolley
Conductor	Clarence Raybould
Costumes and design	Christina Walshe

Dido and Aeneas (18, 23 August)

Dido	Gladys Fisher
Belinda	Marjorie Ffrangcon-Davies
Attendant	Audrey Hudson
Sorceress	Herbert Anderton
Evil Spirit	Carrie Windmill
Witches	Agnes Thomas, Elizabeth Preston, Iris Yeoman
Sailor	Kathleen Dillon
Aeneas	Arthur Jordan
Slaves	Flossie Jolley, May Lemon
Pianist/conductor	Clarence Raybould
Costumes and design	Christina Walshe

The Immortal Hour (14, 12, 24, 25 August)

Dalua	George Painter/Herbert Langley
Etain	Irene Lemon
Eochaidh	Herbert Langley/Frederic Austin
Manus	Bernard Lemon
Maive	Agnes Thomas
Spirit Voice	Muriel Boughton
Old Bard	Bernard Lemon
Midir	Frank Mullings/Arthur Jordan
Dances led by	Kathleen Dillon
Pianist/conductor	Rutland Boughton
Costumes and design	Christina Walshe

Christmas Festival, 28–30 Dcember 1915

Bethlehem (four performances)

Mary	Irene Lemon
Gabriel	Christina Walshe
Joseph	Bernard Lemon
Shepherds: Jem	David Scott
Sym	Tom Filbert
Dave	Percy Holley
Wisemen: Zarathustra	Herbert Anderton
Nubar	David Scott
Merlin	Robert Billingham
Chief Slave	May Lemon
Two Angels	Connie Hinde, Louie Blurton
Two Women	Edith Percy, Elsie Squire
Believer	Agnes Thomas
Unbeliever	Rutland Boughton
Calchas	Percy Holley
Herod	Frank Mullings
Herodias	Irish Yeoman
Pianist/conductor	Charles Kennedy Scott
Costumes and design	Christina Walshe

Whitsuntide Festival, 12–13 June 1916

Everyman (three performances)

Everyman	Gwen Ffrangcon-Davies
Death	Naomi Florence

Fellowship	Roger Clark
Kindred	Agnes Thomas, Hilda Bartholomew
Cousin	William Wall, Elsie Squire
Goods	Percy Holley
Good Deeds	Kitty Raybould
Knowledge	Miss Waller
Confession	David Scott
Discretion	Miss Dent/Mrs Webb
Strength	J. Merriott
Five Wits	Edith Percy
Beauty	Miss Wright
Producer	Rutland Boughton
Costumes and design	Christina Walshe

Easter Festival, 24–26 April 1926

Iphigenia in Tauris: Gluck (three performances)

Iphigenia	Gladys Fisher
First Priestess	Edith Percy
Second Priestess	Muriel Boughton
Woman	Agnes Thomas
Attendant	Herbert Anderton
Thoas	Bernard Lemon
Sythian	Percy Holley
Orestes	Rutland Boughton
Pylades	Louis Godfrey
Diana	Louie Blurton
Chorus	Elsie Squire, Ada Burnell, Margaret Dent, Hilda Bartholemew, Naomi Florence, Hilda Strange, Mabel Burnell, Doris Boroughton, Charles Hannam, William Lewis, Stanley Merriott, Tom Gilbert, David Scott, John Summerhayes, Arthur Wall
Pianist/conductor	Clarence Raybould
Costumes and design	Christina Walshe

Snow White (three performances)

Snow White	Blanche Blurton
Prince	Kathleen Blurton
Wicked Queen	Flossie Jolley
Butcher	Norman Wright
Wolf	Alan Dunthorn

Dwarfs	Ruby Boughton, Cissie Turner, Gladys Blurton, Edward Vincent, Edward Alves, Cyril Gordon, Douglas Checkley
Huntsmen	Beatrice Wright, Queenie Dunthorn, Freda Avery, Irene Bignell
Flowers, Attendants	Irene Holway, Gladys Keep, May Smith, Winnie Hill, Gertie Campbell, Kathleen Goodson, Marjorie Difford, Ella Porch, Dorothy Carter, Beatrice Baulch, Dorothy Corp
Choreography	Margaret Morris
Costumes and design	Margaret Morris
Pianist/conductor	Rutland Boughton

Summer Festival, 8–26 August 1916

There were fifteen events: twelve operatic performances, two recitals, and one evening devoted to a Scholarship Competition and the Glastonbury 'Travesty' Play.

The Round Table (14, 18, 23 August)

Merlin	Percy Hemming
Arthur	Frederic Austin
Dagonet	Arthur Jordan
Kay	David Scott
Ector	William Bennett
Lancelot	Percy Snowden
Bedivere	Herbert Anderton
Lamorek	Louis Godfrey
Gawaine	Stanley Merriott
Bors	Frank Gloyne
Bishop	William Waite
Lady of the Lake	Muriel Boughton/Mrs Tobias Matthay
Nimue	Gwen Ffrangcon-Davies
Guenevere	Irene Lemon
Bettris	Gwen Ffrangcon-Davies
Taliesin	Bernard Lemon
Butler	Percy Holley
Servants	Agnes Thomas, Hilda Bartholemew, Kitty Raybould
Acolytes	Cyril Sparks, Jack Barker
Chorus	Gladys Fisher, Muriel Boughton, Sylvia York Bowen, Edith Percy, Naomi Florence, Ivy Adams, Tom Waller, J. H. Brooks

| Pianist/conductor | Rutland Boughton |
| Costumes and design | Christina Walshe |

The Sumida River: Clarence Raybould (15, 19 24 August)

Ferryman	Percy Hemming
Traveller	Louis Godfrey
Mother	Irene Lemon
Child	Ruby Boughton
Pianist/conductor	Clarence Raybould
Costumes and design	Christina Walshe

Everyman (15, 19 August)

Everyman	Gwen Ffrangcon-Davies
Death	Naomi Florence
Fellowship	Roger Clark
Kindred	Agnes Thomas
Goods	Percy Holley
Good Deeds	Kitty Raybould
Knowledge	Sylvia York Bowen
Confession	David Scott
Discretion	Muriel Boughton
Strength	Stanley Merriott
Five Wits	Edith Percy
Beauty	Irene Lemon
Producer	Rutland Boughton
Costumes and design	Christina Walshe

Iphigenia in Tauris (16, 21, 25 August)

Iphigeia	Gladys Fisher
First Priestess	Edith Percy
Second Priestess	Muriel Boughton
Woman	Agnes Thomas
Attendant	Frank Gloynes
Thoas	Bernard Lemon
Sythian	Percy Holley
Orestes	Percy Snowden
Pylades	Louis Godfrey
Diana	Sylvia York Bowen
Pianist/conductor	Rutland Boughton
Costumes and design	Christina Walshe

The Immortal Hour (17, 22, 26 August)

Dalua	William Waite
Etain	Irene Lemon
Eochaidh	Frederic Austin
Manus	Percy Holley
Maive	Agnes Thomas
Spirit Voice	Muriel Boughton
Old Bard	Frank Gloynes
Midir	Arthur Jordan
Pianist/conductor	Rutland Boughton
Costumes and design	Christina Walshe

Festival Work: September 1916–January 1919

During Boughton's absence on active service, Christina Walshe kept alight a nucleus of interest, presenting what amounted to miniature festivals at, roughly, the usual times of year. All 1917 and 1918 performances designed and produced by Christina Walshe, unless otherwise stated.

24 January 1917

The Land of Heart's Desire: W. B. Yeats (music by Rutland Boughton)

Bridget	Agnes Thomas
Maurteen	Arthur Wall
Maire	Winifred Hardy
Shaun	Stanley Merriott
Father Hart	Herbert Anderton
Fairy Child	Kitty Raybould

Paddly Pools: Miles Malleson

Tony	Ruby Boughton
Little Old Man	Gladys Blurton
Grandpa	Arthur Wall
Fairies and Friends	Agnes Thomas, Kathleen Blurton, Naomi Florence, Hilda Bartholemew, Queenie Dunthorn, Winifred Hardy, Norman Wright, Trissie Wright, Irene Avery, Freda Avery, Mollie Dunthorn, Stella Hersey, Kitty Raybould, Doris Crocker, Blanche Blurton

24 April 1917

The Land of Heart's Desire

Bridget	Agnes Thomas
Maurteen	Arthur Wall
Maire	Christina Walshe
Shaun	John Bright Clark
Father Hart	Herbert Anderton
Fairy Child	Kathleen Blurton

The Maharani of Arakan: George Calderon

Princess Amina	Margaret Mitchell
Princess Roshenara	Christina Walshe
King Dalia	K. N. Das Gupta
Tung Loo	Arthur Wall
Rahmet Sheikh	John Bright Clark
Attendants, Courtiers	Doris Crocker, Hilda Bartholemew, Winnie Edgehill, Herbert Anderton, Edith Percy, Florence Webb, Kitty Raybould
Producer	K. N. Das Gupta

11, 12, 14 September 1918

Sacrifice: Rabindranath Tagore (music by Rutland Boughton)

Queen Gunavati	Zabelle Boyajian
King Govinda	Arthur Jordan
High Priest	Herbert Anderton
Jaising	Gladys Fisher
Aparna	Gwen Ffrangcon-Davies
Prince Nakshatra	Edith Percy
General Nayan Rai	A. W. Arundale
Chandpal	A. E. Webb
Prince Druva	Montefiore Freeman

Paddly Pools (music by Adele Maddison)

Tony	Ruby Boughton
Grandfather	Agnes Thomas
Little Old Man	Gladys Blurton
Three Friends	Edith Percy, Winifred Hardy, Phyllis Jewson

Short Green Grass	Trissie Wright
Wild Flower	Blanche Blurton
Soul of all the Rabbits	Molly Dunthorn
Soul of all the Trees	Winifred Hardy
Sunset Spirits	Girlie Campbell, Ridley Browning, Queenie Dunthorn, Beatrice Baulch, May Smith

The Robin, the Mouse, and the Sausave: John Bostock (music by Rutland Boughton)

Dame Owl	Agnes Thomas
Children	Details not recorded

Young Heaven: Miles Malleson

Daphne	Gwen Ffrangcon-Davies
Dan	Miles Malleson
Servant	Agnes Thomas

The Dark Lady of the Sonnets G. Bernard Shaw

Shakespeare	Rutland Boughton
The Dark Lady	Phyllis Jewson
Warder	A. W. Armitage

The Immortal Hour, Act 1, Scene 2

Etain	Gwen Ffrangcon-Davies
Eochaidh	Arthur Jordan
Manus	Rutland Boughton
Maive	Agnes Thomas

Easter Festival, 23–28 April 1919

There were twelve events: five operatic performances, five recitals, one lecture, and a performance of the St John Passion.

The Moon Maiden (23, 26 April)

Moon Maiden	Ruby Boughton
Fisherman	Herbert Anderton
Pianist/conductor	Rutland Boughton
Costumes and design	Christina Walshe

The Wickedness of Dancing (23, 26 April)

Puritan Father	Percy Holley
Puritan Mother	Agnes Thomas
Daughter	Dollie Crocker
Lover	Stanley Merriott
Pianist/conductor	Rutland Boughton
Costumes and design	Christina Walshe

Cupid and Death, Scene 1: Matthew Locke, Christopher Gibbons (24 April)

Lady	Gwen Brooke-Webb
Lovers	Dolly Crocker, Stanley Merriott
Nature	Gwen Ffrangcon-Davies
Old Couple	Agnes Thomas, Percy Holley
Cupid	Kathleen Blurton
Death	Alice Bostock
Dances arranges by	Florence Jolley
Pianist/conductor	Rutland Boughton
Costumes and design	Christina Walshe
Producer	Marjorie Gilmour

Summer Festival, 17–28 August 1919

There were twenty-two events: eighteen staged works, three recitals, and one lecture-recital.

Cupid and Death (18, 22, 27 August)

Holst	William Millard
Chamberlain	Steuart Wilson
Despair	Phyllis Jewson
Nature	Gwen Ffrangcon-Davies
Mercury	Sheerman Hand
Cupid	Ruby Boughton
Folly	Florence Jolley
Old Couple	Agnes Thomas, Percy Holley
Lady	Gwen Brooke-Webb
Lovers	Dolly Crocker, Stanley Merriott
Two Apes	Douglas Checkley, Willie Crocker
Madness	Gladys Ward
Death	Naomi Florence
Conductor	Edward J. Dent
Costumes and design	Christina Walshe

A Pot o'Broth: W. B. Yeats (18, 22, 27 August)

Tramp	Percy Holley
Sibby	Kathleen Aldridge
John	Frederick Aldridge
Costumes and design	Christina Walshe

The Moon Maiden (20, 25, 29 August)

Moon Maiden	Gwen Ffrangcon-Davies
Fisherman	Percy Holley
Pianist/conductor	William H. Kerridge
Costumes and design	Christina Walshe

The Sumida River (20, 25, 29 August)

Ferryman	W. Johnstone Douglas
Traveller	Louis Godfrey
Mother	Constance Lermit
Child	Molly Dunthorn
Pianist/conductor	William H. Kerridge
Costumes and design	Christina Walshe

The Immortal Hour (19, 23, 28 August)

Dalua	Clive Carey
Etain	Gwen Ffrangcon-Davies
Eochaidh	Percy Snowden
Manus	Percy Holley
Maive	Agnes Thomas
Spirit Voice	Gladys Ward
Old Bard	William Bennett
Midir	Arthur Jordan
Pianist/conductor	Reginald Paul
Costumes and design	Christina Walshe

The Round Table (21, 26, 30 August)

Merlin	Clive Carey
Arthur	W. Johnstone Douglas
Dagonet	Arthur Jordan
Kay	David Scott
Ector	William Bennett

Lancelot	Percy Snowden
Bedivere	Sheerman Hand
Lamorek	Louis Godfrey
Gawaine	Stanley Merriott
Bishop	Hamilton Harris
Taliesin	Percy Holley
Lady of the Lake	Gladys Ward
Nimue	Gwen Ffrangcon-Davies
Bettris	Margaret Longmans
Pianist/conductor	William H. Kerridge, Reginald Paul, David Scott
Costume and design	Christina Walshe

Bournemouth, 1–2 September 1919

The Immortal Hour (three performances)

Cast as Summer Festival

Christmas Festival, 26–29 December 1919

The Festival was preceded by a visit to Bath and Bournemouth (21–3 December) with performances of *Bethlehem* using the same cast.

Snow White (26, 27, 29 December)

Snow White	Blanche Blurton
Prince	Ruby Boughton
Wicked Queen	Kathleen Blurton
Butcher	Edward Vincent
Wolf	Willie Crocker
Pages	Winnie Ford, Vera Wright
Dwarfs	Estelle Boughton, Betty Checkley, Kathleen Dunthorn, Phyllis Goodson, Catherine Miller, Gwendoline Stacey
Huntsmen	Edward Alves, Freda Avery, Irene Avery, Douglas Checkley
Flowers	Eileen Alexander, Joy Boughton, Estelle Boughton, Betty Checkley, Florence Carter, Molly Dunthorn, Kathleen Dunthorn, May Giles, Phyllis Goodson, Eileen Holloway, Margaret Hawkins, Margery Hillard, Enid Maidment, Olive Simpson, Marie Taylor, Gwendoline Stacey
Pianist/conductor	William H. Kerridge

Choreography   Laura Wilson
Costumes and design   Christina Walshe

Bethlehem (26, 27 29 December)

Mary	Gwen Ffrangcon-Davies
Gabriel	Kathleen Beer
Joseph	William Bennett
Shepherds: Jem	Stanley Merriott
Sym	Colin Tilney
Dave	Percy Holley
Wisemen: Zarathustra	Sheerman Hand
Nubar	Rutland Boughton
Merlin	Percy Holley
Two women	Agnes Thomas, Muriel Boughton
Believer	Eva Joyce
Unbeliever	Colin Tilney
Calchas	William Mortimer
Herod	Barrington Hooper
Herodias	Dorothy D'Orsay
Dancers	Norah Allen, Freda Avery, Ruby Boughton, Dolly Crocker, Kathleen Davis, Florence Dunthorn
Pianist/conductor	William H. Kerridge
Costumes and design	Christina Walshe

Easter Festival, 28 March–7 April 1920

There were eight events: five stage performances, and three recitals.

Venus and Adonis: John Blow (5, 6, April)

Venus	Gladys Moger
Adonis	Clive Carey
Cupid	Ruby Boughton/Fedora Turnbull
Huntsmen	Dorothy D'Orsay, Robin Ford
Shepherdess	Kathleen Davis
Shepherds	Arthur Clark, Sheerman Hand, Robin Ford, William Weaver
Graces	Grace Mclearn, Laura Wilson, Norah Allen, Dorothy D'Orsay, Dolley Crocker, Florence Dunthorn, Edith Percy, Naomi Florence, Agnes Thomas

| Pianist/conductor | Rutland Boughton |
| Costumes and design | Christina Walshe |

Everyman (3, 6 April)

Everyman	Gwen Ffrangcon-Davies
Fellowship	Clive Carey
Cousin	Dorothy D'Orsay
Kindred	Agnes Thomas
Goods	Kathleen Aldridge
Good Deeds	Laura Wilson
Strength	William Millard
Discretion	Norah Allen
Five Wits	Edith Percy
Beauty	Kathleen Davis
Knowledge	Sheerman Hand
Confession	Frederick Aldridge
Producer	Rutland Boughton
Costumes and design	Christina Walshe

Whitsuntide Festival 1920

Cancelled so that the Company might appear in London at the Old Vic (31 May–12 June) in a programme that included *The Immortal Hour* and *Venus and Adonis*, *Music Comes*, and *The Children of Lir*, with a cast virtually identical to the previous Easter Festival and the forthcoming Summer Festival.

Summer Festival, 15 August–11 September 1920

There were forty events: thirty-one stage performances, and nine Chamber Concert—including string quartets by Parry, Bainton, and Elgar, Mozart's Clarinet Quintet, Elgar's Piano Quintet, and Vaughan Williams's *On Wenlock Edge*.

The Birth of Arthur (16, 20, 25 August, 3, 8 September)

Uther	Hamilton Harris
Ulfius	Arthur Jacques
Merlin	Herbert Langley
Igraine	Edith Finch
Brastias	Tom Goodey
Pianist/conductor	Edgar Bainton
Costumes and design	Christina Walshe

The Round Table (17, 21, 26, 31 August, 4, 9 September)

Merlin	Herbert Langley
Arthur	W. Johnstone Douglas
Dagonet	Arthur Jordan
Kay	David Scott
Ector	John Goss
Lancelot	Sheerman Hand
Belivere	Colin Ashdown
Bors	Tom Goodey
Lamorek	Louis Godfrey
Gawaine	Stabley Merriott
Bishop	Hamilton Harris
Taliesin	Percy Holley
Lady of the Lake	Gladys Ward
Nimue	Gwen Ffrangcon-Davies
Guenevere	Winifred Lawson
Bettris	Leonora Allen
Pianist/conductor	Edgar Bainton
Costumes and design	Christina Walshe

Dido and Aeneas (18, 23, 27 August, 1, 6, 10 September)

Dido	Dorothy D'Orsay
Belinda	Gwen Ffrangcon-Davies
Attendant	Kathleen Davis
Aeneas	Arthur Jacques
Sorceress	John Goss
Witches	Mary Groser, Gladys Scorer
Sailor	Arthur Clark
Dancers	Ruby Boughton, Mona Guild
Pianist/conductor	Edgar Bainton
Choreography	Laura Wilson
Costumes and design	Christina Walshe

The Immortal Hour (19, 24, 28 August, 2, 7, 11 September)

Dalua	Herbert Langley
Etain	Gwen Ffrangcon-Davies
Eochaidh	W. Johnstone Douglas
Manus	Colin Ashdown
Maive	Agnes Thomas
Spirit Voice	Dorothy D'Orsay

Old Bard	Herbert Anderton
Midir	Arthur Jordan/Tom Goodey
Pianist/conductor	Rutland Boughton
Costumes and design	Christina Walshe

Music Comes: Napier Miles (18, 23, 27 August, 1, 6, 10 September)

Solo Singer	Tom Goodey
Singers and dancers	Members of the Company
Pianist/conductor	Rutland Boughton
Choreography	Laura Wilson
Costumes and design	Christina Walshe

The Children of Lir: Adele Maddison (18, 23, 27 August, 1, 6, 10 September)

Fianula	Gwen Ffrangcon-Davies
Fiacra	Mona Guild
Hugh	Marjorie Kilburn
Conn	Joy Boughton
Hermit	John Goss
Prince Laignen	Ruby Boughton
Princess Deoca	Laura Wilson
Pianist/conductor	Rutland Boughton
Costumes and design	Christina Wilson

New Year Festival, 2–5 January 1921

There were five events: three stage performances, and two recitals. The New Year Festival was followed by a tour: Bath (6, 7, 8 January), Burnham-on-Sea (12 January), Bristol (13, 14, 15 January). The same cast was used, and one concert was given at Bath (7 January).

Bethlehem (3, 4, 5 January)

Mary	Dorothy Silk/Kathleen Davis
Gabriel	Leonora Allen/Kathleen Beer
Joseph	William Bennett
Shepherds: Jem	Stanley Merriott
Sym	Tom Gilbert
Dave	Frederick Woodhouse
Wisemen: Zarathustra	Sheerman Hand
Nubar	Stanley Merriott
Merlin	Frederick Woodhouse

Two women	Amy Hall, Agnes Thomas
Believer	Irene Child
Unbeliever	David Scott
Calchas	P. Hurst Needham
Herod	Steuart Wilson
Herodias	Dorothy D'Orsay
Dancers	Ruby Boughton, Dolly Crocker, Florence Dunthorn, Marjorie Kilburn
Pianist/conductor	Rutland Boughton
Choreography	Laura Wilson
Costumes and design	Christina Walshe

Easter Festival, 31 March–2 April 1921

There were seven events: six stage performances and one recital. The Easter Festival was followed by a tour: Bournemouth (6–8 April), Bath (11–15 April), and Bristol (19–23 April).

The Immortal Hour (31 March, 1 April)

Dalua	Frederick Woodhouse
Etain	Gwen Ffrangcon-Davies
Eochaidh	Sheerman Hand
Manus	Ben Watts
Maive	Agnes Thomas
Spirit Voice	Dorothy D'Orsay
Old Bard	Claude Blanchard
Midir	Arthur Clark
Pianist/conductor	Rutland Boughton
Costumes and design	Christina Walshe

The Moon Maiden (31 March, 2 April)

Moon Maiden	Gwen Ffrangcon-Davies
Fisherman	Claude Blanchard
Spirits of the Mist	Dolly Crocker, Kathleen Davis, Pamela Henderson, Penelope Spencer, Hilda Small, Kathleen Beer
Pianist/conductor	David Scott
Costumes and design	Christina Walshe

Music Comes (31 March, 2 April)

Solo Singer	Arthur Clark
Singers and dancers	Members of the Company

Pianist/conductor	David Scott
Choreography	Penelope Spencer
Costumes and design	Christina Walshe

Dido and Aeneas (31 March, 2 April)

Dido	Dorothy D'Orsay
Belinda	Gwen Ffrangcon-Davies/Irene Lemon
Attendants	Kathleen Davis, Yvonne Whelan
Sorceress	Frederick Woodhouse
Witches	Agnes Thomas, Yvonne Whelan
Sailor	Arthur Clark
Aeneas	Sheerman Hand
Dancers	Penelope Spencer, Hilda Small
Pianist/conductor	Rutland Boughton
Choreography	Penelope Spencer
Costumes and design	Christina Walshe

Summer Festival, 29 August–3 September 1921

This was a less ambitious Festival, comprising eight mixed events (plays, ballets, chamber music), two recitals, and two public readings.

The Fairy: Laurence Housman (31 August, 3 September)

The Grandmother	Angela Cave
The Mother	Jessie Hall
The Child	Gwendoline Toms
Choreography	Penelope Spencer
Costumes and design	Christina Walshe

Spreading the News: Lady Gregory (31 August, 1, 3, September)

Mrs Tarpey	Kathleen Davis
Magistrate	John Bright Clark
Policeman	Percy Davies
Bartley Fallon	W. Johnstone Douglas
Mrs Fallon	Jessie Hall
Jack Smith	Sheerman Hand
Tim Casey	Copley Clark
Mrs Early	Kathleen Beer
Mrs Tully	Olive Hooper
Mrs Ryan	Hilda Price
Producer	W. Johnstone Douglas

The Moon Maiden (30 August, 1 September)

Moon Maiden	Ruby Boughton
Fisherman	Sheerman Hand
Dancers	Members of the Company
Pianist/conductor	Rutland Boughton
Costumes and design	Christina Walshe

The Chapel in Lyonesse (30 August, 1 September)

Sir Ozana	W. Johnstone Douglas
Sir Bors	Sheerman Hand
Sir Galahad	Copley Clark
Pianist/conductor	Rutland Boughton
Costumes and design	Christina Walshe

The Death of Columbine (1, 2 September)

Two Girls	Kathleen Beer, Beatrice Williams
Columbine	Leonora Allen
Harlequin	W. Johnston Douglas
Pierrot	Sheerman Hand
Pantaloon	Percy Holley
Pianist/conductor	Rutland Boughton
Choreography	Penelope Spencer
Costumes and design	Christina Walshe

All Fools Day: Josephine Baretti (music by Clive Carey; 1, 2, September)

The King	Sheerman Hand
The Queen	Jessie Hall
The Fool	W. Johnstone Douglas
Courtier	Roger Clark
Servants	Leonora Allen, Kathleen Beer, Sylvia Spencer, Ruby Boughton, Kathleen Davis, Gwendoline Toms
Dancer	Penelope Spencer

3 October 1921

Bristol Folk Festival School commences operations.

New Year Festival, 2–4 January 1922

Bethlehem (three performances)

Mary	Kathleen Davis
Gabriel	Kathleen Beer
Joseph	Richard Rurnell
Shepherds: Jem	Sheerman Hand
Sym	Archibald Cooper
Dave	William Heyman
Wisemen: Zarathustra	Sheerman Hand
Nubar	Arthur Clark
Merlin	Edmund Davies
Believer	Olive Franks
Unbeliever	Edwin Farwell
Calchas	Reginald Pitt
Herod	Seymour Dosser
Herodias	Dorothy D'Orsay
Pianist/conductor	Rutland Boughton
Costumes and design	Christina Walshe

Spring Tour, 18–31 March 1922

Bristol, Burnham-on-Sea, Winscombe, Bath, Bristol.

Everyman (eight performances)

Everyman	George Holloway
Fellowship	John Case
Cousin	Dorothy Holloway
Kindred	Margery Phillips
Goods	Sheerman Hand
Good Deeds	Laura Wilson
Knowledge	Christina Walshe
Strength	Arthur Dore
Discretion	Eveline Bailes
Five Wits	Vera Pheysey
Beauty	Gwen Brook-Webb
Confession	Ethel Pheysey
Angels	Olive Franks, Alma Croasdale, Ruby Boughton, Kathleen Beer
Producer	Rutland Boughton
Costumes and design	Christina Walshe

Easter Festival, 17–19 April 1922

Venus and Adonis (three performances)

Venus	Kathleen Davis
Adonis	Sheerman Hand
Cupid	Fay Yeatman
Huntsman	Dorothy D'Orsay
Shepherdess	Beatrice Jollyman
Shepherd	Archibald Cooper
Dancer	Ruby Boughton
Pianist/conductor	Rutland Boughton
Choreography	Vera Owen
Costumes and design	Christina Walshe

Summer Festival, 26 August–2 September 1922

There were eighteen events: eleven stage performances, five lectures, and two recitals.

Alkestis (26, 28, 29, 30, 31 August, 2 September)

Apollo	Arthur Jordan
Thanatos	Frederick Woodhouse
Handmaiden	Kathleen Davis
Citizen	Arthur Clark
Alkestis	Astra Desmond
Child	Catherine Miller
Admetus	Steuart Wilson
Herakles	Clive Carey
Pheres	Joseph Eastman
Youth	Greta Don
Pianist/conductor	Rutland Boughton
Costumes and design	Christina Walshe

Venus and Adonis (29, 30 August, 1 September)

Venus	Kathleen Davis
Adonis	Clive Carey
Cupid	Leonora Allen
Huntsman	Kathleen Beer
Shepherdess	Gladys Ward
Shepherd	Arthur Clark

Graces	Gwen Toms, Penelope Spencer, Pamela Williams
Cupids	Joy Boughton, Maire Boughton, May Checkley, June Voake
Pianist/conductor	Julius Harrison
Costumes and design	Christina Walshe

The Death of Herakles: Euripides (22 August, two performances)

Deianeira	Christina Walshe
Hylos	Leslie Bennett
Iole	Dolly Crocker
Herakles	Sheerman Hand
Nurse	Gladys Ward
An Elder	Roger Clark
Attendant	Angela Cave
Messenger	Joan Pym
Lichas	Tom Waller
Woman	Barbara Drummond
Maidservants	Gwendoline Toms, Vera Pheysey, Barbara Pheysey
Chorus	Gladys Ward, Leonora Allen, Marian Murch, Marian Ridgeway-Kitching, Mary de Latour, Sylvia Spencer, Catherine Miller, Ruby Boughton
Captives	Agnes Swaine, Penelope Spencer, Mollie Dunthorn, Freda Avery, Pamela Williams
Producer	Rutland Boughton
Costumes and design	Christina Walshe

Winter Tour, 6–20 December 1922

Weston-super-Mare, Bristol, Bath, Bournemouth

Bethlehem (thirteen performances)

Mary	Kathleen Davis
Gabriel	Arthur Clark
Joseph	Reginald Purnall
Shepherds: Sym	William Bennett
Jem	Albert Monks
Dave	Frank Heyman
Wisemen: Zarathustra	Edmund Davies
Nubar	Archibald Cooper
Merlin	Stuart Smith
Believer	Olive Franks

Unbeliever	John Burton
Angels	Dorothy Bremer, Ruby Boughton
Two Women	Kathleen Beer, Laura Southwell
Calchas	Ernest Bullock
Herod	Seymour Dosser/Arthur Jordan
Herodias	Dorothy D'Orsay
Pianist/conductor	Rutland Boughton
Choreography	Penelope Spencer
Costumes and design	Christina Walshe

1923

There were no festivals. Boughton was too involved with the London productions of *The Immortal Hour* and *Bethlehem*.

Summer Festival, 9–30 August 1924

There were twenty-four events: eighteen stage performances (those of *Agincourt* also included songs and chamber music as a second half), and six lectures.

Agincourt (25, 26, 27, 28, 30 August)

King Henry V	Charles Hedges
Erpingham	Leslie Bennett
Williams	Leyland White
Bates	Frederick Woodhouse
Pistol	Harry Carter
Gloucester	Frank Phillips
Pianist/conductor	Rutland Boughton
Costumes and design	Christina Walshe

The Queen of Cornwall (21, 22, 26, 28, 30 August)

Watchman	Leyland White
Brangwain	Ruby Boughton
King Mark	Frederick Woodhouse
Sir Andret	Harry Carter
Queen Iseult	Gladys Fisher
Tristram	Frank Phillips
Iseult of Brittany	Elena Goeminne
Damsel	Zillan Hobart
Knight	Owen Strong

Ladies	Winifred Barrows, Lorna Collard, Joyce Collard, Marjorie Strong
Shades of the Dead	Leslie Bennett, James Botter, George England, Charles Hedges, Robert Malaghan, George Richardson, Joyce Bartlett, Averice Dickin, Kathleen Beer, Magdalen Hedges, Olive Hooper, Marian Murch, Agnes Swaine, Beatrice Saxson-Snell, Nina Saxson-Snell
Pianist/conductor	Rutland Boughton
Costumes and design	Christina Walshe

The Little Plays of St Francis: Laurence Housman (music by Rutland Boughton); Series I (19, 23, 27, 29 August)

Our Lady of Poverty

Rinaldo	George Richardson
Uberto	Charles Hedges
Isola	Magdalen Hallyar
Julia	Joyce Collard
Francesco	George Holloway
A Leper	Tom Waller
Pompilo	George England
Giovanni	Leyland White
Margherita	Winifred Barrows
Laura	Lorna Collard
Lucio	Leslie Bennett
Dancers	Leonora Allen, Ruby Boughton, Kathleen Beer, Dolly Crocker

The Builders

Juniper	Frederick Woodhouse
Bernard	Leyland White
Don Silvestro	George Richardson
Leo	Leslie Bennett
Francis	George Holloway
Giacomina	Dorothy Holloway
A Beggar	Robert Malaghan
Pietro Bernadone	George England

Sister Clare

Bernard	Leyland White
Clare	Winifred Barrows

Leo	Leslie Bennett
Francis	George Holloway
Juniper	Frederick Woodhouse
Elias	George Richardson
Angelo	Charles Hedges

The Little Plays of St Francis; Series II (20, 23, 28, 29 August)

Brother Sun

The Soldan	Frederick Woodhouse
Councillor	George Richardson
Emir	James Botter
Captain	George England
Francis	George Holloway
Illuminato	Olive Hooper

Brother Juniper

Prior	George Richardson
Rufus	Leyland White
Angelo	Charles Hedges
Anthony	George England
John	Robert Malaghan
Juniper	Frederick Woodhouse
Jerome	James Botter
Simon	Owen Strong
Francis	George Holloway

The Seraphic Vision

Bernardo	Leyland White
Angelo	Charles Hedges
Leo	Leslie Bennett
Francis	George Holloway
Chorus	Not given, but possibly that of *The Queen of Cornwall*
Musical direction	Rutland Boughton
Producer	Laurence Housman
Costumes and design	Christina Walshe

Spring Tour, 16–18 April 1925

Bath and Bournemouth

The Moon Maiden

Moon Maiden	Kathleen Beer
Fisherman	Leslie Bennett

Agincourt

King Henry V	Charles Hedges
Erpingham	Leslie Bennett
Williams	Leyland White
Bates	Frederick Woodhouse
Pistol	Harvey Braban
Gloucester	Frank Phillips

The Queen of Cornwall

Watchman	Leyland White
Brangwain	Ruby Boughton
King Mark	Frederick Woodhouse
Sir Andret	Charles Hedges
Queen Iseult	Gladys Fisher
Tristram	Frank Phillips
Iseult of Brittany	Gloria Dawson
Damsel	Zillah Hobart
Knight	Owen Strong
Ladies and Chorus	Not given, but presumably as in August

The Immortal Hour

Dalua	Frederick Woodhouse
Etain	Nan Ferguson
Eochaidh	Frank Phillips
Manus	George Richardson
Maive	Gladys Fisher
Spirit Voice	Kathleen Beer
Old Bard	Leslie Bennett
Midir	John Dean
Pianist/conductor	Rutland Boughton
Choreography	Penelope Spencer
Costumes and design	Christina Walshe

Summer Festival, 5 August–5 September 1925

Twenty-five stage performances were given.

The Immortal Hour (1, 3, 4, 5 September)

Dalua	Frederick Woodhouse/Leslie Bennett
Etain	Kathleen Vincent/Annette Blackwell
Eochaidh	Frank Phillips
Spirit Voice	Kathleen Beer
Manus	George Richardson
Maive	Gladys Fisher
Old Bard	Leyland White
Midir	Tom Goodey
Pianist/conductor	A. Davies-Adams
Costumes and design	Christina Walshe

The Queen of Cornwall (29, 29 August, 1, 3 September)

Watchman	Leyland White
Brangwain	Ruby Boughton
King Mark	Frederick Woodhouse
Sir Andret	Charles Hedges
Queen Iseult	Gladys Fisher/Mona Benson
Tristram	Arthur Cranmer
Iseult of Britany	Norah Desmond
Damsel	Mable Ritchie
Chorus	Not known
Pianist/conductor	A. Davies-Adams
Costumes and design	Christina Walshe

The Birth of Arthur (27, 28 August, 2 September)

Uther	Frank Phillips
Ulfius	Geoffrey Dunn
Merlin	Frederick Woodhouse
Igraine	Gladys Fisher
Brastias	Charles Hedges
Chorus	Not known
Pianist/conductor	A. Davies-Adams
Choreography	Penelope Spencer
Costumes and design	Christina Walshe

The Round Table (27, 29, 31 August, 2, 4 September)

Merlin	Frederick Woodhouse
Arthur	Arthur Cranmer
Dagonet	Seymour Dosser
Kay	John Thompson
Ector	Eric Godfrey
Lancelot	Augustus Milner
Bedivere	Frank Phillips
Bors	George Richardson
Gawaine	Charles Hedges
Bishop	Leyland White
Lady of the Lake	Mona Benson
Nimue	Kathleen Vincent
Guenevere	Valerie Russell
Bettris	Kennedy Arundel
Butler	Leslie Bennett
Taliesin	Not known
Cook	Cary Davies
Servers	Beatrice Saxson-Snell, Kathleen Beer
Pianist/conductor	A. Davies-Adams
Costumes and design	Christina Walshe

The Little Plays of St Francis: Series I (5, 6, 7, 8 August)

The Revellers

Francesco	George Holloway
Lucio	Henry Cass
Rinaldo	Frederick Woodhouse
Giovanni	Charles Higgins
Antonia	Ruby Boughton
Rudolfo	Howieson Culff
Umberto	George Richardson
Podesta	Sebastian Shaw
Paolo	Donald Finlay
Town Guard	James Botter, George Caines, Tom Waller
Townsfolk	Olive Hooper, Irene Webb, Agnes Swaine

Fellow Prisoners

Francesco	George Holloway
Lucio	Henry Cass

Giovanni	Charles Higgins
Gaoler	Tom Waller
Prisoner	Owen Strong
Rinaldo	Frederick Woodhouse
Rudolfo	Howiesan Culff
Pompilio	Sebastian Shaw
Paolo	Donald Finlay
Turnkeys	George Caines, James Botter

The Bride Feast

Francesco	George Holloway
Lucio	Henry Cass
Rudolfo	Howieson Culff
Umberto	George Richardson
Baldone	Donald Finlay
Arnolfo	Sebastian Shaw
Antonia	Ruby Boughton
Lucrezia	Irene Webb
Fame	Olive Hooper
Cupid	Mary Checkley
Masquers	Penelope Spencer, Sylvia Spencer, Maire Boughton, Beatrice Saxson-Snell, Nina Saxson-Snell, Agnes Swaine
Attendants	Charles Higgins, George Caines

Our Lady of Poverty

Francesco	George Holloway
Lucio	Henry Cass
Umberto	George Richardson
Rinaldo	Frederick Woodhouse
Paolo	Donald Finlay
A Leper	Howieson Culff
Margherita	Dorothy Holloway
Julia	Olive Hooper
Laura	Irene Webb
Isola	Hilda Morland
Pompilio	Sebastian Shaw
Ladies	Penelope Spencer, Sylvia Spencer, Nina Saxson-Snell, Beatrice Saxson-Snell
Courtiers	Charles Higgins, James Botter, George Caines

The Little Plays of St Francis: Series II (5, 6, 7, 8 August)

The Builders

Francis	George Holloway
Juniper	Frederick Holloway
Bernard	Donald Finlay
Leo	Henry Cass
Giacomina	Dorothy Cass
Dom Silvestro	George Richardson
A Beggar	Howieson Culff
Pietro Bernardo	Charles Higgins

Sister Clare

Francis	George Holloway
Juniper	Frederick Woodhouse
Leo	Henry Cass
Angelo	Charles Higgins
Clare	Olive Hooper
Elias	George Richardson
Bernard	Donald Finlay

Brother Juniper

Francis	George Holloway
Elias	George Richardson
Jerome	James Botter
Angelo	Charles Higgins
Leo	Henry Cass
Bernard	Donald Finlay
Matteo	Stephen Morland
Juniper	Frederick Woodhouse
Rufus	Howieson Culff
John	George Caines
Simon	Owen Strong
Anthony	Tom Waller
Sacristan	Sebastian Shaw

Sister Death

Francis	George Holloway
Elias	George Richardson

Rufus	Howieson Culff
Bernard	Donald Finlay
Jerome	James Botter
1st Citizen	Sebastian Shaw
2nd Citizen	Tom Walker
Juniper	Frederick Woodhouse
Angelo	Charles Higgins
Leo	Henry Cass
John	George Caines
Simon	Owen Strong
Matteo	Stephen Morland
Citizens	Dorothy Holloway, Olive Hooper, Agnes Swaine, Irene Webb, W. Difford
Producer	Laurence Housman
Musical Director	Rutland Boughton
Costumes and design	Christina Walshe

1926

Boughton was now occupied with politics and therefore left the Glastonbury Festivals to Housman. He resumed work with The Glastonbury Festival Players in the winter of 1926.

Summer Festival, 9–14 August 1926

This consisted entirely of Housman plays: two cycles of The Little Plays of St Francis and five performances of *Prunella*.

Prunella: Laurence Houseman and Harley Granville-Barker

Prunella	Dorothy Holloway
Prim	Gladys Ward
Prude	Olive Hooper
Privacy	Winifred Farnwell
Queer	Mary Goodson
Quaint	Winifred Barton
1st Gardener	Stuart Fletcher
2nd Gardener	Roger Clark
3rd Gardener	Eric Selley
Boy	Colin Olsson
Love	Ruby Boughton
Pierrot	George Holloway
Scaramee	Ernest Hernu

Hawk	Leyland White
Kennel	Frederick Woodhouse
Callow	Geoffrey Dunn
Mouth	Charles Higgins
Doll	Vivienne Bennett
Romp	Virginia Coit
Tawdry	Dorothy Reynolds
Coquette	Nora Desmond
Choreography	Vivienne Bennett
Music	Joseph Moorat
Costumes and design	Christina Walshe

The Little Plays of St Francis: The Builds, Sister Clare, Brother Sun, Brother Juniper, Brother Ass

Cast lists have not survived, but they may be assumed to have included most of the previous year's players.

Autumn Tour, September–November 1926

The Little Plays of St Francis: Bristol, Eastbourne, Oxford, Guildford, Chislehurst, London, Manchester, Newcastle-on-Tyne. Precise dates not known.

Winter Tour, 6 December 1926–1 January 1927

The tour covered the West Country and ended at Church House, Westminster: 20 December–1 January.

Bethlehem

Mary	Dorothy Silk/Annette Blackwell
Gabriel	Ruby Boughton
Joseph	Cuthbert Smith
Shepherds: Jem	Frederick Woodhouse
Sym	Geoffrey Dunn
Dave	Kennedy Arundel
Wisemen: Zarathustra	Frederick Woodhouse
Nubar	John Patterson
Merlin	Kennedy Arundel
Angels	Juliet Sladden, Gwen Jeaffreson
Believer	Mayne Fisher
Unbeliever	George Richardson

Calchas	Harold Ching
Herod	Edward Nichol
Herodias	Dorothy D'Orsay
Conductor: on tour	A. Davies-Adams
in London	Rutland Boughton
Costumes and design	Christina Walshe

1927

The Glastonbury Festival Players went into liquidation on 7 July, with debts of £1,358. 18s. 7d., of which £830 was 'Issued Capital' and £528. 18s. 7d. was owing to sundry creditors, to whom a dividend of 10s. 6d. in the pound was paid. This brought the Glastonbury Festivals to an end.

THE LAST FESTIVALS

Summer Festival, Stroud, 10–15 September 1934

The Immortal Hour (five performances)

Dalua	Frederick Woodhouse
Etain	May Moore
Eochaidh	Augustus Milner
Manus	W. Salusbury Baker
Maive	Dorothy D'Orsay
Spirit Voice	Dorothy D'Orsay
Old Bard	Julian Were
Midir	Steuart Wilson
Chorus	Maire Boughton, Kathleen Davis, Shelagh Lyle-Smith, Vera Heard, Miriam Prelooker, Emilie Strudwick, Agnes Swaine, Mona Tatham, Ruby Boughton, Joyce Sutton, Joan Case, Dorothy D'Orsay, William Biggs, Norman Bennett, W. Salusbury Baker, Bernard Hall-Bailey, Vivian Miller, Julian Were, Jeffrey Lambourne, Eric Sidney
Dancers	Rosina Bassett, Helga Burgess
Conductor	Anthony Collins
Producer	Christopher Ede
Set design	Walter Spradbury
Costumes	Christina Walshe

The Lily Maid (four performances)

Sir Bernard	Bernard Hall Bailey
Sir Lavaine	Jeffrey Lambourne
Elaine	Nellie Palliser
Sir Lancelot	Augustus Milner
Sir Gawaine	Steuart Wilson
Hod	Frederick Woodhouse
Chorus	The Round Table Singers: Emilie Strudwick, Joan Case, Eric Sidney, Julian Were
Conductor	Rutland Boughton
Producer	Christopher Ede
Set design	Marion Symons Burton
Costumes	Christina Walshe

Winter Tour, 27 December 1934–2 January 1935

Bath, Stroud, Ross on Wye.

Bethlehem (seven performances)

Mary	Jane Vowles
Gabriel	Ruby Boughton
Joseph	Arthur Cranmer
Shepherds: Jem	Thomas Dance
Sym	Julian Were
Dave	William Biggs
Wisemen: Zarathustra	Thomas Dance
Nubar	William Biggs
Merlin	Julian Were
Two Women	Lilian Hughes, Joyce Davies
Believer	Rachel Bernard
Unbeliever	Reginald Ryan
Calchas	W. Salusbury Baker
Herod	Frank Mullings
Herodias	Constance Pavie
Dancers	Helga Burgess, Rosina Bassett, Lesley Hodson, Nancy Hume
Pianist/conductor	Rutland Boughton
Producer	Christopher Ede
Costumes and design	Christina Walshe

Summer Festival, Bath 9–14 September 1935

The Immortal Hour (seven performances)

Dalua	Frederick Woodhouse
Etain	Elsie Suddaby
Eochaidh	Augustus Milner
Manus	W. Salusbury Baker
Maive	Dorothy D'Orsay
Spirit Voice	Dorothy D'Orsay
Old Bard	Julian Were
Midir	Steuart Wilson
Chorus	Maire Boughton, Hylda Corp, Gladys Fisher, Vera Heard, Miriam Prelooker, Aileen Street, Agnes Swaine, Constance Pavie, Ruth Abbott, Joyce Sutton, Gladys Ward, Norman Bennett, Julian Gardiner, Eric Sydney, Bernard Bailey, Vivian Miller, Frank Rendall, Julian Were, W. Salusbury Baker
Dancers	Helga Burgess, Rosina Bassett, Lesley Hodson, Cynthia Nelson, Barbara Ward
Conductor	Boyd Neel
Producer	Christopher Ede
Set design	Walter Spradbury
Costumes	Christina Walshe

The Ever Young (four performances)

Caeria	Astra Desmond
Dagda	Frederick Woodhouse
Boyanna	Kathleen Davis
Aengus	Tom Goodey
Lugh	Steuart Wilson
Bride	Elsie Suddaby
Dana	Dorothy D'Orsay
Chorus and dancers	As for *The Immortal Hour*
Conductor	Rutland Boughton
Producer	Christopher Ede
Set design	Walter Spradbury
Costumes	Christina Walshe

APPENDIX D

Discography

Though there have been a great many recordings of the 'Faery Song' from *The Immortal Hour*, it has seemed best to limit the present list to such records as have either historic or artistic significance.

EARLY RECORDINGS (78 RPM)

The Immortal Hour

Concise Version on four sides (12") sung by the cast of the 1932 Queen's Theatre production, including Gwen Ffrangcon-Davies (Etain), Bruce Flegg (Midir), Arthur Cranmer (Dalua), and William Johnstone Douglas (Eochaidh). Conductor: Ernest Irving.
   Columbia DX 346/347

'Luring Song' and 'Song of Creation', sung by William Heseltine (Midir in the 1922, 1923, 1924, and 1926 London productions).
   Columbia A 2976/77

'Faery Chorus' and 'Love Duet', orchestral version. The London Symphony Orchestra conducted by Rutland Boughton.
   Columbia AX 24/25; reissued as LP on **OPAL 801**

Bethlehem

Four Carols: 'Earth today rejoices'; 'The Holly and the Ivy'; 'In the Ending of the Year'; 'There was a star in Bethlehem'. The English Singers.
   HMV 309

Carol: 'O come, all ye faithful'. The English Singers.
   HMV 307

'The Virgin's Lullaby'. Elsie Suddaby, recital.
   HMV E 366

Lorna Doone

'Lorna's Song'. Sung by Victoria Hopper, as in Basil Dean's film, with the ATP Studio Orchestra conducted by Ernest Irving.
  HMV B 8249

MODERN RECORDINGS (LP AND CD)

The Immortal Hour

First complete recording, with Anne Dawson (Etain), Maldwyn Davies (Midir), David Wilson-Johnson (Eochaidh), Roderick Kennedy (Dalua), the Geoffrey Mitchell Choir and the English Chamber Orchestra conducted by Alan G. Melville.
  Hyperion A 66101/2, CDA 66101/2

Act 2, closing scene. Jeanette Wilson (Etain), David Skewes (Midir), Stuart Conroy (Eochaidh), with the Opera Viva Chorus and the Kensington Symphony Orchestra conducted by Leslie Head. Side 3 of a two-disk Limited Edition of excerpts from nineteenth- and early twentieth-century British opera.
  OV 101/2

Symphony No. 3 in B minor, Oboe Concerto No. 1 in C

Premiere recordings. Sarah Francis (oboe), with the Royal Philharmonic Orchestra conducted by Vernon Handley.
  Hyperion CDA 66343

The Hyperion recordings are the first fruits of a programme begun by the Rutland Boughton Music Trust in 1983, and whose next project (1993) will be a recording of *Bethlehem*, also with Hyperion.

BIBLIOGRAPHY

Books

Bishop, G. W., *Barry Jackson and the London Theatre* (Arthur Barker, London, 1933), 15–20, 36–8, 47, 57–8, 134, 147.

Brook, Donald, *Composers' Gallery* (Rockliffe, London, 1946), 29–34.

Carey, Hugh, *Duet for Two Voices* (Cambridge University Press, Cambridge, 1979), 85.

Cumberland, Gerald, *Set Down in Malice* (Grant Richards, London, 1918), 259–61.

— *Written in Friendship* (Grant Richards, London, 1923), 56–60.

Grew, Sydney, *Our Favourite Musicians* (Peter Davies, London, 1924), 207–23.

Hurd, Michael, *Immortal Hour: The Life and Period of Rutland Boughton* (Routledge & Kegan Paul, London, 1962).

— 'Rutland Boughton's Arthurian Cycle', in Debra N. Mancoff (ed.), *The Arthurian Revival* (Garland Publishing, Inc., New York and London, 1992), 205–29.

Lovell, Percy, *Quaker Inheritance* (The Bannisdale Press, London, 1970), 145–54, 180–3, 218, 238, 243.

Shaw, G. Bernard, *The Perfect Wagnerite* (4th edn. 1922; Collected edn., 1947; Constable, London, 1947).

Sheldon, A. J., *Notes on* The Immortal Hour (Cornish Brothers, Birmingham, 1922).

Trewin, J. C., *The Birmingham Repertory Theatre, 1913–1963* (Barrie & Rockliffe, London, 1963), 15, 61–2, 66–7, 72, 77–80, 86, 106–7, 173, 179, 185.

Wilson, Margaret, *English Singer: The Life of Steuart Wilson* (Duckworth, London, 1947), 77–9, 153–4.

Magazine Articles

Antcliffe, Herbert, *Musical Quarterly*, Jan. 1918: 117–27.

Bennett, Rodney, *Bookman*, Jan. 1926: 238–40.

Bond, F. H., *Musical Standard*, 9 Aug. 1911: 126–7.

Bush, Alan, *Music and Life*, Summer 1960: 4–7.

Cumberland, Gerald, *Musical Opinion*, Aug. 1909: 774.

— *Music Bulletin*, Mar. 1924: 74.

Hurd, Michael, *Musical Times*, Oct. 1963: 700–1.
— *Musical Times*, Aug. 1984: 435–7.
Ould, Hermon, *Musical News and Herald*, 15 May 1927: 200–2.
Russell, Thomas, *Music and Life*, Summer 1960: 3.
Scholes, Percy, *Music Student*, Oct. 1916: 55–61
Stevens, Bernard, *New Reasoner*, Spring 1959: 74–81.
Wortham, H.E., *Music and Youth*, Oct. 1927: 224–6.

INDEX OF WORKS

Italicized references indicate a music example

GENERAL INDEX